Brain Science

Series Editors

Hideyuki Okano
School of Medicine
Keio University
Shinjuku-ku, Tokyo, Japan

Atsushi Iriki
Laboratory for Symbolic Cognitive Development
RIKEN BSI
Wako-shi, Saitama, Japan

The “Brain Science Series” provides critical and comprehensive discussions of the most significant areas of neuroscience research, written by scientific leaders and authorities. Each volume in the series presents the latest and most complete information on its respective subject, making it an unrivalled reference source.To analyze brain functions that have been preserved through evolution, a reductive approach based on gene manipulation using models such as genetically modified rodents were mainly used. At the same time, to analyze high-order brain functions that are uniquely acquired only in non-human primates and humans, analyses of complex behaviors using a psychological approach were mainly adopted. The Brain Science Series focuses on clarifying the operating principles of the neural circuitry governing the human mind, as well as the accompanying molecular mechanisms.

More information about this series at http://www.springer.com/series/13473

Shigeru Watanabe • Michel A. Hofman
Toru Shimizu
Editors

Evolution of the Brain, Cognition, and Emotion in Vertebrates

Springer

Editors
Shigeru Watanabe
Keio University
Minato-ku
Tokyo
Japan

Toru Shimizu
Psychology
University of South Florida
Tampa
Florida
USA

Michel A. Hofman
Department of Neuropsychiatry
Netherlands Institute for Neuroscience
Amsterdam Zuidoost
Noord-Holland
The Netherlands

Brain Science
ISBN 978-4-431-56557-4   ISBN 978-4-431-56559-8 (eBook)
DOI 10.1007/978-4-431-56559-8

Library of Congress Control Number: 2017951870

Cover picture credit: Alexandr Mitiuc / fotolia

Printed on acid-free paper

This Springer imprint is published by Springer Nature
The registered company is Springer Japan KK
The registered company address is: Chiyoda First Bldg. East, 3-8-1 Nishi-Kanda, Chiyoda-ku, Tokyo 101-0065, Japan

Dedicated to Cynthia Shimizu

With kind encouragement of coeditors, Drs. Shigeru Watanabe and Michel Hofman, this book is dedicated to my late wife Cynthia.

When we found out about Cynthia's unexpected terminal illness, this book project was still in the early stages. While her friends and family were not ready to accept her impending death, she was calm, and her main concern was my well-being rather than her own. In the remaining short months, she spent much of her conscious time preparing me psychologically and practically for a life without her. In return, I promised her that I would be alright. Two years after her death, the book project has come to fruition, thanks to our patient publisher, coeditors, and contributors and Dr. Harry Jerison. To me, this book is proof of my intention to honor my promise to Cynthia.

Toru Shimizu

Cynthia Shimizu (1953–2015)

Foreword

This fine collection of essays covers topics that are important for understanding how the brain evolved. In 1859, Charles Darwin wrote "On the Origin of Species" filled with data on the diversity of living species. We continue to appreciate this diversity. Even more than Darwin, we have learned to appreciate the stories told by fossils. Beginning with Cuvier in 1804, we discovered and expanded the fossil evidence of the brain's evolution revealed in endocranial casts. The authors of the essays cover these topics, extending them with data on neuromorphology, brain–behavior relationships, and, of course, fossil endocasts in many species. Among the morphological dimensions covered in this book are issues of network circuitry and their evolutionary diversification. Among the behavioral dimensions are differences in learning and performance in closely related and not so closely related species. The book also contains more specialized issues, such as social behavior, tool use, and emotionality.

Each essay has succeeded in covering its topic broadly and in depth. From Kei Yamamoto and Solal Bloch (Chap. 1), we learn important details on the variation in living fish and its significance for evolutionary analysis. They offer new perspectives on the likely derivation of various behavior patterns. Toru Shimizu and colleagues (Chap. 2) review the variation and likely evolution of the brain in birds with new material that enables them to introduce useful hypotheses on the evolution of behavior. Jon Kaas (Chap. 3), similarly, reviews our knowledge of the origins of mammalian brains, with helpful insight into the variety of their organization, especially in primates. Christine Charvet, Chet Sherwood, and Emi Takahashi (Chap. 4) contribute a marvelous piece on the microevolution of brain circuitry.

The collection tells us where we are now and has many suggestions about the future of our field. A few words about my personal special interest in two of them: Kazuo Okanoya (Chap. 10) presents new data on the origins of sound-generating mechanisms, and Shigeru Miyagawa (Chap. 11) discusses his "parallel" model of language. My own doctoral research was on auditory cortex in monkeys, and I am now concerned with something close to Miyagawa's models (Jerison 2012). I look forward to integrating their new work on speech and language and my own.

Two chapters provide general theoretical approaches to brain evolution and cognition: Gerhard Roth and Ursula Dicke (Chap. 6) and Michel Hofman (Chap. 7). Other chapters contribute more detailed views of the role of cognition and other brain–behavior connections that have evolved. Andrew Iwaniuk (Chap. 5) provides a broad evolutionary perspective in all vertebrate species. In Chap. 8, Yumiko Yamazaki and Atsushi Iriki review tool use, especially in primates, in neural and cognitive adaptations. Behavioral variations among great apes in social attention and its measurement are the subject of the insightful Chap. 9 by Fumihiro Kano and Josep Call. There are, finally, excellent chapters on social and emotional evolution by Edmund Rolls (Chap. 12) and by Shigeru Watanabe and Yutaka Kosaki (Chap. 13).

The book will be of wide interest to students, scholars, and a variety of experts who are interested in keeping track of the new insights and discoveries that are emerging about the evolution of the vertebrate brain in relation to cognition and emotion. It provides a great background for our topic and reviews recent and older literature. The seasoned theoretical statements by these experts are bound to be helpful, and their discussions of new data suggest possibilities for new experiments and observations.

Reference

Jerison HJ (2012) Digitized fossil brains: neocorticalization. Biolinguistics 6: 385–392

Harry J. Jerison
Department of Psychiatry and Biobehavioral Sciences
University of California,
760 Westwood Plaza, Los Angeles,
CA, 90095, USA

Contents

Part I
Evolution of the Brain

Chapter 1
Overview of Brain Evolution: Lobe-Finned Fish vs. Ray-Finned Fish

Kei Yamamoto and Solal Bloch

Abstract Bony jawed vertebrates (Osteichthyes) are divided into two groups, ray-finned fish (Actinopterygii) and lobe-finned fish (Sarcopterygii). The tetrapod (including mammals and birds) is a group of lobe-finned "fish" specialized in terrestrial life. Although the overall brain organization is conserved, significant differences exist within each brain region between Actinopterygii and Sarcopterygii. This chapter introduces a new view of the vertebrate brain organization; more particularly, we propose revised subdivisions in the anterior forebrain, which was revealed through comparative analyses between Actinopterygii and Sarcopterygii.

Behaviorally, some teleosts (a group of Actinopterygii) demonstrate higher order cognitive functions such as tool use or transitive inference. Moreover the "fish brains" are not necessarily simpler than tetrapod brains, and some teleost species have an enlarged pallium (dorsal telencephalon) as it is the case in amniotes. Nonetheless, the anatomical organization of the teleost brain is very different from that of the tetrapods, and there are many inconsistencies when we consider that these functional similarities are inherited from the common ancestor. It is possible that the nervous system is highly plastic during evolution, and more convergent evolution has taken place than is currently thought.

Keywords Forebrain • Pallium • Fish • Teleost • Tetrapod • Phylogeny • Evolution Convergent • Homology

---

K. Yamamoto (✉) • S. Bloch
Paris-Saclay Institute of Neuroscience (UMR 9197), CNRS, Université Paris-Sud, Université Paris-Saclay, Gif-sur-Yvette 91190, France
e-mail: kei.yamamoto@cnrs.fr

S. Watanabe et al. (eds.), *Evolution of the Brain, Cognition, and Emotion in Vertebrates*, Brain Science, DOI 10.1007/978-4-431-56559-8_1

1.1 Introduction: Phylogenetic Relationships of Vertebrates

The brain is an enlarged region of the central nervous system (CNS) located in the head of bilaterian animals, which receives and processes information conveyed by sensory organs. The brain has evolved in different animal groups in bilateria, such as insects, annelids, and vertebrates.

Figure 1.1 shows a simplified phylogenetic tree of vertebrates. The living vertebrates are divided into gnathostomes (jawed vertebrates) and cyclostomes-a group of agnathans (jawless vertebrates). Cyclostomes are the only group of living agnathans and consist of much fewer species (± 100 species, such as the lamprey and hagfish) than living gnathostomes (± 55,000 species). The gnathostomes consist of two large groups, Osteichthyes (bony fish, ± 54,000 species) and Chondrichthyes (cartilaginous fish, ± 1000 species). The group of Osteichthyes contains Sarcopterygii (lobe-finned fish, ± 27,000 species) and Actinopterygii (ray-finned fish, ± 27,000 species; see Bally-Cuif and Vernier 2010).

It is important to point out that phylogenetic trees can be biased based on their layout and focus, which can be misleading for nonspecialists. For example, the overall vertebrate phylogeny is the same in the phylogenetic trees in Fig. 1.1. Figure 1.1a emphasizes the Osteichthyes and shows details of some sets of species included in this group, in particular the tetrapods. This does not mean that there are more animal groups in tetrapods. Alternatively, Fig. 1.1b shows more details of the Chondrichthyes.

Additionally, it is worth noticing that the term "fish" is generally used for aquatic animals without four limbs, but some "fish" such as lungfish are phylogenetically closer to tetrapods than to Actinopterygii. The tetrapod is indeed a group of lobe-finned "fish" specialized in terrestrial life (Fig. 1.1a).

This chapter aims at providing an overview of the brain of Osteichthyes. Due to space limitations, we will focus on the forebrain organization of teleosts and tetrapods, highlighting differences in the basic brain organization between Actinopterygii and Sarcopterygii. From a mammalian-centric point of view, the teleost brain is sometimes used as an example representing a primitive type of brain. However, mammals and teleosts belong to two different groups of Osteichthyes, and as their phylogenetic relationship indicates, the teleost brain does not represent the ancestral form of the mammalian brain.

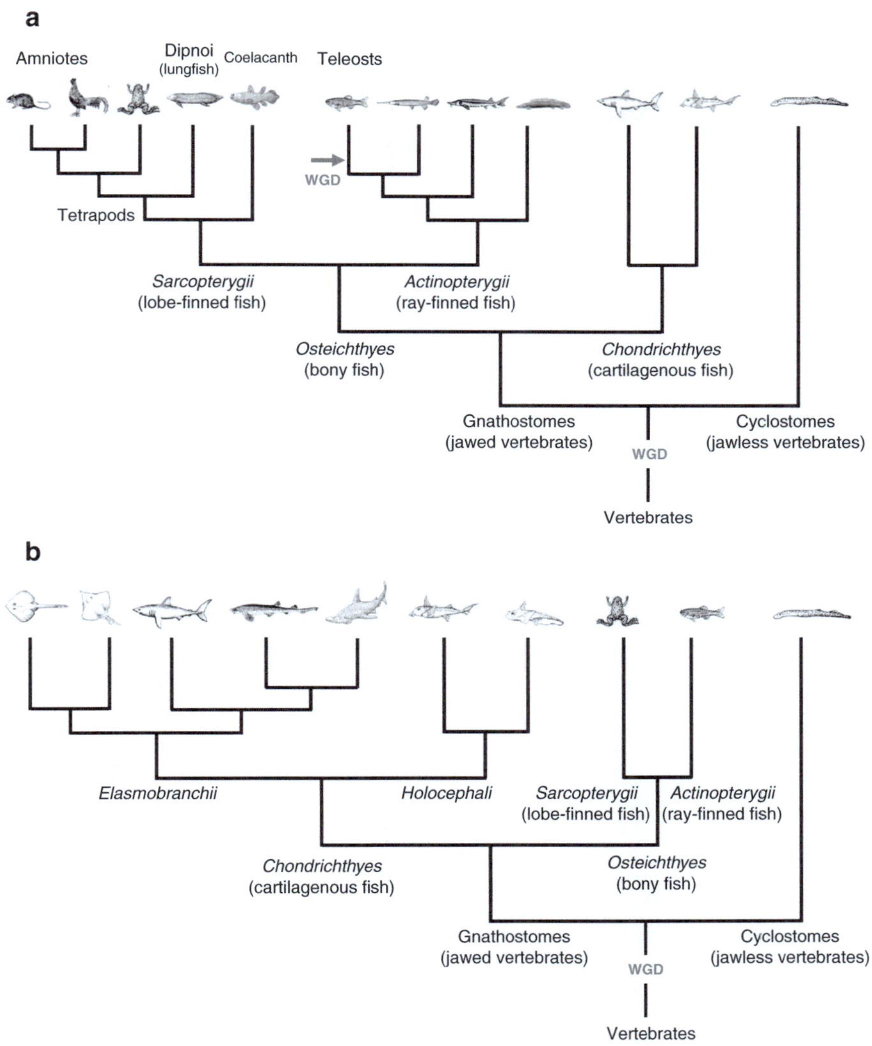

Fig. 1.1 Two different versions of the vertebrate phylogenetic tree. (**a**) A phylogenetic tree focusing on the evolution of Osteichthyes (bony fish). Based on recent findings, it is hypothesized that two rounds of whole genome duplication (WGD) occurred before the gnathostomes-cyclostomes split (Smith et al. 2013). The teleost lineage went through an additional WGD. (**b**) A tree focusing on the evolution of Chondrichthyes (cartilaginous fish). This kind of phylogenetic tree representation is not commonly displayed, simply because we are used to the mammalian-centered point of view

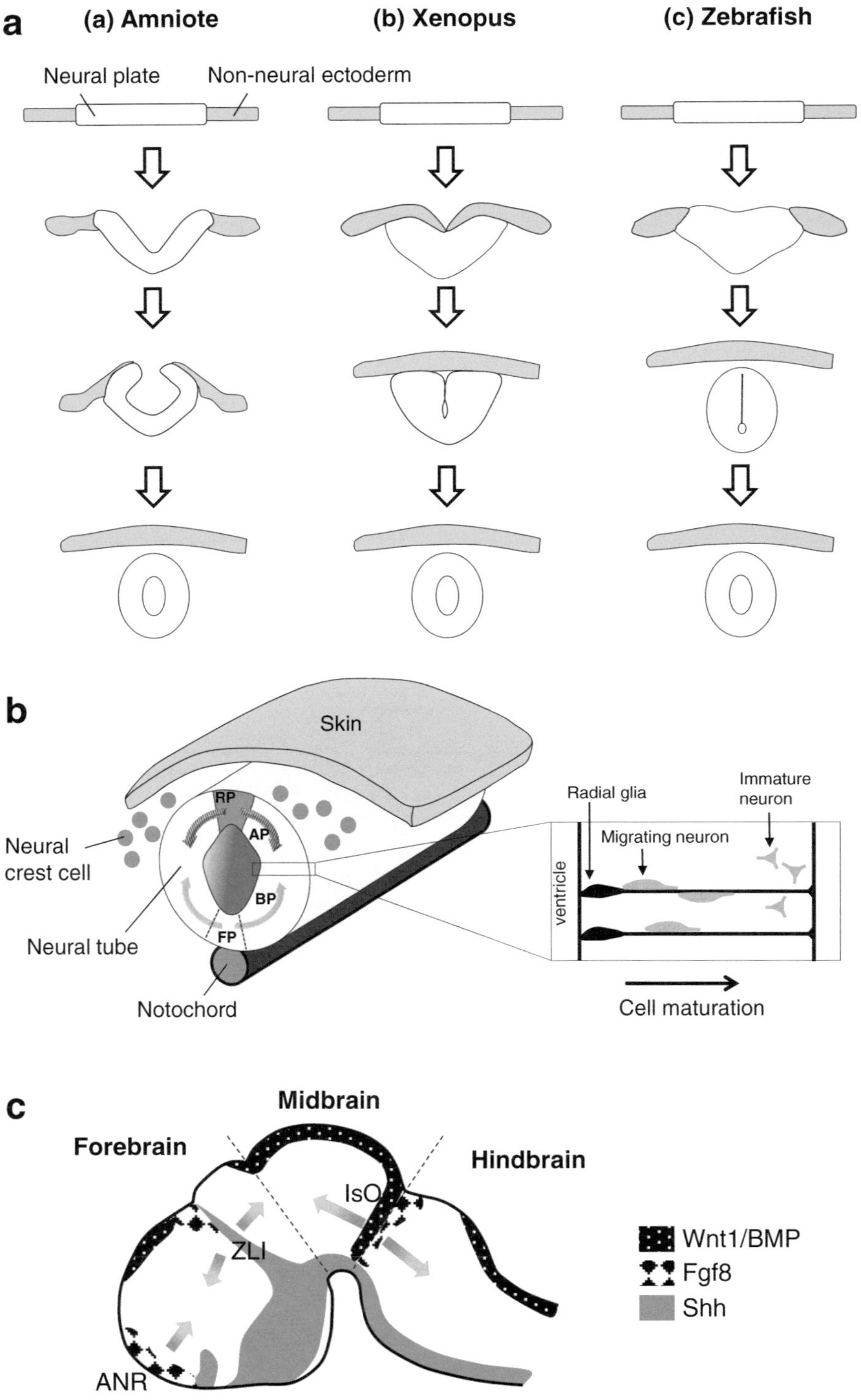
a
(a) Amniote
(b) Xenopus
(c) Zebrafish
Neural plate
Non-neural ectoderm
b
Skin
RP
AP
BP
FP
Neural crest cell
Neural tube
Notochord
Radial glia
Immature neuron
Migrating neuron
ventricle
Cell maturation
c
Midbrain
Forebrain
Hindbrain
IsO
ZLI
ANR
Wnt1/BMP
Fgf8
Shh

1.2 Regionalization of the Brain

1.2.1 Brain Morphogenesis at Early Stages

In all vertebrates, the brain develops from a neural tube, a simple epithelium surrounding a central lumen. The formation of the neural tube from the neural plate (neurulation) and its patterning are induced by morphogens, signaling molecules that are secreted from small cell populations called "signaling centers" or "organizers." Different organizers are set up successively over time during early development, with the "primary organizer" being fundamental for the primary neural induction. Although the mode of neurulation varies across species (Fig. 1.2a; Harrington et al. 2009), signaling molecules involved are shared among species, as is the case with bone morphogenetic proteins (BMP) secreted from the nonneural ectoderm and sonic hedgehog (Shh) secreted from the notochord. After the formation of the neural tube, BMP secreted from the roof plate and Shh secreted from the floor plate establish dorsoventral axis within the neural tube. The dorsal half influenced by BMP is called the alar plate, whereas the ventral half influenced by Shh is called the basal plate (Fig. 1.2b).

Morphogens secreted from "secondary organizers," local signaling centers formed later in the neuroepithelium, further refine the anteroposterior and dorsoventral patterning of the neural tube. The isthmic organizer (IsO) secretes FGF8 and Wnt signals delineating the boundary between the midbrain and hindbrain. In the forebrain, the *zona limitans intrathalamica* (ZLI) secretes Shh, and the anterior neural ridge (ANR) secretes Fgf8 and Wnt antagonists (Fig. 1.2c; Echevarría et al. 2003; Vieira et al. 2010; Cavodeassi and Houart 2012). These secondary organizers are formed and positioned according to the differential expression of genes in adjacent territories. For example, differential expression of *Otx* and *Gbx* genes in the anterior and posterior parts of the brain is crucial for the positioning of the isthmic organizer, and the expression of *Fezf2* and *Irx* genes specifies the position of the ZLI. The gradient distribution of the signaling molecules secreted from the organizers is critical for the determination of regional identity within the neural tube. Interactions between the morphogens and the expression of transcription factors

Fig. 1.2 Neural tube development at early stages. (**a**) Variation in the neurulation process in amniotes such as mouse and chicken (*a*), Xenopus (*b*), and zebrafish (*c*). Although the neural tube is commonly formed from a neuroepithelial neural plate, there is morphological variation in their formation. The schematic drawing is modified from Harrington et al. (2009). (**b**) Neural tube stage. Once the neural tube is formed, neural crest cells located at the boundary of neural/non-neural ectoderm delaminate and migrate. Roof plate (RP) and floor plate (FP) secrete BMP and Shh, respectively, which refine the dorsal-ventral axis within the neural tube. The ventricular zone of the neural tube is a site of cell proliferation, and generated neurons migrate into the mantle following radial glial fibers. (**c**) A lateral view of a representative embryonic brain (rostral to the *left*) showing the location of secondary organizers secreting morphogens. The schematic drawing is modified from Suárez et al. (2014). *ANR* anterior neural ridge, *AP* alar plate, *BP* basal plate, *FP* floor plate, *IsO* isthmic organizer, *RP* roof plate, *ZLI zona limitans intrathalamica* control, in a temporally and spatially regulated manner, the growth of the neuroepithelium and the fate of the neural progenitors.

It has been accepted for a long time that the neural tube gives rise to three primary vesicles by differential proliferation of neuroepithelial territories. These are the forebrain (prosencephalon), the midbrain (mesencephalon), and the hindbrain (rhombencephalon) that is continuous with the spinal cord (the brain and spinal cord together are the CNS). Although a recent study has suggested that the existence of these initial three vesicles is not universal to all vertebrates (Ishikawa et al. 2012), most studies of brain development have been based on this trichotomy. The cerebrum, optic tectum, and cerebellum in the mature brain are considered to be derived from the forebrain, midbrain, and hindbrain, respectively.

1.2.2 Diversity of Brain Morphology in Adults

There is a large diversity in the morphology of mature brains, with different parts of the brain being more or less enlarged depending on the species. The human brain is an example in which the cortical part of the forebrain (cerebral hemispheres) is extremely enlarged (Fig. 1.3a). The term “cerebrum,” which derives from a Latin

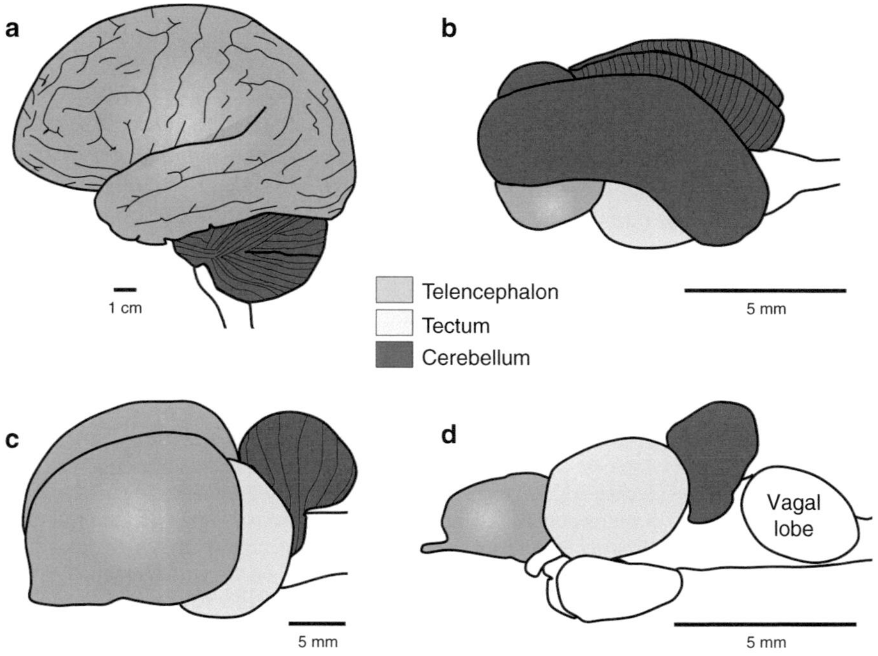

Fig. 1.3 Variation in relative size of brain regions. The entire brain of human (**a**), mormyrid (**b**), pigeon (**c**), and goldfish (**d**) are shown from a lateral view (rostral to the *left*). Each animal has differentially developed specific parts of the brain, which is probably a consequence of the adaptation. (**a**–**c**) are modified from Ruiz i Altaba et al. (2002) and (**d**) is from Yamamoto and Ito (2005)

word for the "brain," is used to refer to the telencephalon, as it covers almost the entire brain in humans. The "cerebellum" (located in the hindbrain) means the "little brain," although in some animals, the "cerebellum" is not small. For example, mormyrids (electric fishes) possess a very large cerebellum covering almost the entire brain (Fig. 1.3b). In mammals, the major role of the cerebellum is to control body balance, coordination, and smooth/rapid movements, while in the mormyrid, it plays specialized roles in electroreception.

In some species of birds and teleosts, the midbrain roof called the tectum (Latin for "roof") is particularly developed. It is often called the "optic tectum" since it receives massive inputs from the eyes (Fig. 1.3c). The large size of the optic tectum is correlated with the high visual capacities in many species of birds and teleosts. The homologous structure in mammals, the superior colliculus (Latin for "mound"), is much smaller and covered by the cerebrum. Indeed, the mammalian ancestors were adapted to nocturnal life, and olfaction became the dominant sensory system. This is still the case for many living mammals including mice, although primates have secondarily evolved elaborated visual systems.

In many vertebrates, the cerebrum (forebrain), optic tectum (midbrain), and cerebellum (hindbrain) are three prominent dorsal bulges. In goldfish, there is an additional dorsal bulge (called the vagal lobe; Fig. 1.3d) in the brainstem. The term "brainstem" is used to refer roughly to the brain regions excluding the three dorsal bulges mentioned above. It contains cranial nervous and autonomic (visceral) sensory-motor centers critical for survival (respiration, heart-rate control, etc.). The vagal lobe in goldfish is an enlarged and highly laminated structure, which is a central termination for gustatory afferents. The elaboration of the vagal lobe would be associated with their feeding behavior: The goldfish vacuums food with other particles and sort it out in the mouth (Finger 2008).

These outlined differences in brain structures among vertebrate groups are likely a consequence of evolutionary adaptation of each animal group to different environments.

1.2.3 Different Models of Forebrain Regionalization

How a simple neural tube develops into an elaborate brain has long been a major question for neuroanatomists and developmental biologists. Due to the difficulty of establishing a universal scheme of brain development throughout vertebrates, several models have been proposed and modified over the years.

Based on classical embryology, the first three vesicles (the forebrain, midbrain, and hindbrain) are further divided into five vesicles. The hindbrain is subdivided into the myelencephalon (containing the medulla oblongata) caudally and the metencephalon (containing the cerebellum and pons) rostrally. The midbrain is considered to remain one division by itself. The forebrain is subdivided into the diencephalon caudally and the telencephalon rostrally. The diencephalon is further divided into the dorsal thalamus and the ventral hypothalamus, and the telencephalon is further divided into the dorsal pallium and the ventral subpallium (Fig. 1.4a).

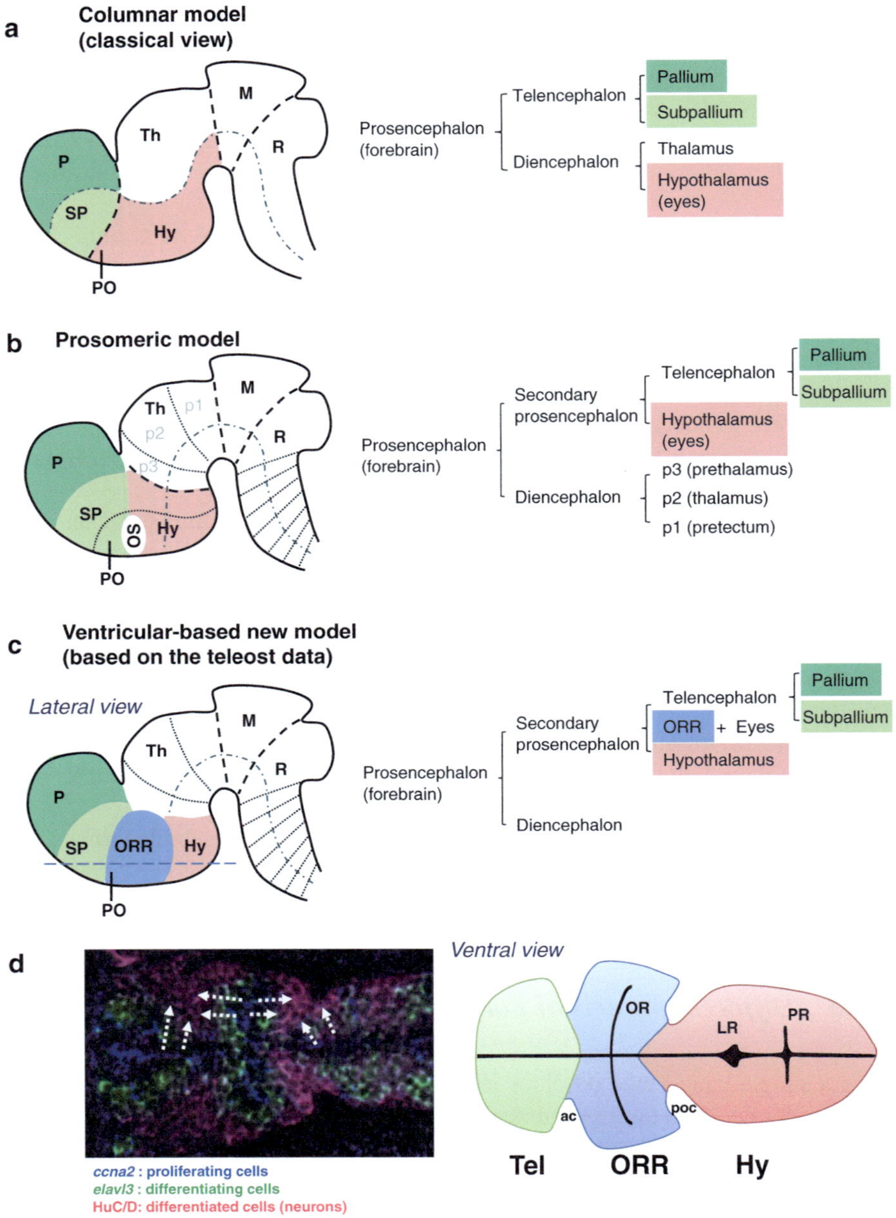
a
Columnar model
(classical view)
P
SP
Th
M
R
Hy
PO
Prosencephalon
(forebrain)
Telencephalon
Pallium
Subpallium
Diencephalon
Thalamus
Hypothalamus
(eyes)
b
Prosomeric model
Th
p1
p2
p3
M
R
P
SP
OS
Hy
PO
Prosencephalon
(forebrain)
Secondary
prosencephalon
Telencephalon
Pallium
Subpallium
Hypothalamus
(eyes)
Diencephalon
p3 (prethalamus)
p2 (thalamus)
p1 (pretectum)
c
Ventricular-based new model
(based on the teleost data)
Lateral view
Th
M
R
P
SP
ORR
Hy
PO
Prosencephalon
(forebrain)
Secondary
prosencephalon
Telencephalon
Pallium
Subpallium
ORR + Eyes
Hypothalamus
Diencephalon
d
ccna2 : proliferating cells
elavl3 : differentiating cells
HuC/D: differentiated cells (neurons)
Ventral view
OR
LR
PR
ac
poc
Tel
ORR
Hy

More recently, a neuromeric model of the brain development has been elaborated and accepted by many developmental biologists (Fig. 1.4b). Neuromeres are defined as transversal divisions which appear transiently in the developing neural tube, with the neuromeres in the rhombencephalon (rhombomeres) being well established. The rhombomeric segmentation is clearly observable and shaped by specific genetic and cellular mechanisms (Keynes and Lumsden 1990; Kiecker and Lumsden 2005), and each segment is named r1, r2, r3..., from rostral to caudal. The cerebellum is a bulge at the roof of r1, and nerve fibers of different cranial nerves (the sensory and motor innervation to the face) are organized along the rhombomeres.

The same idea was applied to the prosencephalon, and neuromeres in this region, termed prosomeres, were proposed based on morphological hallmarks and gene expression patterns (Puelles and Rubenstein 2003). In this view, the forebrain is subdivided into the posterior "diencephalon" whose development is influenced by the notochord and the anterior "secondary prosencephalon" (Fig. 1.4b). The diencephalon is further divided into three prosomeres termed p1 (pretectum), p2 (thalamus), and p3 (prethalamus), from caudal to rostral. The so-called dorsal thalamus corresponds to p2, and the ventral thalamus corresponds to p3. In the mature amniote brain, the dorsal thalamus is extremely enlarged and occupies the majority of the diencephalon. Due to this enlargement of the p2 and the cephalic flexure along amniote brain development, the rostral p3 is pushed ventrally, thus ultimately resulting in the p3 being located ventral to the p2. Although the hypothalamus was originally defined as a region which resides ventral to the thalamus (as the name

Fig. 1.4 Different models for the subdivision of the anterior forebrain. (**a**–**c**) show representative vertebrate brains from a lateral view (rostral to the *left*) demonstrating three different models. (**a**) The columnar model in which the hypothalamus is considered to be the ventral half of the diencephalon. (**b**) The prosomeric model, which was originally proposed by Puelles and Rubenstein in the early 1990s, and has been modified over time. In their models, the hypothalamus is proposed to be the ventral half of the most anterior part of the forebrain (secondary prosencephalon). (**c**) A new model proposed by Affaticati et al. (2015), in which the secondary prosencephalon is divided into three parts, the telencephalon, the hypothalamus, and the optic recess region (ORR). The *blue dotted* line indicates the level of the ventral view shown in (**d**). (**d**) A ventral view of zebrafish embryonic brains illustrating the morphogenesis along the ventricular organization, based on which the new model (shown in **c**) is proposed. The *white dotted arrows* in the *left* image indicate the direction of cell maturation, from proliferation to differentiation. The abutting Hu-positive mature neurons form the regional boundaries. Note that the presence of two hypothalamic ventricles (LR and PR in the *right* diagram) is specific to teleosts, while other vertebrate groups have only one hypothalamic ventricle. *ac* anterior commissure, *Hy* hypothalamus, *LR* lateral recess, *M* mesencephalon, *OR* optic recess, *ORR* optic recess region, *OS* optic stalk, *P* pallium, *p1* prosomere 1, *p2* prosomere 2, *p3* prosomere 3, *PO* preoptic area, *poc* postoptic commissure, *PR* posterior recess, *R* rhombencephalon, *SP* subpallium, *Tel* telencephalon, *Th* thalamus

"hypo"-thalamus indicates), it is now considered to be the most anterior part of the neural tube, occupying the ventral part of the secondary prosencephalon. In the embryonic brain, the subpallium is located anterior (instead of ventral) to the pallium, and during the course of development, the pallium expands thus engulfing the subpallium.

This prosomeric model was established mostly based on the development of the mouse and chicken brains and it has subsequently been applied to other vertebrate species. The assumption is that the prosencephalon can be subdivided into common longitudinal/transversal segmentations throughout vertebrates (Fig. 1.4b). Gene expression patterns are often used to delineate subregions of the brain (genoarchitecture) and to identify homologous brain regions. For example, the expression of *Dlx* genes was used as a marker of the subpallium, and the expression of *Otp* was used as a marker of the supraoptoparaventricular region (SPV; a part of the "alar hypothalamus"), and these genes were used to delineate the telencephalic/hypothalamic border in the Xenopus brain (Domínguez et al. 2013).

However, when this model is applied to the teleost brain, borders delineated by gene expression do not always coincide with morphogenetic borders. A recent study analyzing the teleost (zebrafish) secondary prosencephalon proposed a change to the model in which the subdivision of the anterior forebrain (secondary prosencephalon) is subdivided into three regions: the telencephalon, the hypothalamus, and the optic recess region (ORR) that is continuous with the eyes (Figs. 1.4c and 1.5; Affaticati et al. 2015). This new view takes into consideration the morphogenetic process radially organized around the ventricles, and the ORR is defined as the region which develops around the optic recess (Fig. 1.4d). Importantly, the regional boundaries in this model are delineated by abutting differentiated neurons originated from different ventricular zones, which do not necessarily fit the boundaries of gene expressions (Figs. 1.4d and 1.5a, b). In amniotes, this ORR region would have been difficult to be identified due to its relatively small size compared to the enlarged telencephalon.

The identification of the ORR as a third morphogenetic unit between the telencephalon and the hypothalamus resolves a previously unexplained inconsistency about regional identification, which will be outlined in the following section.

1.2.4 Implication of the New Model on the Anatomical Definition of the Anterior Forebrain

The ORR is flanked by the anterior commissure and the postoptic commissure, which are present throughout vertebrates (Suárez et al. 2014). In the developing brain, the corresponding area has been identified as the "optic stalk (OS)" both in mouse and zebrafish (Shimamura et al. 1995; Wilson and Houart 2004). Until now, the OS was not considered to be a distinct brain region (as represented by the blank between the telencephalon and the hypothalamus in Fig. 1.4b), but reexamination of this area suggests that the OS is indeed the ORR. Specifically, the gene expression patterns of the mammalian OS are similar to those in the zebrafish ORR (Marcus et al. 1999; Roy et al. 2013). Furthermore, studies of eye development show that the ORR develops around the optic recess after the optic vesicles evaginate to form the optic cups (Picker et al. 2009; Ivanovitch et al. 2013), suggesting that the ORR is a part of the eye field.

The new model also clarifies the homology of the *Otp*-dependent neuroendocrine cell population. In both amniotes and teleosts, several categories of neuroendocrine cells regulating pituitary functions are located in the territory expressing the transcription factor *Otp*. Based on the neurochemical and gene expression data, they are suggested to be homologous, but in amniotes, the area containing these neuroendocrine cells has been identified as the hypothalamus, whereas in teleosts, they are located in the preoptic area (PO; Herget et al. 2014; Biran et al. 2015). In the new framework proposed by Affaticati et al. (2015), these *Otp*-expressing neuroendocrine cells are located in the ORR both in amniotes and in teleosts (Fig. 1.5c, d; New model).

Indeed, in the mature teleost brain, the ORR corresponds to the PO. A structure named the PO is also present in tetrapods, however, the tetrapod PO does not exactly correspond to the same domain as the teleost PO (Fig. 1.5c, d). When the new model is applied to the tetrapod brain, the ORR includes domains that have been considered to be the "subpallial PO" and a part of the alar hypothalamus, thus reconciling this previous discrepancy.

Interestingly, this modification of regional boundaries has led to the shrinking of the "hypothalamus proper" in tetrapods, most notably in placental mammals. The idea that the hypothalamus of placental mammals is comparably smaller is not surprising, since the mammalian hypothalamus lacks the cerebrospinal fluid (CSF)-contacting cell population which is present in the hypothalamic region of all other vertebrates. The CSF-contacting cells are located along the periventricular zones and extend processes to contact the CSF. They are abundant in the hypothalami of Actinopterygii, Chondrichthyes (cartilaginous fishes), and cyclostomes (jawless vertebrates), but much less numerous in tetrapods, and are completely lost in placental mammals (Smeets and Reiner 1994; Vígh et al. 2004). Thus it is possible that there are some important hypothalamic functions which have never been addressed because these cells do not exist in mammals.

In teleosts, the CSF-contacting cell populations are organized around two ventricular systems: the lateral recess and the posterior recess (Fig. 1.4d), while there is only one ventricle in the hypothalamus of tetrapods. Due to such large differences in basic organization, one-to-one homology of the hypothalamic cell populations between tetrapods and teleosts requires careful verification.

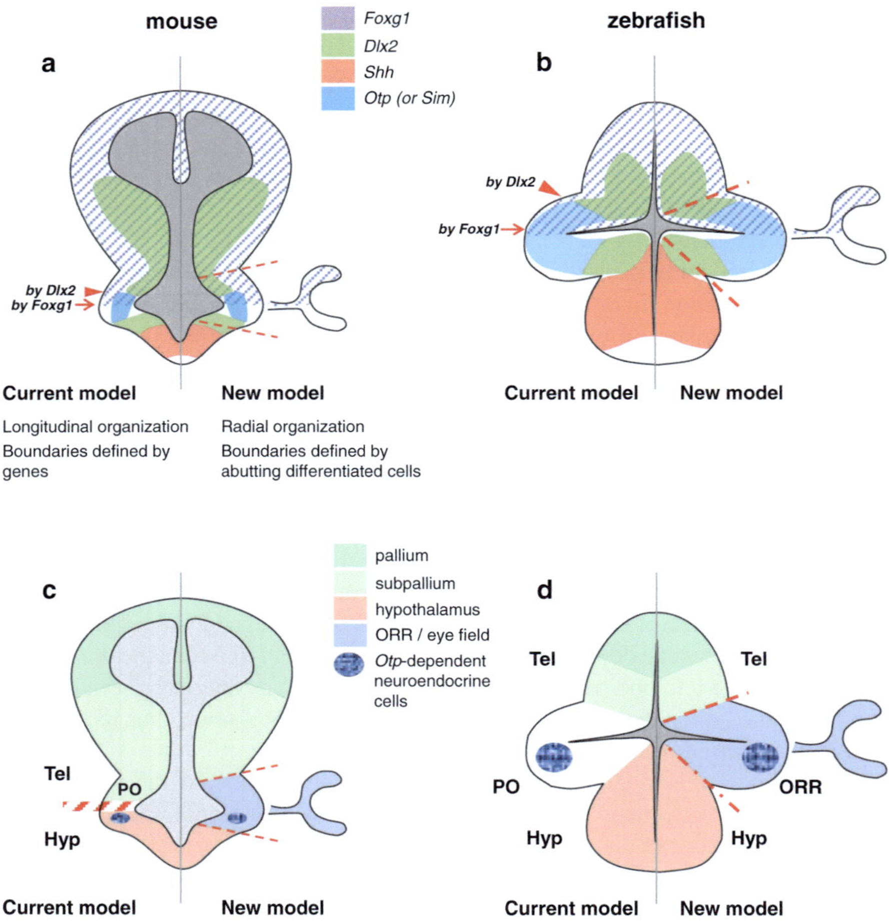

Fig. 1.5 A new model solving the discrepancies of homology in the current model. A frontal view of the anterior forebrain in mouse (**a** and **c**) and zebrafish (**b** and **d**). In each brain section, the *left* half shows the regional boundary proposed in the current prosomeric model, and the *right* half shows the one in the new model proposed by Affaticati et al. (2015). (**a**, **b**) The color code represents gene expression data in mouse (**a**) and zebrafish (**b**). In the current view (*left* half of the brain), the ventral limit of the telencephalon is often delineated by the expression of *Dlx2* (the *arrow head*) or *Foxg1* (the *arrow*), but the two borders do not coincide (which is more prominent in **b** than in **a**). In the proposed model, regional boundaries are delineated by abutting differentiated neurons, and they do not necessarily correspond to the limit delineated by gene expression. (**c**, **d**) Proposed regional identity in mouse (**c**) and zebrafish (**d**). In the current model established in amniotes (*left* half of the brain in **c**), the preoptic area (PO) is considered to be a part of the subpallium due to the expression of *Foxg1* or *Dlx* genes, and *Otp*-dependent neuroendocrine cells are considered to be located in the hypothalamus. In teleosts, the area called the PO contains the *otp*-dependent neuroendocrine cells. In the new model, it corresponds to the optic recess region (ORR). Considering the *Otp*-positive area in mouse as the ORR solves the discrepancy of the homology of neuroendocrine cell population between amniotes and teleosts

1.3 Basic Organization of the Pallium

1.3.1 Enlargement of the Pallium in "Intelligent" Species

Humans have an enlarged neocortex (a part of the pallium with six-layered cytoarchitecture), and thus the evolution of cognition has been associated with the evolution of the cortical structure. However, some birds like corvids and parrots show cognitive capacities equivalent to those of primates, although their pallia do not have layered cytoarchitecture similar to mammals. Their abilities have been described as "insight-related cognition," which includes object permanence, theory of mind, mental time travel, and using/manufacturing tools (Kirsch et al. 2008).

While it is less widely known, some teleost species, such as cichlids and wrasses (Labridae), also exhibit similar cognitive abilities. For example, a tool use-like behavior has been observed in the sixbar wrasse (*Thalassoma hardwicke*; Pasko 2010). Male cleaner wrasses (*Labroides dimidiatus*) are capable of adjusting the strength of punishment according to the behaviors of partner females (Raihani et al. 2012). Studies in cichlids (*Astatotilapia burtoni*) have revealed that they can exhibit logical thinking such as transitive inference in which known relationships are used to deduce unknown ones (e.g., "if A is larger than B and B is larger than C, then A is larger than C"; Grosenick et al. 2007).

There is a remarkable trend that the species displaying such "higher-order cognitions" possess an enlarged pallium (Fig. 1.6). Thus the evolution of the pallium would be largely responsible for the evolution of cognition. For this reason, this chapter mainly discusses the evolution of the forebrain, with a particular focus on the pallium.

In line with this theory, amphibian pallia are relatively simple with a tube-like structure, and there is no amphibian species known to show such higher cognitive abilities. Since the lungfish and coelacanth (sister groups of tetrapods; see Fig. 1.1a) also have a pallium similar to amphibians, the simple tube-like pallium may be close to the situation of the stem Sarcopterygii.

1.3.2 The Pallium as an Integration Center

Accumulation of hodological (connectivity) and behavioral studies has revealed that the functional organization of the avian pallium is similar to the mammalian cortex, although it is not organized in layers. A detailed description of the mammalian and avian pallia is beyond the focus of this chapter since these structures will be the focus of following chapters of this book. Yet we will briefly describe their common functions to provide some overview on the general functions of the pallium (Fig. 1.7a). It is important to keep in mind that most studies in nonmammals are interpreted by comparing with mammalian data.

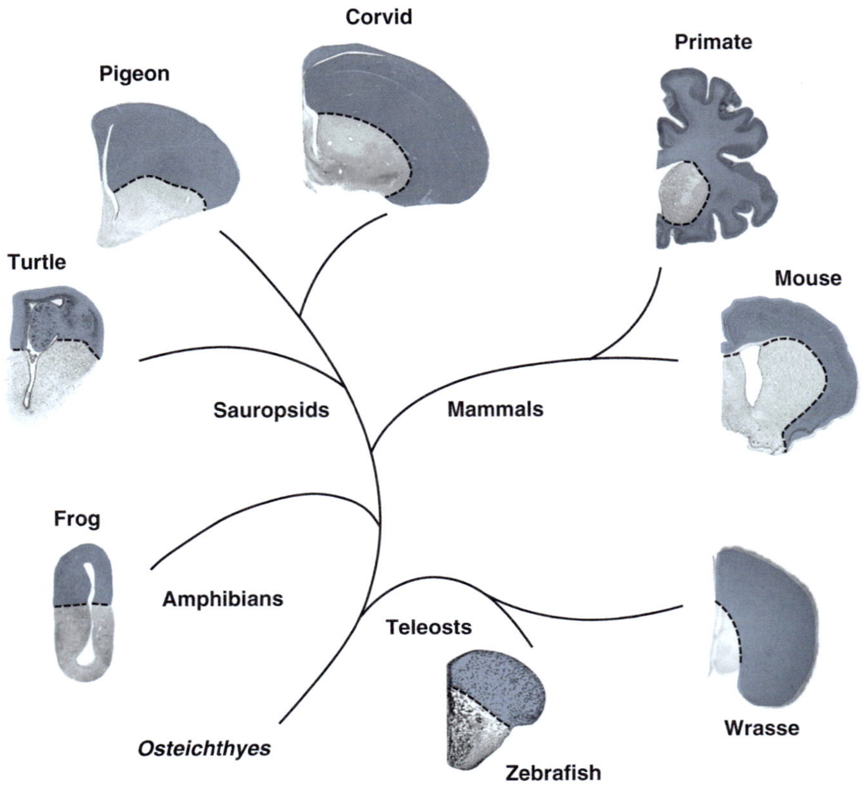

Fig. 1.6 Variation in relative size of the pallium within each animal group. Frontal sections of the telencephalon in some species of mammals, sauropsids, amphibians, and teleosts are shown. The *dotted lines* indicate the boundary between the pallium and the subpallium. The basic organization of the telencephalon is shared in the same animal group, but the relative size of the pallium (*shaded* dorsal part) compared to the subpallium (*non-shaded* ventral part) varies. For example, extreme enlargement of the pallium is found in species of primate, corvid, and wrasse

Both in mammals and birds, modal-specific sensory information is conveyed to the pallium via specific thalamic nuclei (Butler 1994a, b), with the exception of olfactory inputs, in which the piriform cortex receives direct projections from the main olfactory bulb. All these sensory information is processed in the secondary or higher-order areas.

Autonomic/visceral-related brain areas are collectively called the limbic system, and the amygdala and the hippocampus are often considered to be part of it. In mammals, the amygdala is thought to be a key component of the emotional response, based on its involvement in fear conditioning and aggressive behaviors. A well-studied mechanism underlying such amygdala functions is an association between unconditioned and conditioned stimuli. In birds, an amygdala-like pallial area is found (Zeier and Karten 1971; Cohen 1975), but the homology is controversial. The dorsomedial edge of the pallium is identified as the hippocampus both in mammals and birds, based on several hodological and functional similarities: the connections with the hypothalamus and the septum (Krayniak and Siegel 1978a, b; Casini et al. 1986), the

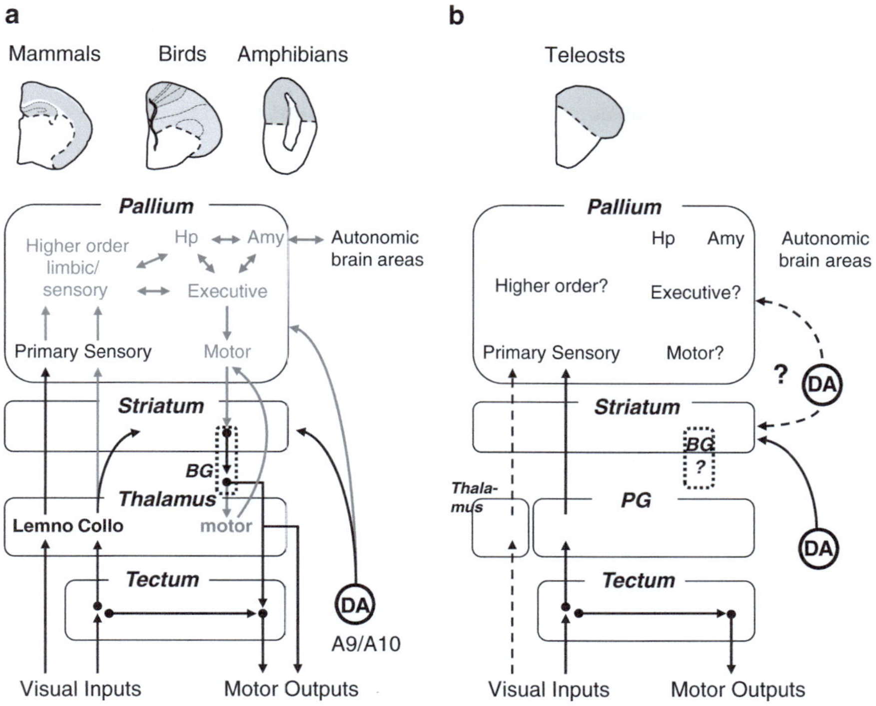

Fig. 1.7 Similar functional connectivity in different groups of vertebrates. (**a**) A simplified diagram showing inputs and outputs related to the pallium. The *black color* indicates the connectivity or brain areas also shared in amphibians, while the *gray color* indicates those not found in amphibians. Here visual pathways are shown as an example of modal-specific sensory afferents to the pallium. Two parallel afferents are found, the collopathway ("collo" represents the colliculus, which is relayed via the midbrain roof) and lemnopathway ("lemno" represents lemniscal inputs, but the term is generally used to refer non-collicular inputs). Dopaminergic (DA) projection from the mesencephalon (A9/A10) to the striatum is critical for the basal ganglia (BG) function in tetrapods. Note that many pallial components found in amniotes are not found in amphibians. (**b**) A diagram showing teleost data comparable with **a**. In teleosts, the presence of the two parallel pathways is not clear in many species, and the majority of sensory projections to the pallium are relayed via the preglomerular nucleus (PG) instead of the thalamus. There is no DA neuron in the mesencephalon, and it is suggested that some forebrain DA cells project to the telencephalon. The BG organization has not been investigated in teleosts

existence of long-term potentiation (LTP; Margrie et al. 1998), and the involvement in spatial learning and memory (Sherry et al. 1989; Bingman and Yates 1992).

In the mammalian brain, the prefrontal cortex is a center for executive functions. Birds also possess an executive area, which is interconnected with and regulates various pallial areas, such as the sensory, limbic, and motor-like areas (Güntürkün 2005a, b). Many animals possess emotions such as aggression, but only some species have the ability to control these emotions. The evolution of connectivity between limbic areas and the executive area is critical for emotional control.

Although the projection patterns of sensory inputs and motor outputs outside of the pallium in amphibians are relatively similar to those found in mammals and birds, the functional organization of pallial components in amphibians are less developed than in amniotes (Fig. 1.7a). The majority of the sensory inputs terminate in the striatum (subpallium), and it is thus possible that the sensory motor integration is mainly performed at the level of the striatum instead of in the pallium.

In contrast to amphibians, the teleost pallium receives prominent sensory inputs of different modalities (visual, auditory, olfactory, lateral line, etc.; Fig. 1.7b; Yamamoto et al. 2007). Although they are often compared to the primary sensory areas in mammals (Ito and Yamamoto 2009), it is unlikely that these similar pathways are inherited from the common ancestor of Osteichthyes. A prominent difference between teleost and tetrapod sensory pathways is that the major sensory relay nuclei are not located in the dorsal thalamus but in another nuclear complex located ventrally named the preglomerular nucleus. Additionally, the importance of the optic tectum for visual processing in teleosts is much higher than in mammals. This is most prominent at early larval stages when telencephalic afferents are not fully developed. For example, a motor choice depending on "prey or predators" is performed using relatively simple tectal circuitry, based on the size of the object covering the retinal field (Del Bene et al. 2010). Such visual-motor processing at the level of the tectum is comparable to the circuitry involved in saccade (unconscious adjustment of eye movement following the detection of motion) in mammals, and is completely distinct from voluntary movements involving the pallium.

In addition to sensory inputs, hippocampal-like and amygdala-like functions have been reported in goldfish (Rodriguez et al. 2002; Portavella et al. 2004), but little is known about how information is processed within the pallium (Fig. 1.7b).

Interestingly, a recent study suggests that zebrafish possess a pallial area involved in context-dependent retrieval of remote memory (Aoki et al. 2013), which is an executive-like function. Some teleost species possess behavior/cognitive repertories as elaborate as some amniotes; thus, it is conceivable that an executive area might be present in teleosts.

1.3.3 *Anatomical Organization of the Pallium and Questions of Homology*

Since the mammalian pallium can be divided into four cytoarchitectonically distinct parts, consisting of the tri-layered hippocampal formation, piriform (olfactory) cortex, the six-layered neocortex (also known as isocortex), and non-layered nuclear

Fig. 1.8 A different organization of the pallium between the Actinopterygii and Sarcopterygii. (**a**) Schematic drawing of the telencephalon in Holmgren (1925), the Sarcopterygii on the *left* and the Actinopterygii on the *right*. It has long been believed that the pallial organization in Actinopterygii is an inverted version of that in Sarcopterygii. Based on this theory, the piriform pallium (pp; indicated in *gray*) originates from the ventral end of the pallium in both Sarcopterygii and Actinopterygii. (**b**) Schematic diagram modified from Dirian et al. (2014) showing the development of the pallium in zebrafish. The lateral and posterior zones of the teleost pallium (Dl and Dp; indicated in *gray*) are formed from the dorsal tip of the pallium, and neurogenesis of this region starts much later than the medial part of the pallium. *Dl* lateral zone of the dorsal telencephalic area, *Dp* posterior zone of the dorsal telencephalic area, *gp* general pallium, *hp.* hippocampal pallium, *pp.* piriform pallium

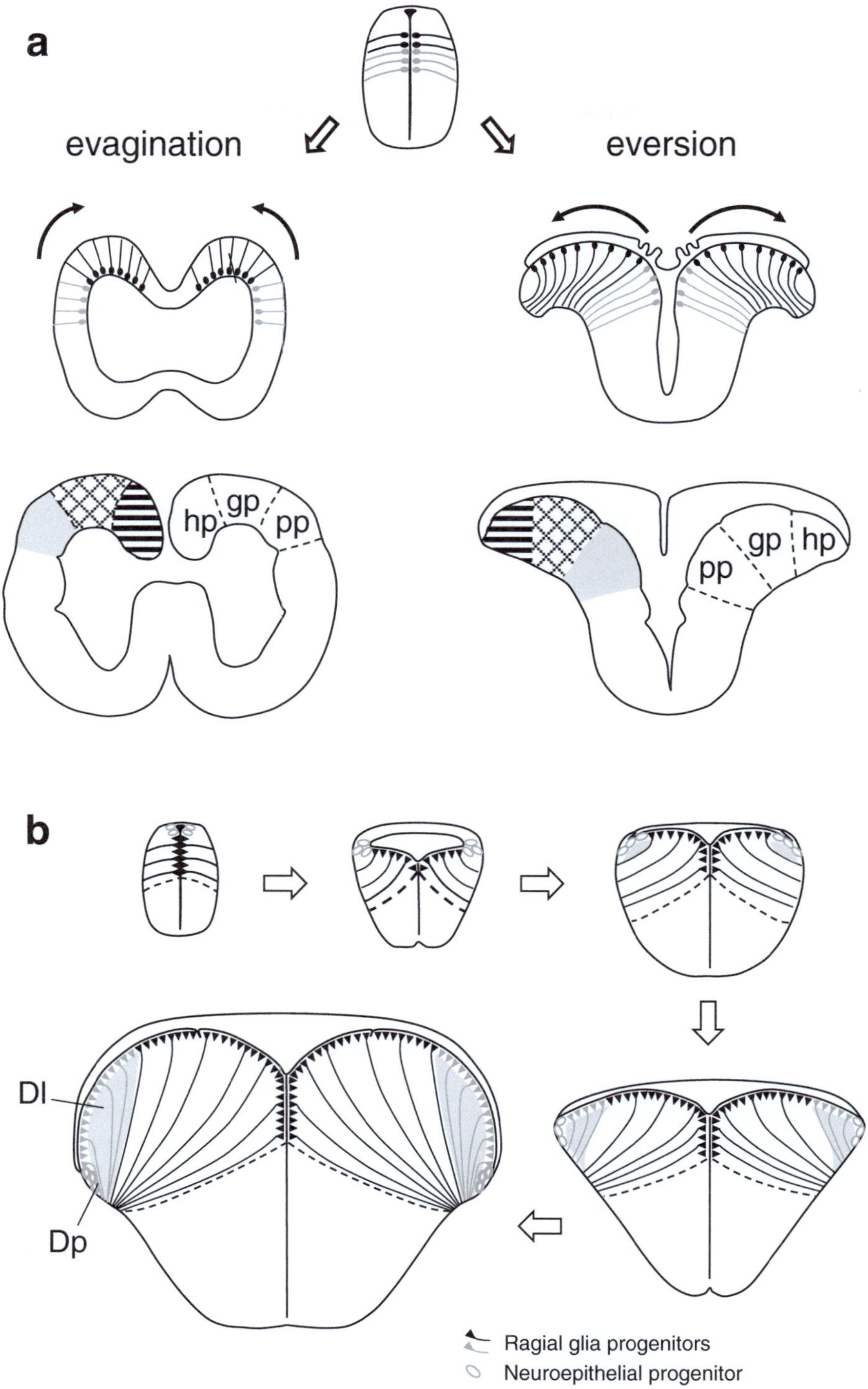
a
evagination
eversion
hp
gp
pp
pp
gp
hp
b
Dl
Dp
Ragial glia progenitors
Neuroepithelial progenitor

complex including the amygdala, comparative neuroanatomists have been examining non-mammalian pallial regions to identify the equivalent of each of these pallial components. Particularly, it has been a long-standing question to identify the neocortex homolog in non-mammalian species as it relates to the evolution of cognition. Due to the morphological diversity of pallia in vertebrates, the homology of the pallial subdivisions is still controversial among comparative neuroanatomists.

1.3.3.1 Three Classical Subdivisions: The Hippocampal, General, and Piriform Pallia

Based on observations of the pallium in various vertebrate species, Holmgren identified three pallial subdivisions: the hippocampal pallium (hp), general pallium (gp), and piriform pallium (pp; Fig. 1.8a; Holmgren 1922, 1925). The piriform pallium contains a superficial cortical structure (the piriform cortex) and a deep nuclear structure (the piriform lobe) containing the claustrum/amygdala complex. In recent literature, the hp, gp, and pp are more commonly called the medial pallium (MP), dorsal pallium (DP), and lateral pallium (LP), reflecting their topology in the pallium of Sarcopterygii. Regardless of which terms are applied, it has been accepted that the presence of the three subdivisions is the "morphotype" (a model comprising the characteristics believed to have been present in common ancestors) of the vertebrate pallium (Northcutt 1995).

Notably, the amphibian pallium has a simple tube-like structure, which is subdivided into the medial, dorsal, and lateral pallia, the same terminology applied to the vertebrate morphotype. In this chapter, the amphibian structures will be abbreviated with lowercase letters "mp, dp, and lp" to distinguish from the vertebrate morphotype subdivisions. In the time of Holmgren (1925), they were considered to have a one-to-one correspondency. However hodological studies have revealed that the olfactory projections extend to dp, and the "cortical-like" projection patterns (inputs from the thalamus and distal output projections) are found more in mp than dp (Fig. 1.9a; amphibian). Thus the proposed hypothesis of homology is quite confusing, with the mp in amphibians being homologous to the MP/DP in mammals, and the dp/lp in amphibians being homologous to the LP in mammals (Bruce and Braford 2009).

In most vertebrates including Sarcopterygii, the telencephalon develops via a process termed evagination. In this process, the central lumen of the neural tube enlarges to form two lateral ventricles (Fig. 1.8a; left). In Actinopterygii, the lateral ventricles are not formed, since the roof of the neural tube elongates outward wrapping over the pallium. This way of morphogenesis has been called "eversion" (Fig. 1.8a; right). In the classical eversion theory, the medio-lateral organization of the pallium in Actinopterygii is inverted from that in Sarcopterygii (compare the position of hp and pp in Fig. 1.8a).

Based on topological location, the nomenclature of the teleost pallial subdivisions is as follows (with the "D" referring to the dorsal telencephalic area): Dl is the lateral zone of the pallium, Dm is the medial zone, Dc is the central zone, Dd is the dorsal zone, and Dp is the posterolateral zone (Figs. 1.8b and 1.9a teleost; Wullimann et al. 1996). A set of studies in several teleost species have demonstrated that the

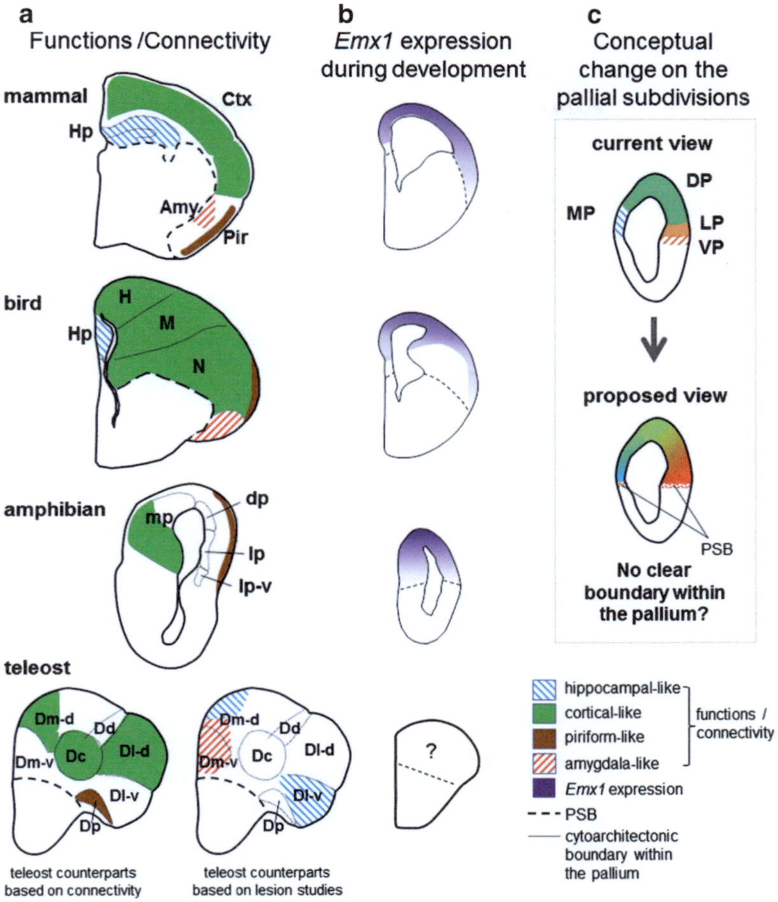

Fig. 1.9 Comparison of functional, hodological, and gene expression data in cytoarchitectonically different pallia. (**a**) Schematic drawing of frontal sections of mature brains showing pallial areas with functional/hodological properties similar to the mammalian hippocampus, neocortex, piriform cortex, and amygdala. The pallial homology has classically been proposed using these sets of data. (**b**) Schematic drawing of frontal sections of developing brains showing the expression of *Emx1* in mouse, chick, and frog. A large part of the ventral DVR in chick is *Emx1* negative, but otherwise, the expression is a dorsal to ventral gradient. The *emx1* gene is known to be expressed in the teleost pallium, but an *emx1*-negative area in the developing pallium has never been demonstrated. (**c**) Proposed modification of the concept of pallial subdivisions. We propose that there is no distinct subdivision within the pallium. The topology is not a critical factor for determining the pallial properties, and any part of the pallium has potential to generate hippocampal-like, cortical-like, piriform-like, and amygdala-like characteristics during evolution. *Amy* amygdala, *Ctx* neocortex, *Dc* central zone of the dorsal telencephalon, *Dd* dorsal zone of the dorsal telencephalon, *Dm-d* medial zone of the dorsal telencephalon (dorsal part), *Dm-v* medial zone of the dorsal telencephalon (ventral part), *Dl-d* lateral zone of the dorsal telencephalon (dorsal part), *Dl-v* lateral zone of the dorsal telencephalon (ventral part), *DP* dorsal pallium, *dp* dorsal pallium (amphibian structure), *Dp* posterior zone of the dorsal telencephalon, *H* hyperpallium, *Hp* hippocampus, *LP* lateral pallium, *lp* lateral pallium (amphibian structure), *lp-v* ventral part of the lateral pallium (amphibian structure), *M* mesopallium, *MP* medial pallium, *mp* medial pallium (amphibian structure), *N* nidopallium, *Pir* piriform cortex, *PSB* pallial-subpallial boundary

sensory-recipient areas are largely distributed throughout the pallium. For example, visual inputs terminate in the Dl, auditory inputs in the Dm, and olfactory inputs in the Dp (Fig. 1.9a; teleost left; Yamamoto et al. 2007).

Other sets of studies in goldfish have indicated that a hippocampal-like function resides in Dl (involved in spatial learning; Rodriguez et al. 2002) and an amygdaloid-like function resides in Dm (involved in aversive learning; Portavella et al. 2004; Fig. 1.9a; teleost right). Although a medio-dorsal part of the pallium is also involved in spatial learning (Saito and Watanabe 2006; Fig. 1.9a; teleost right), a simplified hypothesis that the Dl is hippocampal and the Dm is amygdaloid has been widely accepted, because it fits better to the eversion theory (Mueller et al. 2011).

However, a recent cell lineage study in zebrafish suggests that the developmental process of the teleost pallium does not go through the "eversion" process as previously believed. Cell lineage tracing demonstrated that the lateral part of the zebrafish pallium containing both Dl and Dp originates from the dorsal tip of the embryonic pallium (Fig. 1.8b; gray zones) at a later time point than the medial part (Dirian et al. 2014). If Dp (the primary olfactory area in teleosts) is regionally homologous to the piriform cortex in mammals, it is expected to originate from the ventral part of the pallium close to the subpallium (Fig. 1.8a; gray zones). Overall, this suggests that the eversion theory, and accordingly the hypothesis of pallial homology between Sarcopterygii and Actinopterygii, needs to be reevaluated.

It is also important to point out that Dl and Dm are well developed in certain teleost species and can be further subdivided (Fig. 1.9a; teleost). They both contain sensory-recipient (cortical-like) areas and hippocampal-like/amygdala-like areas, respectively.

To summarize, although the amphibian and teleost pallia both have three to four cytoarchitectonic subdivisions, they do not fit the hypothetical MP/DP/LP morphotype divisions in a simple one-to-one manner.

1.3.3.2 The Fourth Subdivision: The Ventral Pallium

Based on the absence of the expression of the transcription factor *Emx1* within the LP, another subdivision, the ventral pallium (VP), was later added to the three classical ones (Fernandez et al. 1998; Puelles et al. 2000). Notably, *Emx1*-negative or poor expression is a common feature at the pallial-subpallial boundary (PSB) throughout tetrapods (Fig. 1.9b). In the mouse, VP includes a large part of the amygdala nuclei and the ventral part of the piriform cortex.

It is still controversial whether the absence of *Emx1* expression is sufficient to define a subregion throughout vertebrates. Firstly, besides this absence of a single gene, there is no other VP marker commonly expressed in different vertebrate groups. For example, although *Dbx1* is highly expressed in the *Emx1*-negative territory in mammals, it is not the case in the avian telencephalon. Secondly, as it is the case for many transcription factors expressed in the cortex, the *Emx1* expression is gradient, without a clear-cut border. Although the VP was initially defined as an *Emx1*-negative area, a subsequent study found transient expression of *Emx1* in this region (Gorski et al. 2002). Thus the VP needs to be redefined as an "*Emx1*-poor" area. Thirdly, there is no data showing a *Tbx1*-positive/*Emx1*-negative domain in the pallium outside of tetrapods. For instance, in teleosts, the Dm is proposed to be the VP based on its amygdala-like functions, but there is no evidence that the *emx1* expression fits this hypothesis.

The reduced expression of *Emx1* at the pallium-subpallium border (PSB) most likely has a functional significance during development in the tetrapod pallium, but this may not be the case in the teleost pallium. In the mouse, *Emx1* and *Emx2* play critical roles in pallial development, but in the zebrafish, knocking down *emx1* and *emx2* does not have a significant effect. Instead, *emx3* plays much more important roles in the development of the zebrafish pallium (Viktorin et al. 2009). The *Emx3* gene has been lost in many tetrapod species except a few species such as Xenopus and opossum. This indicates that the role of distinct *Emx* genes may be different in tetrapods and teleosts, and thus *Emx1* expression may not be the best criteria to define the morphotype of the pallium.

1.3.3.3 Do the Four Subdivisions Really Exist?

The presence of the VP is a controversial issue among avian researchers (Butler et al. 2011; Dugas-Ford and Ragsdale 2015). This is because the majority of the avian nidopallium is functionally similar to the mammalian neocortex (DP like), but it is negative for *Emx1* (VP like). One could claim that the functional similarity or the connectivity should not be used as criteria for homology because it can change. However, it should be noted that the definition of the classical three subdivisions, the hippocampal pallium, the general pallium, and the piriform pallium, reflects their functional properties. Other than the mammalian cortex (in which hippocampus and piriform cortices are three layered while neocortex is six layered), hodological and functional data have been the criteria to identify the MP/DP/LP. Thus if the use of functional or hodological properties to identify homology is questioned, we would need to doubt the presence of the MP/DP/LP subdivisions.

It may be worth reconsidering whether the MP (hp), DP (gp), LP (pp), and VP are true subdivisions present throughout vertebrates (Fig. 1.9c). It is possible that any pallial region can potentially evolve functional properties similar to the mammalian neocortex.

1.4 Evolvability of the Nervous Systems: Lessons from the Dopamine Systems

To understand brain evolution correctly by using genetic data in vertebrates, it must be put in the context of gene duplication and gene loss. This is especially important when we compare the data between mammals and teleosts, as exemplified above by the different functional importance of *Emx* genes.

1.4.1 Gene Duplication and Gene Loss

Gene duplication is one of the important sources of novelty during evolution. The presence of redundant gene copies facilitates the occurrence of a mutation in one of the duplicated genes. Often, the mutation is disadvantageous and the mutated gene is lost over time. When the mutation is advantageous, it is kept resulting in a new function different from the ancestral one (neofunctionalization). It is also common for a modification to occur in regulatory elements controlling the expression of the gene, and the two duplicated genes (paralogs) maintain the same functions but are expressed in a differential manner (e.g., different timing or different part of the body), a phenomenon termed subfunctionalization. Note that genes are referred as "paralogous" when homologous sequences are generated by a gene duplication event. By contrast, genes are "orthologous" when homologous sequences present in different species are originating from a gene already present in their common ancestor (Fig. 1.10).

Many genes are present as single copies in genomes of amphioxus and ascidians, whereas mammalian genomes generally contain three or more orthologs. Such observations led to the two-round (2R) hypothesis which speculates that gnathostomes (jawed vertebrates) experienced two rounds of whole genome duplication (WGD; Lundin 1993; Holland et al. 1994; Hughes 1999). The timing of the duplications is not yet conclusive, but a recent study analyzing the lamprey genome indicates that two WGDs likely occurred before the divergence of the gnathostome and cyclostome lineages (Fig. 1.1; Smith et al. 2013). The teleost lineage went through an additional round of WGD (Jaillon et al. 2004). As mentioned above, many paralogous genes tend to be lost secondarily, but some species such as zebrafish have kept many of the paralogs.

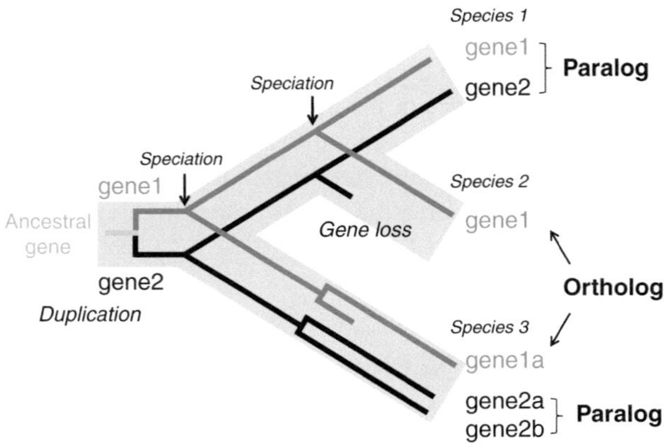

Fig. 1.10 Evolutionary history of genes. When an ancestral gene is duplicated giving rise to gene 1 and gene 2, these duplicated genes are paralogs, while the gene 1 copies found in different species are orthologs. Duplicated genes can be lost in certain lineages or can be further duplicated. It is known that the teleost lineage went through an additional whole genome duplication and thus tends to possess additional copies compared to other vertebrate groups

A series of studies suggests that the evolution of dopamine (DA) systems has been heavily influenced by gene duplications and gene losses, giving rise to a large diversity among different groups of vertebrates (Yamamoto et al. 2010, 2011, 2013, 2015).

1.4.2 A Large Diversity of Dopamine Systems

DA is one of the most intensively studied modulatory neurotransmitters due to its implication in many brain functions, such as neuroendocrine regulation, motor programming, motivational behaviors, learning, and executive functions. The majority of the knowledge on DA systems comes from mammalian data. Since the basic organization is often considered to be relatively conserved, data obtained from non-mammalian species have been interpreted based on a one-to-one comparison with mammalian data.

However, detailed observations indicate that DA plays a critical role in behavioral/physiological repertories which are similar between different animal groups but not necessarily inherited from the common ancestors. One example is the role of DA in the executive functions (Güntürkün 2005a, b). Executive functions similar to the ones managed by the mammalian prefrontal cortex are performed in an area of the avian pallium called the nidopallium caudolateral (NCL). These two structures are likely to have evolved independently.

DA control of reproduction is another example of convergent evolution. Inhibitory effects of DA on the anterior pituitary, counteracting the stimulatory effects of GnRH, are found in mammals, birds, amphibians, and teleosts (Dufour et al. 2005), but only in some species within each of these groups. Comparisons between sheep (a well-studied mammalian species) and zebrafish suggest that the DA cell populations involved in this function are not likely to be homologous (Fontaine et al. 2015). In addition, the mode of DA transmission is different: In amniotes, DA is released in the median eminence and transmitted via the portal blood system to the pituitary, while in teleosts, DA neurons directly innervate the pituitary (Ball 1981).

Indeed, phylogenetic analyses have demonstrated that the vertebrate DA system is more diversified than previously thought, due to gene duplications and gene losses (Fig. 1.11a). The mammalian DA system (especially in placental mammals) is an exceptional case, as mammals have lost many genes involved in DA neurotransmission which are commonly present in other jawed vertebrate species. For example, placental mammals have lost one of the two paralogs encoding tyrosine hydroxylase (*TH1* and *TH2*), the rate-limiting enzyme for DA synthesis. The loss of the *TH2* may have accompanied the lack of CSF-contacting neurons expressing *TH2*, which occupy a large portion of the hypothalamus in some animal groups such as teleosts (Yamamoto et al. 2010, 2011; Yamamoto and Vernier 2011).

In addition to having lost genes involved in DA transmission, mammals have also lost many genes encoding DA receptors. Classically five DA receptors, D1 to D5, were identified based on mammalian studies. D1 and D5 are classified as the D1 family, which is coupled to the Gs family of G proteins and activates adenylate cyclase.

D2, D3, and D4 are classified as the D2 family which is coupled to the Gi/o family of G proteins and does not activate adenylate cyclase. However, more DA receptor genes have been found in non-mammals, and recent data strongly suggest that the ancestral Osteichthyes possessed four D1-family dopamine receptors (D1, D5, D6, and D7) and five D2-family dopamine receptors (D2, D3, D4, D8, and D9; Demchyshyn et al. 1995; Yamamoto et al. 2013, 2015).

For an in-depth understanding of DA contribution to brain functions, these differences should be taken into account. Notably, the mammalian and teleostean systems represent two extremes in Osteichthyes. In contrast to the numerous gene losses in mammals, teleosts have gone through an additional genome duplication. Interestingly, teleosts lack mesencephalic DA cells which modulate various important telencephalic functions in amniotes (Fig. 1.11a), further demonstrating that the comparison of data from these two groups requires a careful approach. Comparative analyses including other animal groups would be helpful.

1.4.3 To What Extent Is the Basal Ganglia Circuitry Conserved?

Due to its implication in Parkinson's disease, the role of DA neurotransmission is especially well studied in relation to the organization of the basal ganglia (BG). A simplified schema outlining the BG circuitry is shown in Fig. 1.11b. Pharmacological studies in mammals have demonstrated that DA agonists induce hyperkinesia and DA antagonists cause hypokinesia. In mammals, substance P (SP)-containing GABAergic medial spiny neurons abundantly express mRNA encoding the D1 dopamine receptor, whereas enkephalin (ENK)-containing

Fig. 1.11 A large diversity of DA systems. (**a**) Some of the comparable DA cell populations are plotted in the sagittal brain sections (rostral to the *left*) of chicken (bird), mouse (mammal), Xenopus (amphibian), and zebrafish (teleost). The presence of D1 and D2 dopamine receptor genes in the genome of each animal is also shown. The subregions of the secondary prosencephalon are color coded (as shown in Fig. 1.3c) in order to visualize the homologous brain areas in mature brains. Zebrafish have abundant *TH2*-expressing CSF-contacting neurons (red diamonds) around the two hypothalamic ventricles, while mammals have lost the CSF-contacting neurons in the hypothalamic region. Mammals have lost almost half of the ancestral DA receptor genes, while zebrafish have additional copies due to the teleost-specific genome duplication. Teleosts lack mesencephalic DA cell populations which are a major source of DA to the telencephalon in tetrapods (*green dots* and *arrows*). In contrast, they have abundant DA cells in the telencephalon, which seem to have intra-telencephalic projections. (**b**) A simplified diagram showing the direct and indirect basal ganglia (BG) circuitry which is suggested to be conserved in vertebrates. *ENK* enkephalin, *GPe* external globus pallidus, *GPi* internal globus pallidus, *M* mesencephalon, *P* prosencephalon, *SNr* substantia nigra pars reticulata, *SP* substance P, *STN* subthalamic nucleus

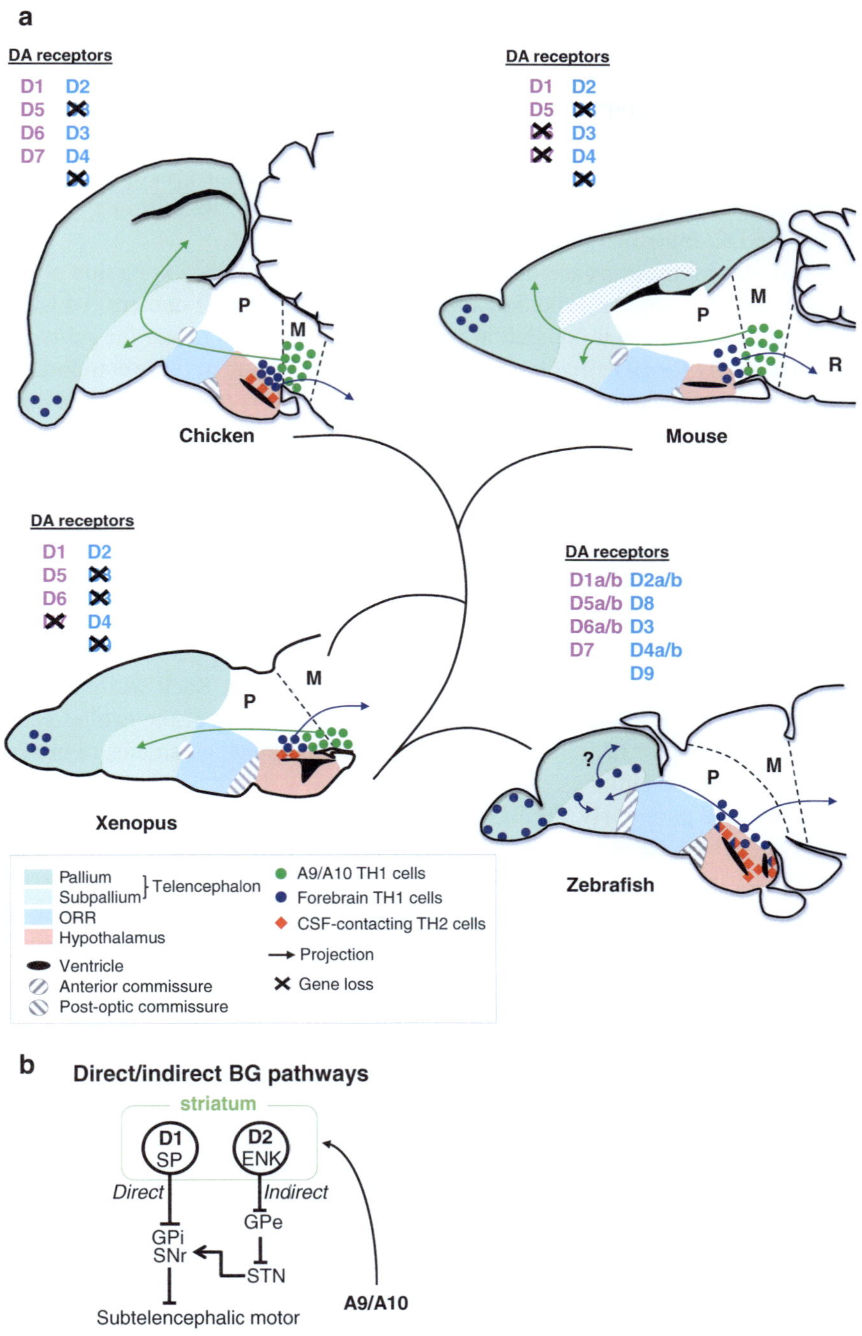
a
DA receptors
D1 D2
D5
D6 D3
D7 D4
P
M
Chicken
DA receptors
D1 D2
D5
D3
D4
P
M
R
Mouse
DA receptors
D1 D2
D5
D6
D4
P
M
Xenopus
DA receptors
D1a/b D2a/b
D5a/b D8
D6a/b D3
D7 D4a/b
D9
?
P
M
Zebrafish
Pallium
Subpallium
Telencephalon
ORR
Hypothalamus
Ventricle
Anterior commissure
Post-optic commissure
A9/A10 TH1 cells
Forebrain TH1 cells
CSF-contacting TH2 cells
Projection
Gene loss
b
Direct/indirect BG pathways
striatum
D1
SP
D2
ENK
Direct
Indirect
GPe
GPi
SNr
STN
Subtelencephalic motor
A9/A10

GABAergic medial spiny neurons abundantly express mRNA encoding the D2 dopamine receptor. They project to two different pallidal populations, constituting the direct and indirect BG pathways. It has been hypothesized that voluntary movements are regulated by the balanced activity of the direct and indirect BG pathways (Albin et al. 1989; DeLong 1990): the direct pathway promotes desired movements by activating the SP neurons via the D1 dopamine receptor, whereas the indirect pathway suppresses unwanted movements by inhibiting the ENK cells via the D2 dopamine receptor.

For this reason, the presence of DA inputs to the striatum, the expression of SP, ENK, D1, and D2 receptors, and the presence of the subthalamic nucleus (which is the only glutamatergic cell population within the BG circuitry) were the principle criteria to identify the BG circuitry. Based on the identification of these components in the bird, amphibian, and lamprey, the BG organization is considered to be conserved throughout vertebrates (Reiner et al. 1998; Maier et al. 2010; Stephenson-Jones et al. 2011).

However, considering the diversity of DA systems, it would be worth verifying the extent to which the BG circuitry is conserved. Indeed, the exclusive dichotomy of D1 and D2 receptor expression in SP and ENK cells is an oversimplification (Surmeier et al. 1996). A significant percentage of striatal neurons co-express both D1 and D2 receptors, and other DA receptors such as D3 or D4 are also expressed in SP cells. Thus, in non-mammals, it is possible that DA receptor subtypes which are not present in mammals are involved in the BG function. Such verification is important not only in terms of the fundamental understanding of the evolutionary perspective but also for taking full advantage of non-mammalian studies to understand pathological conditions.

As mentioned previously, the BG organization is found in the lamprey (cyclostome). Since the separation of the cyclostomes and gnathostomes occurred at an early stage of vertebrate evolution, it is considered that the mechanisms underlying the similarity are well conserved (Stephenson-Jones et al. 2011). However, to confirm the conservation throughout the vertebrates, studies in Actinopterygii and Chondrichthyes are needed. Especially in teleosts, there is no mesencephalic DA cell population (A9/A10 in mammals), which are a major source of DA to the telencephalon in amniotes. Since some diencephalic DA cells project to the striatum in zebrafish, these diencephalic DA cells have been proposed to be A9/A10-like (Rink and Wullimann 2001). However, later studies suggested that they are more similar to A11, projecting posteriorly to the spinal cord (Ryu et al. 2007; Fig. 1.11a). Based on behavioral observations, it would be surprising if teleosts do not possess cells equivalent to the tetrapod A9/A10. As is the case for the inhibitory effect of DA on reproduction, it is possible that non-homologous DA cell populations play a role equivalent to the tetrapod A9/A10. One candidate is the telencephalic DA cell population, as these DA cells project locally to the telencephalon (Fig. 1.11a, zebrafish; Bloch and Yamamoto, unpublished observation). More studies are needed to understand how similar behavioral outputs are controlled by highly diversified brains.

1.5 Conclusion

According to a view inherited from the *scala naturae*, brain complexity increases linearly with the phylogenetical "modernity" of the vertebrate species. Later studies have suggested that non-mammalian vertebrates have some functional and hodological similarities comparable to mammals. These findings have led to the idea that the basic brain organization is relatively conserved.

However, more detailed analyses revealed many inconsistencies when we consider that these similarities are inherited from the common ancestor. It is possible that the nervous system is highly plastic during evolution and more convergent evolution has taken place than is currently thought. The similar functional organizations may have been built using genetic regulatory networks inherited from a common ancestor, but they might have evolved independently in different lineages.

Acknowledgments We would like to thank Drs. Ann Butler, Naoyuki Yamamoto, Laura Bruce, Cliff Ragsdale, and Onur Güntürkün for the discussion which helped a lot for writing this chapter. We also thank laboratories of Drs. Laure Bally-Cuif, Sylvie Rétaux, and Koichi Kawakami, for sharing their unpublished data. Finally we thank Drs. Michaël Demarque, Catherine Pasqualini, Philippe Vernier, Alessandro Alunni, Maryline Blin, and Shauna Katz for the critical reading.

References

Affaticati P, Yamamoto K, Rizzi B, Bureau C, Peyriéras N, Pasqualini C, Demarque M, Vernier P (2015) Identification of the optic recess region as a morphogenetic entity in the zebrafish forebrain. Sci Rep 5:8738

Albin RL, Young AB, Penney JB (1989) The functional anatomy of basal ganglia disorders. Trends Neurosci 12:366–375

Aoki T, Kinoshita M, Aoki R, Agetsuma M, Aizawa H, Yamazaki M, Takahoko M, Amo R, Arata A, Higashijima S, Tsuboi T, Okamoto H (2013) Imaging of neural ensemble for the retrieval of a learned behavioral program. Neuron 78:881–894

Ball JN (1981) Hypothalamic control of the pars distalis in fishes, amphibians, and reptiles. Gen Comp Endocrinol 44:135–170

Bally-Cuif L, Vernier P (2010) Organization and physiology of the zebrafish nervous system. In: Perry SF, Ekker M, Farrell AP, Brauner CJ (eds) Zebrafish. Elsevier, Amsterdam, pp 25–80

Del Bene F, Wyart C, Robles E, Tran A, Looger L, Scott EK, Isacoff EY, Baier H (2010) Filtering of visual information in the tectum by an identified neural circuit. Science 330:669–673

Bingman VP, Yates G (1992) Hippocampal lesions impair navigational learning in experienced homing pigeons. Behav Neurosci 106:229–232

Biran J, Tahor M, Wircer E, Levkowitz G (2015) Role of developmental factors in hypothalamic function. Front Neuroanat 9:47

Bruce LL, Braford MR (2009) Evolution of the limbic system. In: Squire LR (ed) Encyclopedia of neuroscience, vol 4. Academic, Oxford, pp 43–55

Butler AB, Reiner A, Karten HJ (2011) Evolution of the amniote pallium and the origins of mammalian neocortex. Ann N Y Acad Sci 1225:14–27

Butler AB (1994a) The evolution of the dorsal thalamus of jawed vertebrates, including mammals: cladistic analysis and a new hypothesis. Brain Res Brain Res Rev 19:29–65

Butler AB (1994b) The evolution of the dorsal pallium in the telencephalon of amniotes: cladistic analysis and a new hypothesis. Brain Res Brain Res Rev 19:66–101
Casini G, Bingman VP, Bagnoli P (1986) Connections of the pigeon dorsomedial forebrain studied with WGA-HRP and 3H-proline. J Comp Neurol 245:454–470
Cavodeassi F, Houart C (2012) Brain regionalization: of signaling centers and boundaries. Dev Neurobiol 72:218–233
Cohen DH (1975) Involvement of the avian amygdalar homologue (archistriatum posterior and mediale) in defensively conditioned heart rate change. J Comp Neurol 160:13–35
DeLong MR (1990) Primate models of movement disorders of basal ganglia origin. Trends Neurosci 13:281–285
Demchyshyn LL, Sugamori KS, Lee FJ, Hamadanizadeh SA, Niznik HB (1995) The dopamine D1D receptor. Cloning and characterization of three pharmacologically distinct D1-like receptors from Gallus domesticus. J Biol Chem 270:4005–4012
Dirian L, Galant S, Coolen M, Chen W, Bedu S, Houart C, Bally-Cuif L, Foucher I (2014) Spatial regionalization and heterochrony in the formation of adult pallial neural stem cells. Dev Cell 30:123–136
Domínguez L, Morona R, González A, Moreno N (2013) Characterization of the hypothalamus of Xenopus laevis during development. I. The alar regions. J Comp Neurol 521:725–759
Dufour S, Weltzien F-A, Sebert M-E, Le Belle N, Vidal B, Vernier P, Pasqualini C (2005) Dopaminergic inhibition of reproduction in teleost fishes: ecophysiological and evolutionary implications. Ann N Y Acad Sci 1040:9–21
Dugas-Ford J, Ragsdale CW (2015) Levels of homology and the problem of neocortex. Annu Rev Neurosci 38:351–368
Echevarría D, Vieira C, Gimeno L, Martínez S (2003) Neuroepithelial secondary organizers and cell fate specification in the developing brain. Brain Res Brain Res Rev 43:179–191
Fernandez AS, Pieau C, Reperant J, Boncinelli E, Wassef M (1998) Expression of the Emx-1 and Dlx-1 homeobox genes define three molecularly distinct domains in the telencephalon of mouse, chick, turtle and frog embryos: implications for the evolution of telencephalic subdivisions in amniotes. Development 125:2099–2111
Finger TE (2008) Sorting food from stones: the vagal taste system in Goldfish, Carassius auratus. J Comp Physiol A Neuroethol Sens Neural Behav Physiol 194:135–143
Fontaine R, Affaticati P, Bureau C, Colin I, Demarque M, Dufour S, Vernier P, Yamamoto K, Pasqualini C (2015) Dopaminergic neurons controlling anterior pituitary functions: anatomy and ontogenesis in zebrafish. Endocrinology 156:2934–2948
Gorski JA, Talley T, Qiu M, Puelles L, Rubenstein JL, Jones KR (2002) Cortical excitatory neurons and glia, but not GABAergic neurons, are produced in the Emx1-expressing lineage. J Neurosci 22:6309–6314
Grosenick L, Clement TS, Fernald RD (2007) Fish can infer social rank by observation alone. Nature 445:429–432
Güntürkün O (2005a) The avian “prefrontal cortex” and cognition. Curr Opin Neurobiol 15:686–693
Güntürkün O (2005b) Avian and mammalian “prefrontal cortices”: limited degrees of freedom in the evolution of the neural mechanisms of goal-state maintenance. Brain Res Bull 66:311–316
Harrington MJ, Hong E, Brewster R (2009) Comparative analysis of neurulation: first impressions do not count. Mol Reprod Dev 76:954–965
Herget U, Wolf A, Wullimann MF, Ryu S (2014) Molecular neuroanatomy and chemoarchitecture of the neurosecretory preoptic-hypothalamic area in zebrafish larvae. J Comp Neurol 522(7):1542–1564
Holland PW, Garcia-Fernàndez J, Williams NA, Sidow A (1994) Gene duplications and the origins of vertebrate development. Development 1994(Suppl):125–133
Holmgren N (1922) Points of view concerning forebrain morphology in lower vertebrates. J Comp Neurol 34:391–459
Holmgren N (1925) Points of view concerning forebrain morphology in higher vertebrates. Acta Zool 6:413–477
Hughes AL (1999) Phylogenies of developmentally important proteins do not support the hypothesis of two rounds of genome duplication early in vertebrate history. J Mol Evol 48:565–576

Ishikawa Y, Yamamoto N, Yoshimoto M, Ito H (2012) The primary brain vesicles revisited: are the three primary vesicles (forebrain/midbrain/hindbrain) universal in vertebrates? Brain Behav Evol 79:75–83

Ito H, Yamamoto N (2009) Non-laminar cerebral cortex in teleost fishes? Biol Lett 5:117–121

Ivanovitch K, Cavodeassi F, Wilson SW (2013) Precocious acquisition of neuroepithelial character in the eye field underlies the onset of eye morphogenesis. Dev Cell 27:293–305

Jaillon O, Aury J-MM, Brunet F, Petit J-LL, Stange-Thomann N, Mauceli E, Bouneau L, Fischer C, Ozouf-Costaz C, Bernot A, Nicaud S, Jaffe D, Fisher S, Lutfalla G, Dossat C, Segurens B, Dasilva C, Salanoubat M, Levy M, Boudet N, Castellano S, Anthouard V, Jubin C, Castelli V, Katinka M, Vacherie B, Biémont C, Skalli Z, Cattolico L, Poulain J, De Berardinis V, Cruaud C, Duprat S, Brottier P, Coutanceau J-PP, Gouzy J, Parra G, Lardier G, Chapple C, McKernan KJ, McEwan P, Bosak S, Kellis M, Volff J-NN, Guigó R, Zody MC, Mesirov J, Lindblad-Toh K, Birren B, Nusbaum C, Kahn D, Robinson-Rechavi M, Laudet V, Schachter V, Quétier F, Saurin W, Scarpelli C, Wincker P, Lander ES, Weissenbach J, Roest Crollius H (2004) Genome duplication in the teleost fish Tetraodon nigroviridis reveals the early vertebrate proto-karyotype. Nature 431:946–957

Keynes R, Lumsden A (1990) Segmentation and the origin of regional diversity in the vertebrate central nervous system. Neuron 4:1–9

Kiecker C, Lumsden A (2005) Compartments and their boundaries in vertebrate brain development. Nat Rev Neurosci 6:553–564

Kirsch JA, Güntürkün O, Rose J (2008) Insight without cortex: lessons from the avian brain. Conscious Cogn 17:475–483

Krayniak PF, Siegel A (1978a) Efferent connections of the septal area in the pigeon. Brain Behav Evol 15:389–404

Krayniak PF, Siegel A (1978b) Efferent connections of the hippocampus and adjacent regions in the pigeon. Brain Behav Evol 15:372–388

Lundin LG (1993) Evolution of the vertebrate genome as reflected in paralogous chromosomal regions in man and the house mouse. Genomics 16:1–19

Maier S, Walkowiak W, Luksch H, Endepols H (2010) An indirect basal ganglia pathway in anuran amphibians? J Chem Neuroanat 40:21–35

Marcus RC, Shimamura K, Sretavan D, Lai E, Rubenstein JL, Mason CA (1999) Domains of regulatory gene expression and the developing optic chiasm: correspondence with retinal axon paths and candidate signaling cells. J Comp Neurol 403:346–358

Margrie TW, Rostas JA, Sah P (1998) Long-term potentiation of synaptic transmission in the avian hippocampus. J Neurosci 18:1207–1216

Mueller T, Dong Z, Berberoglu MA, Guo S (2011) The dorsal pallium in zebrafish, Danio rerio (Cyprinidae, Teleostei). Brain Res 1381:95–105

Northcutt RG (1995) The forebrain of gnathostomes: in search of a morphotype. Brain Behav Evol 46:275–318

Pasko L (2010) Tool-like behavior in the sixbar wrasse, Thalassoma hardwicke (Bennett, 1830). Zoo Biol 29:767–773

Picker A, Cavodeassi F, Machate A, Bernauer S, Hans S, Abe G, Kawakami K, Wilson SW, Brand M (2009) Dynamic coupling of pattern formation and morphogenesis in the developing vertebrate retina. PLoS Biol 7:e1000214

Portavella M, Torres B, Salas C (2004) Avoidance response in goldfish: emotional and temporal involvement of medial and lateral telencephalic pallium. J Neurosci 24:2335–2342

Puelles L, Kuwana E, Puelles E, Bulfone A, Shimamura K, Keleher J, Smiga S, Rubenstein JL (2000) Pallial and subpallial derivatives in the embryonic chick and mouse telencephalon, traced by the expression of the genes Dlx-2, Emx-1, Nkx-2.1, Pax-6, and Tbr-1. J Comp Neurol 424:409–438

Puelles L, Rubenstein JL (2003) Forebrain gene expression domains and the evolving prosomeric model. Trends Neurosci 26:469–476

Raihani NJ, Pinto AI, Grutter AS, Wismer S, Bshary R (2012) Male cleaner wrasses adjust punishment of female partners according to the stakes. Proc Biol Sci 279:365–370

Reiner A, Medina L, Veenman CL (1998) Structural and functional evolution of the basal ganglia in vertebrates. Brain Res Brain Res Rev 28:235–285

Rink E, Wullimann MF (2001) The teleostean (zebrafish) dopaminergic system ascending to the subpallium (striatum) is located in the basal diencephalon (posterior tuberculum). Brain Res 889:316–330
Rodriguez F, Lopez JC, Vargas JP, Gomez Y, Broglio C, Salas C (2002) Conservation of spatial memory function in the pallial forebrain of reptiles and ray-finned fishes. J Neurosci 22:2894–2903
Roy A, de Melo J, Chaturvedi D, Thein T, Cabrera-Socorro A, Houart C, Meyer G, Blackshaw S, Tole S (2013) LHX2 is necessary for the maintenance of optic identity and for the progression of optic morphogenesis. J Neurosci 33:6877–6884
Ruiz i Altaba A, Palma V, Dahmane N (2002) Hedgehog-Gli signalling and the growth of the brain. Nat Rev Neurosci 3:24–33
Ryu S, Mahler J, Acampora D, Holzschuh J, Erhardt S, Omodei D, Simeone A, Driever W (2007) Orthopedia homeodomain protein is essential for diencephalic dopaminergic neuron development. Curr Biol 17:873–880
Saito K, Watanabe S (2006) Deficits in acquisition of spatial learning after dorsomedial telencephalon lesions in goldfish. Behav Brain Res 172:187–194
Sherry DF, Vaccarino AL, Buckenham K, Herz RS (1989) The hippocampal complex of food-storing birds. Brain Behav Evol 34:308–317
Shimamura K, Hartigan DJ, Martinez S, Puelles L, Rubenstein JL (1995) Longitudinal organization of the anterior neural plate and neural tube. Development 121:3923–3933
Smeets WJAJ, Reiner A (1994) Phylogeny and development of catecholamine systems in the CNS of vertebrates. Cambridge University Press, Cambridge
Smith JJ, Kuraku S, Holt C, Sauka-Spengler T, Jiang N, Campbell MS, Yandell MD, Manousaki T, Meyer A, Bloom OE, Morgan JR, Buxbaum JD, Sachidanandam R, Sims C, Garruss AS, Cook M, Krumlauf R, Wiedemann LM, Sower SA, Decatur WA, Hall JA, Amemiya CT, Saha NR, Buckley KM, Rast JP, Das S, Hirano M, McCurley N, Guo P, Rohner N, Tabin CJ, Piccinelli P, Elgar G, Ruffier M, Aken BL, Searle SM, Muffato M, Pignatelli M, Herrero J, Jones M, Brown CT, Chung-Davidson Y-WW, Nanlohy KG, Libants SV, Yeh C-YY, McCauley DW, Langeland JA, Pancer Z, Fritzsch B, de Jong PJ, Zhu B, Fulton LL, Theising B, Flicek P, Bronner ME, Warren WC, Clifton SW, Wilson RK, Li W (2013) Sequencing of the sea lamprey (Petromyzon marinus) genome provides insights into vertebrate evolution. Nat Genet 45(415–21):421e1–421e2
Stephenson-Jones M, Samuelsson E, Ericsson J, Robertson B, Grillner S (2011) Evolutionary conservation of the basal ganglia as a common vertebrate mechanism for action selection. Curr Biol 21:1081–1091
Surmeier DJ, Song WJ, Yan Z (1996) Coordinated expression of dopamine receptors in neostriatal medium spiny neurons. J Neurosci 16:6579–6591
Suárez R, Gobius I, Richards LJ (2014) Evolution and development of interhemispheric connections in the vertebrate forebrain. Front Hum Neurosci 8:497
Vieira C, Pombero A, García-Lopez R, Gimeno L, Echevarria D, Martínez S (2010) Molecular mechanisms controlling brain development: an overview of neuroepithelial secondary organizers. Int J Dev Biol 54:7–20
Viktorin G, Chiuchitu C, Rissler M, Varga ZMM, Westerfield M (2009) Emx3 is required for the differentiation of dorsal telencephalic neurons. Dev Dyn 238:1984–1998
Vígh B, Manzano e Silva MJ, Frank CL, Vincze C, Czirok SJ, Szabó A, Lukáts A, Szél A (2004) The system of cerebrospinal fluid-contacting neurons. Its supposed role in the nonsynaptic signal transmission of the brain. Histol Histopathol 19:607–628
Wilson SW, Houart C (2004) Early steps in the development of the forebrain. Dev Cell 6:167–181
Wullimann MF, Rupp B, Reichert H (1996) Neuroanatomy of the zebrafish brain: a topological atlas. Birkhäuser Verlag, Basel
Yamamoto K, Ruuskanen JO, Wullimann MF, Vernier P (2010) Two tyrosine hydroxylase genes in vertebrates new dopaminergic territories revealed in the zebrafish brain. Mol Cell Neurosci 43:394–402

Yamamoto K, Vernier P (2011) The evolution of dopamine systems in chordates. Front Neuroanat 5:21

Yamamoto K, Fontaine R, Pasqualini C, Vernier P (2015) Classification of dopamine receptor genes in vertebrates: nine subtypes in osteichthyes. Brain Behav Evol 86:164–175

Yamamoto K, Mirabeau O, Bureau C, Blin M, Michon-Coudouel S, Demarque M, Vernier P (2013) Evolution of dopamine receptor genes of the D1 class in vertebrates. Mol Biol Evol 30:833–843

Yamamoto K, Ruuskanen JO, Wullimann MF, Vernier P (2011) Differential expression of dopaminergic cell markers in the adult zebrafish forebrain. J Comp Neurol 519:576–598

Yamamoto N, Ishikawa Y, Yoshimoto M, Xue H-GG, Bahaxar N, Sawai N, Yang C-YY, Ozawa H, Ito H (2007) A new interpretation on the homology of the teleostean telencephalon based on hodology and a new eversion model. Brain Behav Evol 69:96–104

Yamamoto N, Ito H (2005) Fiber connections of the anterior preglomerular nucleus in cyprinids with notes on telencephalic connections of the preglomerular complex. J Comp Neurol 491:212–233

Zeier H, Karten HJ (1971) The archistriatum of the pigeon: organization of afferent and efferent connections. Brain Res 31:313–326

Chapter 2
The Origins of the Bird Brain: Multiple Pulses of Cerebral Expansion in Evolution

Toru Shimizu, Kazutaka Shinozuka, Ahmet K. Uysal, and S. Leilani Kellogg

Abstract Birds demonstrate extraordinary cognitive and emotional capabilities. The majority of these performances are most likely supported by their developed cerebrum. Birds, as well as mammals, have a much larger cerebrum compared to reptiles, given a similar body size. Since the common ancestral reptiles of birds and mammals had a relatively small brain according to paleobiological evidence, birds and mammals must have evolved to expand their brains independently after they diverged into different lineages. In the lineage leading to modern birds, brain expansion occurred multiple times, possibly in response to different selective pressures. This chapter includes focused discussions on three major pulses regarding brain evolution of the bird lineage. In each discussion, possible important selection factors to trigger the brain expansion are proposed. First, a discussion is on the emergence of amniotes (the common ancestor of reptiles, birds, and mammals) in the Paleozoic Era. Adaptation to terrestrial habitats and increased parental investment might play essential roles in brain expansion. Second, a discussion focuses on how theropod dinosaurs in the bird stem lineage evolved their brains in the Mesozoic Era. In the bird stem lineage, predatory behavior and body miniaturization were probably associated with the development of the brain. Finally, we discuss the evolutionary process of cerebrum expansion in modern birds during the Cenozoic Era. Acquisition of powered flight and endothermic metabolism are proposed as the main contributing factors of cerebral expansion in modern birds.

Keywords Amniotes • Reptiles • Dinosaurs • Cognition • Flight • Metabolism Endothermy

---

T. Shimizu, Ph.D. (✉) • A.K. Uysal • S. Leilani Kellogg
Department of Psychology, University of South Florida, 4202 E. Fowler Ave., PCD 4118G, Tampa, FL 33620, USA
e-mail: shimizu@usf.edu

K. Shinozuka
RIKEN Brain Science Institute, 2-1 Hirosawa, Wako, Saitama, Japan

S. Watanabe et al. (eds.), *Evolution of the Brain, Cognition, and Emotion in Vertebrates*, Brain Science, DOI 10.1007/978-4-431-56559-8_2

2.1 Introduction

Mental capacities of nonhuman animals have fascinated us since the beginning of time. We know and hear many intriguing reports and anecdotes about surprisingly high "intelligence" and rich "emotion" of animals. Some stories are rather products of imagination due to our anthropomorphic bias. However, rigorous scientific research has also proven that various kinds of "intelligence" and "emotion" truly exist in nonhuman animals (Shettleworth 2012; Wasserman and Zentall 2006). Among different "smart" animals, birds have been one of the most popular subjects of extensive studies (Ackerman 2016; Marzluff and Angell 2013; Heinrich 2009). Just like mammals, birds have shown superb abilities in parental care, communication, long-term memory, associative learning, observational learning, concept formation, and toolmaking and use. Some researchers even suggest that birds demonstrate theory of mind (Emery and Clayton 2004, 2005) and support or console each other (Fraser and Bugnyar 2010, 2012). Based on these lines of evidence, the similarities of mental capacities between some avian species and primates have been pointed out (Emery and Clayton 2004; Güntürkün and Bugnyar 2016; Navarrete et al. 2016).

These abilities in birds, as well as in mammals, are substantially associated with their enlarged and differentiated brains, the cerebrum in particular (Cross et al. 2013; Shanahan et al. 2013). Both birds and mammals have much larger brain as a whole, the cerebrum specifically, compared to any other vertebrates (Northcutt 1981). Relative to similar-sized reptiles and fishes, the brain size difference is at least a factor of 10. As discussed extensively in other chapters in this volume, there are significant differences in the cerebral organization between birds and mammals. The avian brain lacks a laminated cortex, which plays essential roles in perception, cognition, and motor control in the mammalian brain and is highly expanded especially in the primate brain. Despite the lack of a laminated cortex, the fact that birds have the capability of accomplishing similar cognitive feats suggests that the enlarged cerebrum of birds and mammals has, to some degree, equivalent functions. In contrast to the cerebrum expansion, the brainstem is relatively conserved in terms of size, structure, and function among birds, mammals, and reptiles, although its role in cognitive and affective functions should not be underestimated.

When and why did the avian cerebrum become enlarged? For the last 300 million years, the cerebrum expansion occurred at least three times in the lineage leading to modern birds. First, it occurred when the common ancestors of all amniotes (reptiles, birds, and mammals) emerged during the Carboniferous Period in the Paleozoic Era over 300 million years ago (MYA). Almost immediately, lineages of reptiles and mammals were diverged from the ancestral amniotes. About 200 MYA, the lineage of stem birds was then separated from the non-avian reptilian group. The second cerebral expansion occurred during the transition from theropod dinosaurs to ancestral birds in the Mesozoic Era. Many Mesozoic birds became extinct with other dinosaurs by 66 MYA, and direct ancestors of modern bird species appeared and flourished in the Cenozoic Era. The third and most extensive expansion of the cerebrum occurred then. In this chapter, we will discuss these three events of cerebrum expansion—ancestral amniotes, dinosaurs, and modern birds—to understand the origin of avian brain capacity.

2.2 From Water to Land: Emergence of Amniotes

2.2.1 Timeline

Reptiles, birds, and mammals are together categorized in a group called amniotes because only amniotes have an amnion, an extraembryonic membrane. The amniotic membrane protects embryos in a stable fluid environment within a shell, enabling them to survive on a dry land. Anamniotes, such as fishes and amphibians, lay eggs in the water and have an aquatic larval stage, while amniotes no longer need to stay close to a body of water. The first amniotes evolved from ancestral amphibians over 300 MYA during the Carboniferous Period in the Paleozoic Era (Fig. 2.1). The emergence of amniotes is probably one of the most seminal events in vertebrate evolution. This is the beginning of the diverse and successful terrestrial

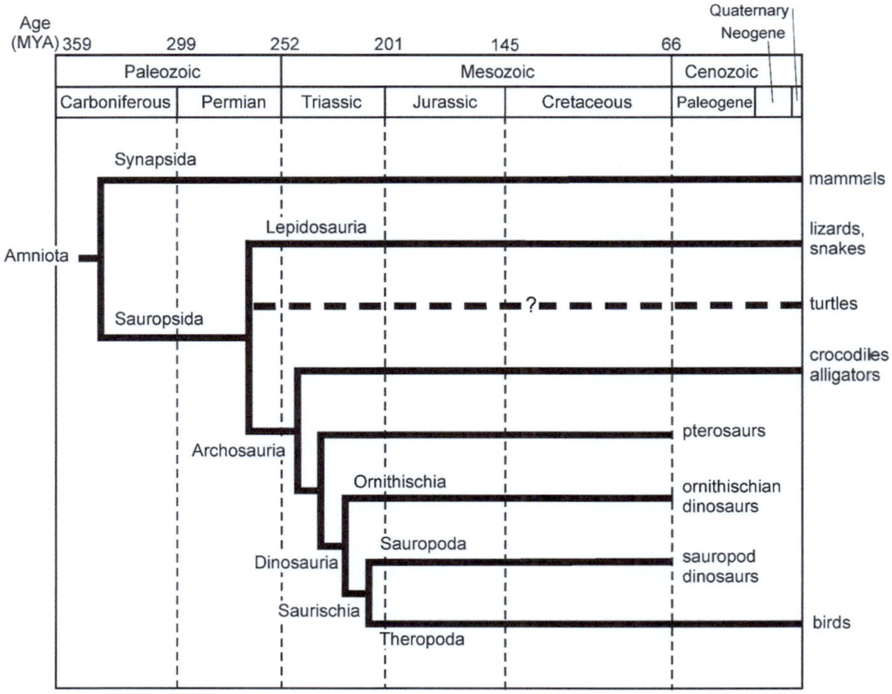

Fig. 2.1 Timeline of amniote evolution. Phylogeny based on Carroll (1988)

animals that ultimately lived in a wide variety of physical and biological niches. Almost immediately after the first amniotes appeared, the lineage was divided into sauropsids ("lizard faces" including anapsids and diapsids) and theropsids ("beast faces," also called synapsids). Sauropsids ultimately led to diverse groups of reptiles and birds. The lineage of theropsids eventually gave rise to mammals in the Late Triassic (235–201 MYA).

2.2.2 Early Amniotes

What did the early amniotes look like? *Diadectes*, which had some characteristics of both reptiles and amphibians, were terrestrial animals during the Early Permian (290–272 MYA) (Carroll 1988). They had a proportionally large skull and a barrel-shaped body of 1.5–3 m long, unlike relatively small extant amphibians. Their dental features suggest that they were probably herbivores.

Truly indisputable amniotes include *Hylonomus* and *Paleothyris* during the Late Carboniferous (Carroll 1988). Unlike large Permian amphibians, they were lizard-like creatures of a small body (about 20 cm), slender limbs, and sharp teeth. They probably possessed developed stretch receptors in the muscle for improved musculoskeletal coordination, as do extant lizards. Based on these physical characteristics, we can assume that these animals were most likely agile hunters and fed on insects (e.g., spiders, dragonflies) and small animals (e.g., amphibians) on land.

2.2.3 Brain Development

To study the brains of extinct animals is a challenge. Detailed neuroanatomical analysis is impossible. Soft tissues like brains are rarely fossilized. However, cranial endocasts can be generated from fossil braincases. By examining endocasts, the general size, shape, and surface morphology of the brain can be postulated. Endocasts may be formed naturally through fossilization. They can also be produced artificially by filling a molding material into the endocranium. While traditional molding has a risk of harming the fossils, recent advances in imaging technology allow researchers to digitally generate virtual endocasts using high-resolution computed tomography techniques. Although endocasts are quite informative, they need to be evaluated with caution. The braincase holds not only the brain itself but also cerebrospinal fluid (CSF) surrounding the brain. The CSF can take up a significant portion of the cranial volume, and thus endocasts may not represent the exact shape and surface characteristics of the brain. In addition, the brain size itself does not indicate the number of neurons and the complexity of connections. A recent study showed that the forebrains of large-brained parrots and corvids have a higher density of neurons than the primate brains of the same size (Olkowicz et al. 2016). Thus, the capacity of neural computation per unit mass may be different even if the brain sizes are comparable.

The general organization of vertebrate brains is essentially the same among all living species. Based on endocasts, this seems plesiomorphic for extinct vertebrates as well. Vertebrate brains include three major subdivisions along the rostral-caudal axis: forebrain (or prosencephalon), midbrain (or mesencephalon), and hindbrain (or rhombencephalon) (Butler and Hodos 1996). The forebrain consists of the cerebrum and thalamus, which are closely interconnected and often serve functions in synchrony. As a result, their development is highly correlated to each other. At the most rostral part of the cerebrum, olfactory bulbs are visibly large in many vertebrates. The midbrain includes the optic tectum (superior colliculus in mammals), torus semicircularis (inferior colliculus in mammals), and tegmentum. The optic tectum is particularly well-developed and elaborated in amniotes. The caudal-most region, hindbrain, includes the pons, medulla oblongata, and cerebellum. The midbrain and hindbrain are together categorized as the brainstem.

The innovation of extraembryonic membranes, especially the amnion, led to dramatic changes in the brains of amniotes (e.g., *Hylonomus* and *Paleothyris*) from those of Permian amphibians (e.g., *Diadectes*). Amniote brains, the forebrain in particular, are larger than anamniote brains given comparable body sizes (Jerison 1973; Northcutt 1981).

According to the endocast of *Diadectes*, they had a relatively small, narrow, and elongated brain (Hopson and Gans 1979). Such characteristics suggest that the brain of *Diadectes* was probably similar to that of living amphibians. In general, amphibians have a rather diminutive cylindrical brain, which includes relatively large olfactory bulbs, small cerebral hemispheres which are in contact with optic lobes, and a small cerebellum. The relative volumes of their entire brain, as well as the forebrain, are similar to those of bony fishes of similar body size (Jerison 1973; Northcutt 1981).

No data are currently available about the brains of *Hylonomus* and *Paleothyris*. However, it is reasonable to assume that their brains were more similar to lizards (reptiles) than amphibians in order to carry out their active locomotion for insectivorous diets (Fig. 2.2). If so, their brains, forebrains in particular, were enlarged and developed to some degree, compared to *Diadectes*. Forebrains of living reptiles are generally about one to two times larger than amphibians of similar body sizes (Northcutt 1981). As discussed in a chapter by Jarvis in this book, this enlargement may be due to an expansion of the cerebrum, a pallial region called the dorsal ventricular ridge (DVR) in particular (Jarvis et al. 2005; Reiner et al. 2004). Directly caudal to the cerebral hemispheres, reptiles have optic lobes, which are in turn adjacent to an expanded cerebellum.

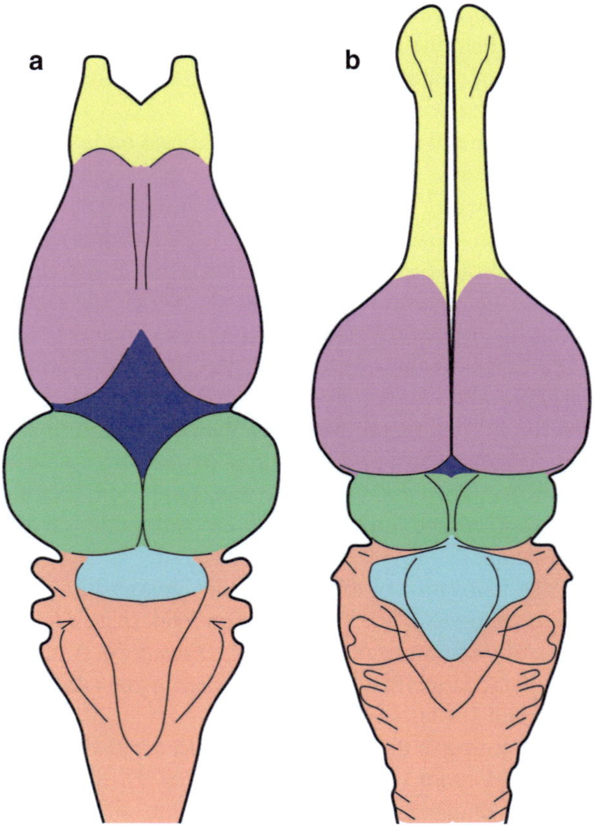

Fig. 2.2 The top views of brains of a living amphibian (**a**, frog) and reptile (**b**, alligator). Each color represents a different brain subdivisions: olfactory bulbs (*yellow*), cerebrum (*magenta*), thalamus (*dark blue*), optic lobe (*green*), cerebellum (*light blue*), and brainstem (*orange*). Brains are not scaled to size

2.2.4 Selective Factors

2.2.4.1 Terrestrial Habitat

One important consequence of the emergence of embryonic membranes is that amniotes evolved to truly and freely explore the terrestrial environment. The amniote brains must now have a different neural system to obtain sensory signals that transmitted through the surrounding air, instead of fluid. Compared to the aquatic habitat, a new motor system also became necessary to control different kinds of movements on land. Because early amniotes (e.g., *Hylonomus* and *Paleothyris*) were capable of agile locomotion, it is likely that brain areas for motor systems needed to be developed and enlarged. For example, similar to the reptilian brain, the early amniote brain probably had more developed and differentiated basal ganglia (subpallium) in the cerebrum, compared to those of amphibians.

The basal ganglia are known to play a major role in motor control in all vertebrates (Reiner et al. 1998). They receive dopaminergic projections from the midbrain tegmentum and influence motor functions through outputs to the tectum and/

or the pallium in the cerebrum. In living amphibians, both the basal ganglia and midbrain dopaminergic areas are relatively small and not clearly differentiated. The amphibian tectum is the primary output for motor control, while projections to the pallial regions are limited. In contrast, the reptilian (as well as avian and mammalian) basal ganglia are highly enlarged and differentiated into striatal and pallidal regions. Similarly, the midbrain dopaminergic cell populations include the substantia nigra and ventral tegmental area. As for the output of the basal ganglia, the amniote basal ganglia send projections to both the tectum and pallium to control movements.

2.2.4.2 Parental Investment

Ancestral amphibians, just like their extant counterparts, had external fertilization, laid numerous eggs in the water while newly emerged amniotes fertilized internally, and laid fewer eggs (or kept them within the body). As a result, amniotes had increased parental investment compared to anamniotes. Parental investment can be defined as any investment (e.g., time and energy) by the parent for the survival and reproductive success of offspring (Trivers 1972). Such significant changes in parental investment were associated with various social aspects of sexual selection, including male-male competition and mate choice. In extant amniotes, these complex social behaviors often necessitate the development of the forebrain (a pallial region in particular) to process species-specific sensory signals, make appropriate choices and decisions, and control complex social behaviors. While the pallium in all tetrapods receives projections from thalamic sensory nuclei to execute such functions, projections are more extensive in amniotes than anamniotes, and both the pallium and thalamus of amniotes are much more enlarged and elaborate (Butler and Hodos 1996). We can assume that early amniotes had a larger and more developed pallial region than anamniotes to deal with these biological problems.

2.3 From Large to Small: Dinosaurs in the Bird Stem Lineage

2.3.1 Timeline

During the Mesozoic Era (about 252 to 66 MYA), sauropsids were greatly proliferated ("the Age of Reptiles"). Sauropsids diverged into archosaurs ("ruling lizards") and lepidosaurs ("scaly lizards") by the Early Triassic Period. The former led to crocodiles, dinosaurs, pterosaurs, and birds, while the latter led to lizards and snakes. Dinosaurs ("terrible lizards") were one of the most successful vertebrates over 150 million years in the Mesozoic Era. They appeared in the Late Triassic (about 225 MYA) and then became dominant in the terrestrial, aquatic, and aerial niches of the Jurassic Period. Huxley (1868) first proposed and later Ostrom (1973) reinforced the hypothesis that the stem lineage of birds appeared from a group of theropod dinosaurs. Although the issue has been debated extensively, most of paleontologists support this hypothesis today. By the end of the Cretaceous (about 66 MYA), all the dinosaurs except the lineage to modern birds rather abruptly disappeared.

2.3.2 *Dinosaurs in the Bird Stem Lineage*

Dinosaurs are categorized into two groups: saurischians ("lizard-hipped") and ornithischians ("bird-hipped") based on pelvic structure. Interestingly, birds were originated from a "lizard-hipped" branch of dinosaurs, rather than "bird-hipped" ornithischian dinosaurs. Saurischians further include carnivorous theropods ("beast-footed") and herbivorous sauropods ("lizard-footed"). The earliest birds were considered to be derived from bipedal theropods during the Jurassic Period. The exact evolutionary steps from theropods to modern birds are not completely understood and continuously revised due to new findings. Here, five major clades of the bird stem lineage—Tetanurae, Coelurosauria, Maniraptora, Paraves, and Avialae—are discussed. The relationships of these clades are shown in Fig. 2.3.

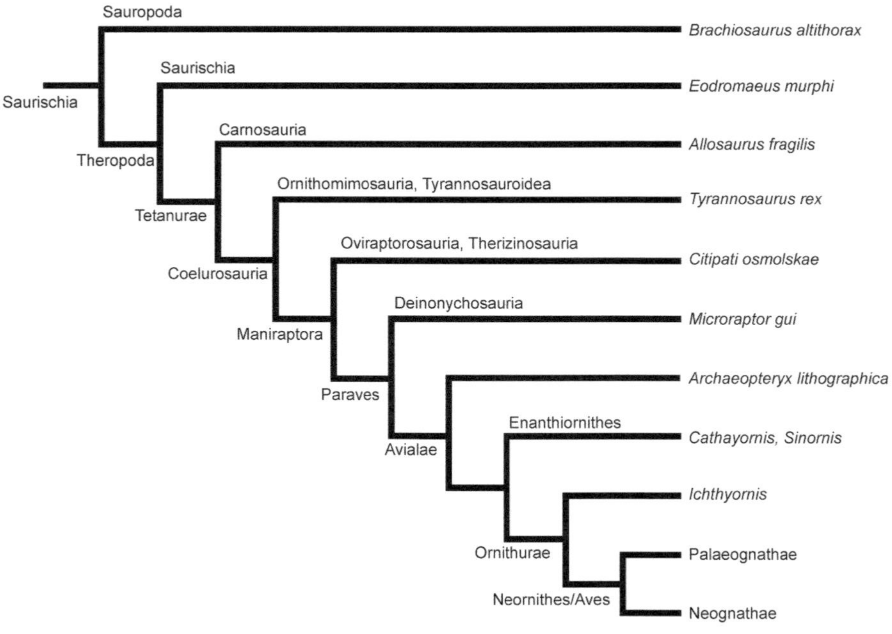

Fig. 2.3 Phylogeny of theropods, focusing on the bird stem lineage. Phylogeny based on Gauthier and de Queiroz (2001) and Turner et al. (2012)

Tetanurae ("stiff tails"): Dinosaurs in this large theropod group are diverse in terms of size and appeared during the Early or Middle Jurassic Period. It includes two subgroups—Carnosauria and Coelurosauria—primarily based on size and proportion differences. In general, coelurosaurs are smaller than carnosaurs, which include large *Allosaurus fragilis* (about 8.5 m long, Glut 1997) and *Carcharodontosaurus saharicus* (over 12 m long, Sereno et al. 1996). However, enormous tyrannosaurids (e.g., *Tyrannosaurus rex*, about 12 m long, Hutchinson et al. 2011) belong to Coelurosauria.

Coelurosauria ("hollow-tailed lizards"): This clade consists of all theropods more closely related to birds than to carnosaurs. These relatively small theropods are recognized from the Late Jurassic and Early Cretaceous. Many feathered dinosaurs have been discovered in China from the Middle-Upper Jurassic and Lower Cretaceous. Most of these feathered dinosaurs turn out to be coelurosaurs (Xu and Guo 2009; Xu et al. 2012). The major coelurosaurian groups include Maniraptora, Ornithomimidae, and Tyrannosauridae.

Maniraptora ("hand snatchers"): All dinosaurs closer to birds than to ornithomimids are members of this group. It contains the subgroups Paraves, Oviraptorosauria, and Therizinosauria. Fossil records of maniraptorans appeared during the Jurassic Period. They show skeletal characteristics, which were essential steps for the evolution of gliding and/or powered flight. For example, maniraptorans are the only dinosaurs that have elongated forearms (which become wings in birds) and a sternum (where flight muscles attach, but see Zheng et al. 2014).

Paraves ("near bird"): This group includes all dinosaurs more closely related to birds than to oviraptorosaurs. It has two major subgroups: Avialae and Deinonychosauria. The latter further includes the dromaeosaurids (e.g., *Microraptor gui, Deinonychus antirrhopus*) and troodontids (e.g., *Zanabazar junior*). Although the origin and evolution of avian flight is still in debate, at least some paravians were experimenting with flight—parachuting, gliding, and/or hopping from tree to tree. For example, *Microraptor gui,* a basal dromaeosaurid dinosaur found in China, had four wings located on both the forelimb and hind limb (Xu et al. 2003). Using these developed wings, they should have been able to glide. However, powered flight probably did not start till the clade of Avialae.

Avialae ("bird wing"): All dinosaurs closer to birds than to *Deinonychus* are members of this group. Gauthier and de Queiroz (2001) used this term to include all flying [winged] dinosaurs. This group includes Ornithurae (including Aves), Enantiornithes, as well as the famous *Archaeopteryx lithographica*, the exact position of which is still uncertain. *Archaeopteryx* lived in the Late Jurassic (about 150 MYA). The first specimen was discovered in 1861 in Southern Germany, just a few years after the publication of Darwin's *On the Origin of Species*. It is small in body size (up to 50 cm long, 0.8–1 kg body weight) and has both dinosaur-like characteristics (e.g., a jaw with sharp teeth, claws, and a bony tail) and birdlike features (e.g., broad feathered wings) (Callaway 2014). Since the early 1990s, new specimens showing dinosaur-bird transition have been discovered in China. They include other feathered dinosaurs, such as *Anchiornis huxleyi* (160 MYA) and *Aurornis xui* (160 MYA). A recent phylogenetic analysis suggests that *Aurornis xui*, but not *Archaeopteryx*, is the basal-most avialan (Godefroit et al. 2013).

2.3.3 Brain Development

Fossil endocasts suggest that the brains of non-avian theropod dinosaurs are not much different from those of living reptiles in terms of relative size and overall shape (Hopson and Gans 1979; Hopson 1977; Jerison 1969, 1973). Due to scarce fossil records in the bird stem lineage, little is known about the transition process from diminutive theropod brains to highly developed avian brains. However, new fossil findings of maniraptorans from China have started to shed an insight into this issue. These data show that maniraptoran brains (e.g., *Archaeopteryx*) are volumetrically situated intermediately between early theropods and modern birds (Balanoff et al. 2013).

Recent studies suggest that the cerebral expansion in the bird stem lineage occurred at least three times and started early in theropod history—probably after the divergence of coelurosaurs in the Middle Jurassic Period about 180–160 MYA (Balanoff et al. 2013; Larsson et al. 2000). Endocasts of the coelurosaur *Tyrannosaurus* show overall cerebral features that are different from those of two carnosaur species (*Carcharodontosaurus and Allosaurus*). For example, the dorsal roof of the cerebrum is positioned higher than the cerebellum in carnosaurs, while the cerebellum of *Tyrannosaurus* has moved to a more elevated position (Larsson et al. 2000). This coelurosaur characteristic of an elevated cerebellum is also observed in *Archaeopteryx* as well as in modern birds. In terms of endocast volumes, *Tyrannosaurus* is about 50% larger for total brain volume and 100% larger for cerebrum volume than *Carcharodontosaurus.* The expansion of the total brain volume of *Tyrannosaurus* is probably due to a larger proportion of the cerebrum (32.6%) compared to that in *Carcharodontosaurus* (24%).

The next development and enlargement of brains probably occurred in the Late Jurassic about 150 MYA after maniraptorans appeared. The brain volume of avialan *Archaeopteryx* is approximately three times larger than those of reptiles of comparable size (Alonso et al. 2004). Compared to reptiles and ancestral theropods, the brain of *Archaeopteryx* shows some characteristics shared by modern birds (Fig. 2.4), such as reduced olfactory bulbs, expanded cerebral hemispheres, a large cerebellum directly caudal to the cerebrum, and a midbrain (including optic lobes) displaced lateroventrally. The cerebral surface has a slight sign of a longitudinal indentation (vallecula), implying that it has a brain region called the Wulst (Balanoff et al. 2013). However, the brain of *Archaeopteryx* is not an avian brain. If the body mass is about the same, modern birds have much larger brains than *Archaeopteryx.*

Examinations of endocasts of other maniraptorans suggest that the cranial expansion was not unique in *Archaeopteryx*, but probably a generalized phenomenon in maniraptorans (Balanoff et al. 2013). For example, endocast images of oviraptorosaur *Citipati osmolskae* (Fig. 2.4) and an unnamed troodontid show the avian characteristics mentioned above, such as reduced olfactory bulbs, expanded cerebral hemispheres, a large cerebellum directly caudal to the cerebrum, and optic lobes displaced lateroventrally. Volumetric analyses show that the total brain and cerebrum volumes relative to body size were similar among all maniraptorans including *Archaeopteryx* (Balanoff et al. 2013).

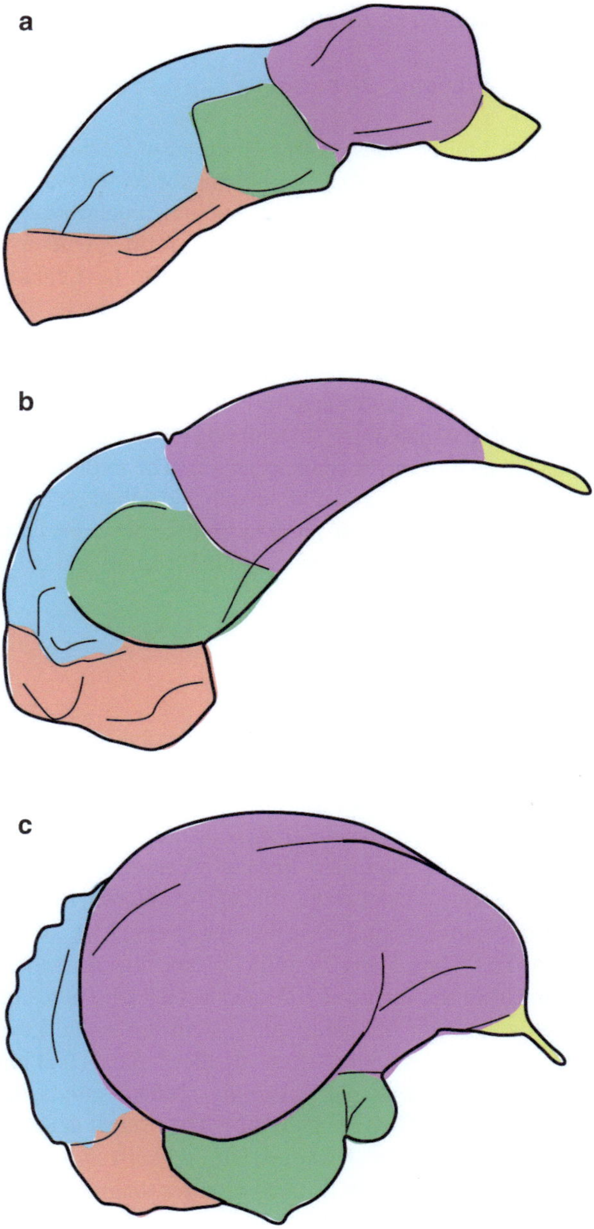

Fig. 2.4 Postulated brain organization based on endocasts (adapted from Balanoff et al. 2013 with permission). (**a**) *Citipati osmolskae* (an oviraptorosaur from the Late Cretaceous), (**b**) *Archaeopteryx lithographica*, and (**c**) *Melanerpes aurifrons* (woodpecker). Each color represents a different brain subdivisions: olfactory bulbs (*yellow*), cerebrum (*magenta*), optic lobe (*green*), cerebellum (*light blue*), and brainstem (*orange*). Endocasts are not scaled to size

Furthermore, the data suggest that at least some members of Paraves have brains that are somehow different from those of other maniraptorans. A principal component analysis was conducted to study the effects of different brain structures (olfactory bulbs, cerebrum, optic lobes, cerebellum, and brainstem) relative to total brain volumes (Balanoff et al. 2013). The results showed a clear volumetric separation between Paraves and Oviraptorosauria. The difference is largely defined by the cerebral expansion in Paraves.

2.3.4 Selective Factors

2.3.4.1 Predatory Behavior

An early brain expansion was observed in Coelurosauria compared to Carnosauria (Larsson et al. 2000). One possible factor for the brain expansion might be related to differences in active predatory behavior in coelurosaurs. Although both coelurosaurs and carnosaurs are carnivorous, the former tend to be smaller than the latter. Compared to carnosaurs, coelurosaurs also have relatively small skulls and longer forelimbs (Carroll 1988). Such physical characteristics suggest that coelurosaurs were more vigilant and agile hunters than carnosaurs and that they needed a developed sensory (visual in particular) system to detect prey, as well as an efficient motor system to control swift activity. The midbrain contains the optic tectum that has a precise visual map of the surrounding environment (Butler and Hodos 1996) and the acoustic tectum (torus semicircularis). The midbrain also sends descending projections to motor areas in the brainstem and spinal cord in living amniotes. It is reasonable to assume that these predators have a developed midbrain system, as well as the cerebellum, to react swiftly to external stimuli and generate quick movements.

2.3.4.2 Body Miniaturization

Subsequently, multiple stages of brain expansion occurred in the bird stem lineage. Another important factor that made continued effects on this lineage is sustained miniaturization of the body size. By analyzing extensive fossil databases, researchers showed that such miniaturization occurred specifically in the bird stem lineage, but not in non-avian dinosaur lineages (Lee et al. 2014b). The ancestral tetanuran is approximately 163 kg about 198 MYA, followed by coelurosaurs (27 kg, 173 MYA), maniraptorans (10 kg, 170 MYA), paravians (3 kg, 167.5 MYA), and avialans (0.8 kg, 163 MYA). Thus, along 50 million years, dinosaurs in the lineage leading to avialans shrunk from an average of 163 kg to just 0.8 kg. These changes are depicted in Fig. 2.5.

As a result of miniaturization, stem birds could evolve to obtain physiological and anatomical changes. These smaller animals had higher metabolic rates, feather elaborations, increased aerial ability, reduced snouts, developed beaks, and enlarged eyes and brains (Lee et al. 2014b). The critical selective pressures for miniaturization are not known. However, arboreal lifestyle might be one important driver (Benton 2014). Living in trees requires smaller bodies, along with sharper claws and keener vision, in addition to elongated forearms for wings.

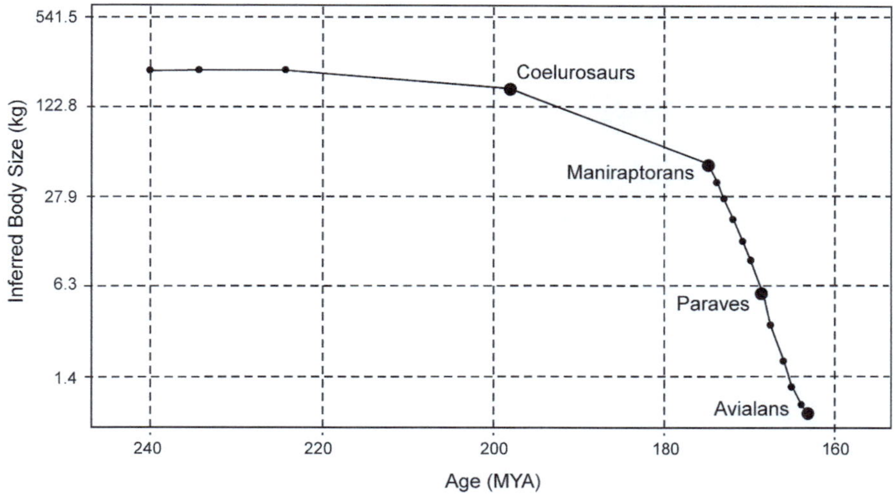

Fig. 2.5 Body miniaturization along the bird stem lineage (adapted from Lee et al. 2014b with permission). Circles represent nodes ("ancestors") along the bird stem lineage. Large circles represent nodes discussed in this chapter

2.4 From Cold to Warm: Flying Birds

2.4.1 Timeline

Modern birds emerged from a subgroup of Avialae called *Ornithurae* ("bird tails"), which includes all decedents of the common ancestor of modern birds as well as extinct toothed birds, such as *Ichthyornis* and Hesperornithes. Both *Ichthyornis* and Hesperornithes lived in the Cretaceous Period (Clarke 2004). *Ichthyornis* resembles modern flying seabirds, such as gulls, with a developed wings and a large sternum. Hesperornithes is a group of flightless, but strong, swimming waterbirds. The crown group—the most recent common ancestor of all living birds and all of its descendants—is categorized in the clade *Neornithes* ("new birds") or *Aves* (sensu Gauthier and de Queiroz 2001). The oldest fossil definitively placed in Neornithes was *Vegavis iaai*, a member of the duck lineage (Clarke et al. 2005). This bird lived in the Late Cretaceous (about 67 MYA) in Antarctica, suggesting that living bird lineages coexisted with other avialans before the Cretaceous/Paleogene (K-Pg) boundary about 66 MYA.

2.4.2 Diversification of Modern Birds

Modern birds are the most successful tetrapods in terms of the number of species. According to the International Ornithological Congress World Bird List (v6.2), there are 10,637 extant avian species, 40 orders, 239 families, and 2289 genera (Gill and Donsker 2016). The timing of this great diversification is still in debate. Some molecular studies indicate a gradual radiation during the Cretaceous Period (Haddrath and Baker 2012; Jetz et al. 2012; Lee et al. 2014a). On the other hand, fossil records of Neornithes from the Cretaceous Period are quite limited (Clarke et al. 2005), suggesting that modern birds underwent extraordinary "big bang" diversification for a short period of time after the mass extinction event (Ericson et al. 2006; Feduccia 1995; Jarvis et al. 2014). A recent extensive whole-genome analyses also support the "big bang" hypothesis that a rapid diversification occurred within a short period of time during the K-Pg transition (36 lineages within 10–15 million years) (Jarvis et al. 2014). If this is the case, modern birds, together with placental mammals (O'Leary et al. 2013), diversified rapidly in the ecological niches where their potential predators and rivals (e.g., non-avian dinosaurs, pterosaurs, and other avialans) no longer existed.

Based on comparative anatomy, they are divided into two subgroups, Palaeognathae ("old jaw") and Neognathae ("new jaw") (Livezey and Zusi 2007). Palaeognathae include flightless ratites, such as Struthioniformes (ostriches), Rheiformes (rheas), Casuariiformes (emus), and Apterygiformes (kiwis). Neognathae consists of Galloanserae (Galliformes, e.g., chickens, turkey; Anseriformes, e.g., ducks, geese) and Neoaves (all other diverse birds). Neoaves further include Columbiformes (pigeons), Passeriformes (songbirds, corvids), Psittaciformes (parrots), Falconiformes (falcons), Strigiformes (owls), Accipitriformes (eagles, hawks), Charadriiformes (gulls, shorebirds), and Apodiformes (hummingbirds).

2.4.3 Brain Development

The brain of modern birds has some fundamental characteristics that were shared with other Mesozoic avialans based on information from endocast data. The avian brain has relatively small olfactory bulbs, large cerebral hemispheres, lateroventrally displaced optic lobes, and an enlarged cerebellum. However, endocasts also clearly show that modern birds evolved to have much larger brains than other avialans (Alonso et al. 2004; Balanoff et al. 2013). For example, when total endocranial volume relative to body mass is compared, birds with the same size have 1/3 to 5 times larger brains than the brain of *Archaeopteryx* (Alonso et al. 2004). In particular, the enlargement of the cerebrum is clearly obvious in modern birds compared to that of *Archaeopteryx*. The extensive cerebral expansion causes a further displacement of optic lobes ventrally to the point that the midbrain beneath the cerebrum is only partially visible from the dorsal viewpoint.

From the endocast information, it is difficult to pinpoint the exact cerebral region that expanded in birds. However, it is most likely that two pallial areas, DVR (Jarvis's chapter) and Wulst, expanded the most in modern birds (Shimizu 2001). When the cerebra of extant birds and reptiles are compared, these two regions are highly developed and enlarged in birds compared to that of reptiles. Other cerebral substructures, such as the basal ganglia (including striatum and pallidum), appear to be similar volumetrically between reptiles and birds. Either the Wulst or DVR is not a unitary entity with a single function. Both structures are critically involved in diverse functions associated with sensory, cognitive, motor, and limbic systems (Shimizu and Watanabe 2012).

The extant avian cerebellum is an extremely differentiated and efficient machine consisting of numerous neurons. It has a number of parallel grooves (folia, lobules) on the cortical surface, a characteristic which is also found in mammals, as well as some elasmobranches and fishes, but not reptiles. These grooves are fissures of the cerebellar cortex, in which layers of numerous neurons are efficiently folded like an accordion. The cerebellum of *Archaeopteryx* appears to be well-developed in terms of size (Alonso et al. 2004; Balanoff et al. 2013). However, the interpretation was challenged because at least a portion of the "cerebellum" might be a torus semicircularis, making the "true" cerebellum much smaller (Kurochkin et al. 2007). Furthermore, no detailed information is available about the surface morphology of the *Archaeopteryx* cerebellum. From endocast data, it is unclear whether non-avian maniraptorans had folia or lobules in the cerebellum. Since cerebella of extant reptiles are lissencephalic, it is possible that *Archaeopteryx* did not have an avian-like elaborated cerebellum. Thus, even if endocasts suggest that the general size and shape of the cerebellum are comparable, the cerebellum of *Archaeopteryx* or any other non-avian maniraptorans might not have the same computational capacity as that of modern birds.

2.4.4 Selective Factors

2.4.4.1 Flight

Of over 10,000 species of Neornithes, most birds fly, while about 60 living species do not (Roots 2006). Powered flight is definitely a signature behavior of modern birds. The exact origin of volant avians has been a contentious issue among paleobiologists for decades. Some non-avian maniraptorans in the Mesozoic Era had feathered wings, with which they were also probably able to leap, glide, parachute, and/or flap for powered flight. Two main hypotheses about the origin of flight have been proposed—ground-up and tree-down. The former hypothesis proposes that small-bodied terrestrial avian ancestors obtained feathers for thermal insulation or sexual display, but not for locomotion. Then the flight stroke evolved for fast ground-running activity. The latter hypothesis argues that avian ancestors lived in an arboreal environment. They climbed up trees using claws and then parachuted or glided between trees using feathered wings. *Archaeopteryx* probably flew, but whether or not they had the capability of powered flight has been debated. For example, a relatively developed cerebellum of *Archaeopteryx* suggests that their neural system is equipped for flight (Balanoff et al. 2013, but see a discussion above). However, they also had a rather flat sternum lacking a well-developed keel. Modern flying birds have sterna with the keel, to which breast muscles (pectoralis and sternocoracoideus) necessary for powerful strokes attach.

Regardless of the exact beginning of the avian powered flight, it is clear that powered flight became ubiquitous among Neornithes in the Cenozoic. In order to have the capability of powered flight, the avian brain needed to undergo radical changes in the neural system—the visuomotor system in particular. While there were major neural changes for the coordination of muscles related to flying locomotion, the most important change involved all systems adapting to high speeds of flight behavior. Birds are the fastest animals in the animal kingdom. Falcons can fly at a speed of over 100 km/h and dive about 400 km/h (Tucker 1998; Tucker et al. 1998). With such a high speed, humans have only blurry views, while birds can detect and analyze their surroundings instantly and maneuver themselves precisely to avoid collisions. Furthermore, flying adds the vertical dimension of the environment which birds must also process. These biological problems forced the bird brain to expand such neural areas for processing abundant visual input efficiently, while controlling or correcting motor activities instantly.

In all amniotes, visual information is sent from the retina to both the midbrain (optic tectum) and thalamus (lateral geniculate nucleus) (Shimizu and Watanabe 2012). Each area further sends projections eventually to the DVR via the tectofugal pathway and the Wulst via the thalamofugal pathway. In birds, the optic tectum is an extremely developed and differentiated structure, containing at least 15 layers of diverse neurons. Similarly, the subsequent cerebral targets, the DVR and the Wulst, are also large developed structures, which in turn send projections directly and indirectly to enlarged motor areas in the brainstem and cerebellum. It is likely that these areas—the optic tectum, lateral geniculate nucleus, DVR, Wulst, and visuomotor areas in the brainstem—evolved to meet the demands associated with becoming animals with powered flight.

2.4.4.2 Endothermy

Today, birds and mammals are only homeothermic endotherms that can internally regulate and maintain their body temperature. In endotherms, the primary source of the heat is a high resting metabolic rate, which is supported by aerobic metabolism. In contrast, reptiles are ectothermic animals that regulate their body temperature using external heat sources, the sun in particular. Since skeletal evidence suggests that the common ancestral amniotes of the avian and mammalian lineages were almost definitely ectothermic, endothermy must have evolved independently in the therapsid-mammalian lineage and the theropod-bird lineage.

When and how endothermy in these lineages evolved is not fully understood. Paleobiological evidence suggests that metabolic changes did not occur suddenly in the lineages of birds and mammals. Rather, the resting metabolic rate and body temperature had increased gradually throughout their stem lineages (Lovegrove 2016). However, in the case of birds, it is most likely that true homeothermic endothermy was acquired only by Cretaceous Ornithurae (Hillenius and Ruben 2004). Strong evidence of, or the lack of, endothermy can be physically observed in the design of the nasal cavity in fossils. Based on the design of the nasal cavity, we can speculate whether or not extinct animals had nasal respiratory turbinates, or conchae, which are curled bony protrusions from the walls of the nasal cavity into the breathing passage. Respiratory turbinates are essential for animals with a high resting metabolic rate (i.e., living birds and mammals) in order to recover a significant portion of water and heat (Fig. 2.6). However, ectothermic animals, such as reptiles, do not need or possess such a mechanism. Fossil analyses suggest that

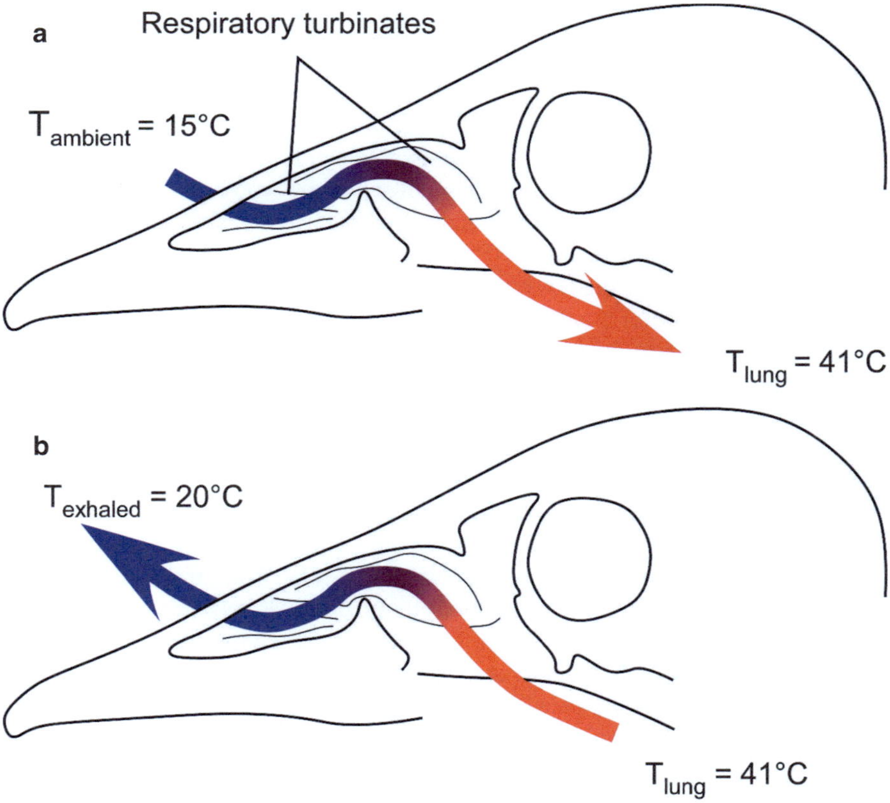

Fig. 2.6 Respiratory turbinates in birds (adapted from Hillenius and Ruben 2004 with permission). When birds inhale (**a**), air passing through the respiratory turbinates is heated and humidified. When birds exhale (**b**), air is cooled down and dehumidified through the respiratory turbinates

theropod dinosaurs, as well as other avialans (e.g., *Archaeopteryx* and Enantiornithes, Martin and Zhou (1997), did not have any indication of the mechanism, while Ornithurae in the Early to Late Cretaceous probably had ones. These observations are consistent with the hypothesis that endothermy probably evolved with modern birds and their immediate ancestors.

The major reason that only ornithurine birds acquired endothermy might be related to their ability of powered flight although there are other relevant abiotic and biotic factors, such as ambient temperature, humidity, CO_2 level, body size, and musculature type. Bennett and Ruben (1979) argue that the evolution of a higher metabolic rate and endothermy are accidental secondary events, preceded by an enhanced aerobic capacity for prolonged locomotor activity. In the case of birds, long-distance powered flight is the sustained performance requiring aerobic metabolism. The capacity for long-distance activity was probably required to compete against rivals in order to increase territory/home range size for resources. In contrast, ectotherms rely on anaerobic metabolism for bursts of intense movements, but such behavior cannot last long due to fatigue associated with the accumulation of lactic acid and the depletion of stored fuel. Once endothermy was obtained, it perhaps enabled birds to have stamina and time necessary to perform behaviors above and beyond long-distance powered flight. These behaviors include sophisticated courtships to attract potential mates, parental care for offspring, patrolling of extensive territories, and other complex behaviors that can be categorized as cognition. Selection of these complex behaviors might result in the cerebrum expansion, especially the DVR and Wulst, which may play important roles in such higher cognitive capacities (Shimizu and Watanabe 2012).

In order to clarify the relationship between the endothermy and cerebral development, it would be useful if we had a detailed information about the metabolism of Pterosauria ("wing lizards"). Although their exact flight capabilities are also still controversial, skeletal features suggest that most pterosaurs could sustain powered flapping flight (Witton 2013). Pterosaurs lived from the Late Triassic to the end of the Cretaceous Period (228 to 66 million years ago). Since pterosaurs and dinosaurs belong to two different clades in archosaurs, pterosaurs evolved volancy independently from the stem bird lineage. In terms of brain development, the two most notable characteristics of the pterosaur brain are its large semicircular canals in the inner ear and the cerebellum, follicular lobes in particular. Both structures are essential for maintaining equilibrium and controlling vestibular coordination associated with aerial movements. However, the brain volume of pterosaurs is intermediate between reptiles and birds (Witmer et al. 2003). Importantly, the cerebrum development is not impressive compared to that of modern birds. Whether pterosaurs are endothermic or ectothermic is not yet resolved (Clarke and Portner 2010). If there is a relationship between endothermy and cerebral expansion, the hypothesis predicts that they had a high metabolic rate and body temperature, but did not reach the level of endothermy as in modern birds.

Finally, the importance of homeothermic endothermy as a necessary step for the cerebrum expansion may be the case for mammals as well. In mammals, the evolution of large and complex brains occurred in three major pulses (Rowe et al. 2011), which may roughly correspond to the evolutionary processes of endothermy in the avian lineage (Lovegrove 2016). According to the examination of nasal respiratory turbinates in synapsids that gave rise to mammals, endothermy was attained relatively early in the lineage even before the earliest eutherian mammals emerged about 160 MYA from the Jurassic of China (Hillenius and Ruben 2004; Luo et al. 2011). In the lineage of synapsids, terrestrial cursoriality might have been the sustained performance requiring aerobic metabolism. Such activities are further associated with olfaction and brain areas for olfaction, such as olfactory bulbs and piriform cortex. Further analysis is warranted to clarify the relationship between the evolution of endothermy and cerebral expansion in mammals.

2.5 Concluding Remarks

In the bird lineage, expansions of the brain have occurred multiple times during the last 300 million years. As seen in Fig. 2.7, discussions in this chapter focused on three possible pulses—at the emergence of amniotes in the Paleozoic, during the theropsid evolution in the Mesozoic, and the most extensive cerebral expansion in the crown birds in the Cenozoic. Thus, it is most likely that truly impressive cognitive and emotional capabilities of birds flourished only after the K-Pg boundary. We discussed several critical selective factors at each pulse of expansion, including terrestrial habitats, parental investment, predatory behaviors, body miniaturization, and increased metabolic rate. The most critical factor for the enlarged cerebrum of

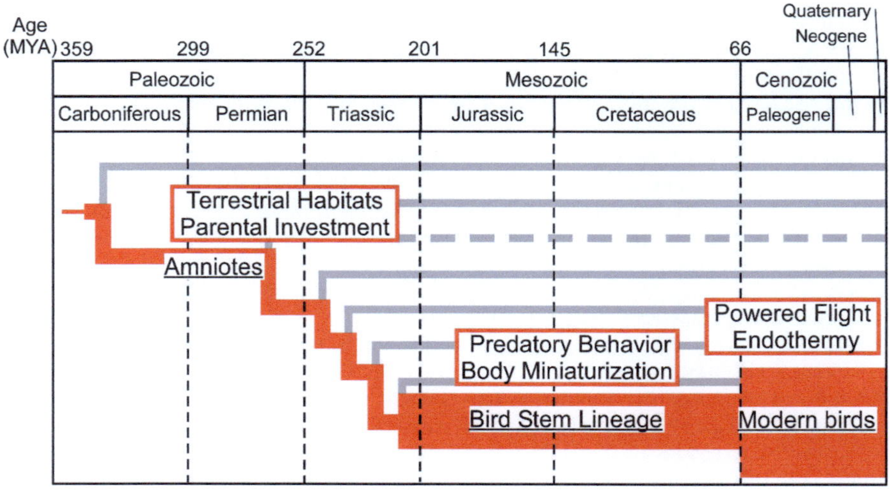

Fig. 2.7 Three pulses of the cerebral expansion along the lineage leading to modern birds. The expansion occurred multiple times—at the emergence of amniotes in the Paleozoic, during the theropsid evolution in the Mesozoic, and the flourishment of modern birds in the Cenozoic. The thickness of the red line schematically represents the degree of cerebral expansion

modern birds might have been the evolution of homeothermic endothermy 66 MYA in order to support long-distance flight. Endurance based on endothermy in turn enabled birds to spend more time and energy to conduct various complex behaviors, which required to develop and enlarge the cerebrum. Without endothermic metabolism, birds might have had a brain like Mesozoic avialans or pterosaurs—having developed visual, vestibular, and motor systems, but a relatively small cerebrum. A large cerebrum is a brain structure for animals that can afford a long duration of sustained activity owing to endothermy. In this respect, the evolution of cognition and emotion in modern birds can be regarded as a fortuitous result of long-range powered flight.

Acknowledgments The authors thank Tadd B. Patton, Michel A. Hofman, Douglas G. Barron, and Lynn B. Martin for critically reading the manuscript and providing helpful suggestions.

References

Ackerman J (2016) The genius of birds. Penguin Press, New York

Alonso PD, Milner AC, Ketcham RA, Cookson MJ, Rowe TB (2004) The avian nature of the brain and inner ear of *Archaeopteryx*. Nature 430(7000):666–669

Balanoff AM, Bever GS, Rowe TB, Norell MA (2013) Evolutionary origins of the avian brain. Nature 501(7465):93–96

Bennett AF, Ruben JA (1979) Endothermy and activity in vertebrates. Science 206(4419):649–654

Benton MJ (2014) How birds became birds. Science 345(6196):508–509

Butler AB, Hodos W (1996) Comparative vertebrate neuroanatomy: evolution and adaptation. Wiley, Hoboken

Callaway E (2014) Rival species recast significance of 'first bird'. Nature 516(7529):18–19

Carroll RL (1988) Vertebrate palaeontology and evolution. W. H. Freeman, New York

Clarke A, Portner HO (2010) Temperature, metabolic power and the evolution of endothermy. Biol Rev Camb Philos Soc 85(4):703–727

Clarke JA (2004) Phylogenetic taxonomy, and systematics of *Ichthyornis* and *Apatornis* (Avialae: Ornithurae). Bull Am Mus Nat Hist 286:1–179

Clarke JA, Tambussi CP, Noriega JI, Erickson GM, Ketcham RA (2005) Definitive fossil evidence for the extant avian radiation in the Cretaceous. Nature 433(7023):305–308

Cross DJ, Marzluff JM, Palmquist I, Minoshima S, Shimizu T, Miyaoka R (2013) Distinct neural circuits underlie assessment of a diversity of natural dangers by American crows. Proc Biol Sci 280(1765):20131046

Emery NJ, Clayton NS (2004) The mentality of crows: convergent evolution of intelligence in corvids and apes. Science 306(5703):1903–1907

Emery NJ, Clayton NS (2005) Evolution of the avian brain and intelligence. Curr Biol 15(23):R946–R950

Ericson PGP, Anderson CL, Britton T, Elzanowski A, Johansson US, Källersjö M, Ohlson JI, Parsons TJ, Zuccon D, Mayr G (2006) Diversification of Neoaves: integration of molecular sequence data and fossils. Biol Lett 2(4):543–547

Feduccia A (1995) Explosive evolution in tertiary birds and mammals. Science 267(5198):637–638

Fraser ON, Bugnyar T (2010) Do ravens show consolation? Responses to distressed others. PLoS One 5(5):e10605

Fraser ON, Bugnyar T (2012) Reciprocity of agonistic support in ravens. Anim Behav 83(1):171–177

Gauthier J, de Queiroz K (2001) Feathered dinosaurs, flying dinosaurs, crown dinosaurs, and the name "Aves". In: Gauthier JA, Gall LF (eds) New perspectives on the origin and early evolution of birds: proceedings of the international symposium in honor of John H. Ostrom. Peabody Museum of Natural History, Yale University, New Haven, pp 7–41

Gill F, Donsker D (2016) IOC World Bird List (v 6.2). doi:10.14344/IOC.ML.6.2

Glut DF (1997) Dinosaurs, the encyclopedia. McFarland, Jefferson

Godefroit P, Cau A, Dong-Yu H, Escuillie F, Wenhao W, Dyke G (2013) A Jurassic avialan dinosaur from China resolves the early phylogenetic history of birds. Nature 498(7454):359–362

Güntürkün O, Bugnyar T (2016) Cognition without cortex. Trends Cogn Sci 20(4):291–303

Haddrath O, Baker AJ (2012) Multiple nuclear genes and retroposons support vicariance and dispersal of the palaeognaths, and an Early Cretaceous origin of modern birds. Proc R Soc Lond B Biol Sci 279(1747):4617–4625

Heinrich B (2009) Mind of the raven. HarperCollins, New York

Hillenius WJ, Ruben JA (2004) The evolution of endothermy in terrestrial vertebrates: Who? When? Why? Physiol Biochem Zool 77(6):1019–1042

Hopson J, Gans C (1979) Paleoneurology. In: Gans C, Northcutt RG, Ulinski P (eds) Biology of the reptilia, vol 9. Academic, New York, pp 39–146

Hopson JA (1977) Relative brain size and behavior in archosaurian reptiles. Ann Rev Ecol Syst 8:429–448

Hutchinson JR, Bates KT, Molnar J, Allen V, Makovicky PJ (2011) A computational analysis of limb and body dimensions in *Tyrannosaurus rex* with implications for locomotion, ontogeny, and growth. PLoS One 6(10):e26037

Huxley TH (1868) On the animals which are most nearly intermediate between birds and reptiles. Ann Mag Nat Hist 4(2):66–75

Jarvis ED, Gunturkun O, Bruce L, Csillag A, Karten H, Kuenzel W, Medina L, Paxinos G, Perkel DJ, Shimizu T, Striedter G, Wild JM, Ball GF, Dugas-Ford J, Durand SE, Hough GE, Husband S, Kubikova L, Lee DW, Mello CV, Powers A, Siang C, Smulders TV, Wada K, White SA, Yamamoto K, Yu J, Reiner A, Butler AB (2005) Avian brains and a new understanding of vertebrate brain evolution. Nat Rev Neurosci 6(2):151–159

Jarvis ED, Mirarab S, Aberer AJ, Li B, Houde P, Li C, Ho SYW, Faircloth BC, Nabholz B, Howard JT, Suh A, Weber CC, da Fonseca RR, Li J, Zhang F, Li H, Zhou L, Narula N, Liu L, Ganapathy G, Boussau B, Bayzid MS, Zavidovych V, Subramanian S, Gabaldón T, Capella-Gutiérrez S, Huerta-Cepas J, Rekepalli B, Munch K, Schierup M, Lindow B, Warren WC, Ray D, Green RE, Bruford MW, Zhan X, Dixon A, Li S, Li N, Huang Y, Derryberry EP, Bertelsen MF, Sheldon FH, Brumfield RT, Mello CV, Lovell PV, Wirthlin M, Schneider MPC, Prosdocimi F, Samaniego JA, Velazquez AMV, Alfaro-Núñez A, Campos PF, Petersen B, Sicheritz-Ponten T, Pas A, Bailey T, Scofield P, Bunce M, Lambert DM, Zhou Q, Perelman P, Driskell AC, Shapiro B, Xiong Z, Zeng Y, Liu S, Li Z, Liu B, Wu K, Xiao J, Yinqi X, Zheng Q, Zhang Y, Yang H, Wang J, Smeds L, Rheindt FE, Braun M, Fjeldsa J, Orlando L, Barker FK, Jønsson KA, Johnson W, Koepfli K-P, O'Brien S, Haussler D, Ryder OA, Rahbek C, Willerslev E, Graves GR, Glenn TC, McCormack J, Burt D, Ellegren H, Alström P, Edwards SV, Stamatakis A, Mindell DP, Cracraft J, Braun EL, Warnow T, Jun W, Gilbert MTP, Zhang G (2014) Whole-genome analyses resolve early branches in the tree of life of modern birds. Science 346(6215):1320–1331

Jerison HJ (1969) Brain evolution and dinosaur brains. Am Nat 103(934):575–588

Jerison HJ (1973) Evolution of the brain and intelligence. Academic, New York

Jetz W, Thomas GH, Joy JB, Hartmann K, Mooers AO (2012) The global diversity of birds in space and time. Nature 491(7424):444–448

Kurochkin EN, Dyke GJ, Saveliev SV, Pervushov EM, Popov EV (2007) A fossil brain from the Cretaceous of European Russia and avian sensory evolution. Biol Lett 3(3):309–313

Larsson HC, Sereno PC, Wilson JA (2000) Forebrain enlargement among nonavian theropod dinosaurs. J Vert Paleontol 20(3):615–618

Lee MSY, Cau A, Naish D, Dyke GJ (2014a) Morphological clocks in palaeontology, and a mid-Cretaceous origin of crown Aves. Syst Biol. doi:10.1093/sysbio/syt110

Lee MSY, Cau A, Naish D, Dyke GJ (2014b) Sustained miniaturization and anatomical innovation in the dinosaurian ancestors of birds. Science 345(6196):562–566

Livezey BC, Zusi RL (2007) Higher-order phylogeny of modern birds (Theropoda, Aves: Neornithes) based on comparative anatomy. II Analysis and discussion. Zool J Linn Soc 149(1):1–95

Lovegrove BG (2016) A phenology of the evolution of endothermy in birds and mammals. Biol Rev 92(2):1213–1240. doi:10.1111/brv.12280

Luo Z-X, Yuan C-X, Meng Q-J, Ji Q (2011) A Jurassic eutherian mammal and divergence of marsupials and placentals. Nature 476(7361):442–445

Martin LD, Zhou Z (1997) *Archaeopteryx*-like skull in Enantiornithine bird. Nature 389(6651):556–556

Marzluff J, Angell T (2013) Gifts of the crow: how perception, emotion, and thought allow smart birds to behave like humans. Simon & Schuster, New York

Navarrete AF, Reader SM, Street SE, Whalen A, Laland KN (2016) The coevolution of innovation and technical intelligence in primates. Philos Trans R Soc Lond B Biol Sci 371(1690):20150186

Northcutt RG (1981) Evolution of the telencephalon in nonmammals. Annu Rev Neurosci 4:301–350

O'Leary MA, Bloch JI, Flynn JJ, Gaudin TJ, Giallombardo A, Giannini NP, Goldberg SL, Kraatz BP, Luo Z-X, Meng J, Ni X, Novacek MJ, Perini FA, Randall ZS, Rougier GW, Sargis EJ, Silcox MT, Simmons NB, Spaulding M, Velazco PM, Weksler M, Wible JR, Cirranello AL (2013) The placental mammal ancestor and the post–K-Pg radiation of placentals. Science 339(6120):662–667

Olkowicz S, Kocourek M, Lučan RK, Porteš M, Fitch WT, Herculano-Houzel S, Němec P (2016) Birds have primate-like numbers of neurons in the forebrain. Proc Natl Acad Sci U S A 113(26):7255–7260. doi:10.1073/pnas.1517131113

Ostrom JH (1973) The ancestry of birds. Nature 242(5393):136–136

Reiner A, Medina L, Veenman CL (1998) Structural and functional evolution of the basal ganglia in vertebrates. Brain Res Brain Res Rev 28(3):235–285

Reiner A, Perkel DJ, Bruce LL, Butler AB, Csillag A, Kuenzel W, Medina L, Paxinos G, Shimizu T, Striedter G, Wild M, Ball GF, Durand S, Gunturkun O, Lee DW, Mello CV, Powers A, White SA, Hough G, Kubikova L, Smulders TV, Wada K, Dugas-Ford J, Husband S, Yamamoto K, Yu J, Siang C, Jarvis ED (2004) Revised nomenclature for avian telencephalon and some related brainstem nuclei. J Comp Neurol 473(3):377–414

Roots C (2006) Flightless birds. Greenwood Press, Santa Barbara

Rowe TB, Macrini TE, Luo Z-X (2011) Fossil evidence on origin of the mammalian brain. Science 332(6032):955–957

Sereno PC, Dutheil DB, Iarochene M, Larsson HCE, Lyon GH, Magwene PM, Sidor CA, Varricchio DJ, Wilson JA (1996) Predatory dinosaurs from the Sahara and Late Cretaceous faunal differentiation. Science 272(5264):986–991

Shanahan M, Bingman VP, Shimizu T, Wild M, Güntürkün O (2013) Large-scale network organization in the avian forebrain: a connectivity matrix and theoretical analysis. Front Comput Neurosci 7:89

Shettleworth SJ (2012) Fundamentals of comparative cognition. Oxford University Press, Oxford

Shimizu T (2001) Evolution of the forebrain in tetrapods. In: Roth G, Wullimann MF (eds) Brain evolution and cognition. Wiley, New York, pp 135–184

Shimizu T, Watanabe S (2012) The avian visual system: overview. In: Lazareva OF, Shimizu T, Wasserman EA (eds) How animals see the world: comparative behavior, biology, and evolution of vision. Oxford University Press, New York, pp 473–482

Trivers R (1972) Parental investment and sexual selection. In: Campbell BG (ed) Sexual selection and the descent of man 1871–1971. Aldine, Chicago, pp 136–179

Tucker VA (1998) Gliding flight: speed and acceleration of ideal falcons during diving and pull out. J Exp Biol 201(3):403–414

Tucker VA, Cade TJ, Tucker AE (1998) Diving speeds and angles of a gyrfalcon (*Falco rusticolus*). J Exp Biol 201(13):2061–2070

Turner AH, Makovicky PJ, Norell M (2012) A review of dromaeosaurid systematics and paravian phylogeny. Bull Am Mus Nat Hist 371:1–206

Wasserman EA, Zentall TR (2006) Comparative cognition: experimental explorations of animal intelligence. Oxford University Press, Oxford

Witmer LM, Chatterjee S, Franzosa J, Rowe T (2003) Neuroanatomy of flying reptiles and implications for flight, posture and behaviour. Nature 425(6961):950–953

Witton MP (2013) Pterosaurs: natural history, evolution, anatomy. Princeton University Press, Princeton

Xu X, Guo Y (2009) The origin and early evolution of feathers: insights from recent paleontological and neontological data. Vertebrat Palasiatic 47(4):311–329

Xu X, Wang K, Zhang K, Ma Q, Xing L, Sullivan C, Hu D, Cheng S, Wang S (2012) A gigantic feathered dinosaur from the lower Cretaceous of China. Nature 484(7392):92–95

Xu X, Zhou Z, Wang X, Kuang X, Zhang F, Du X (2003) Four-winged dinosaurs from China. Nature 421(6921):335–340

Zheng X, O'Connor J, Wang X, Wang M, Zhang X, Zhou Z (2014) On the absence of sternal elements in Anchiornis (Paraves) and Sapeornis (Aves) and the complex early evolution of the avian sternum. Proc Natl Acad Sci U S A 111(38):13900–13905

Chapter 3
The Evolution of Mammalian Brains from Early Mammals to Present-Day Primates

Jon H. Kaas

Abstract The changes in brains as they evolved from early mammals to modern humans involved a great expansion of overall brain size and especially neocortex. Early mammals were small, and their small brains had a proportionately small cap of neocortex divided into approximately 20 functionally distinct areas. Numbers of areas and the size of the cortical sheet increased with the first primates and contained 40–50 cortical areas. Overall the six million years of the evolution of modern humans from early bipedal apes, brains evolved from a great ape size of 400 cc to roughly 1400 cc, with neocortex having an estimated 200 cortical areas occupying 80% of the brain. This cortical mass of 16 billion neurons, together with a high level of hemispheric specialization, appears to be critically involved in mediating the impressive cognitive abilities of modern humans.

Keywords Cortical areas • Thalamus • Marsupials • Monotremes • Prosimians • Monkeys • Apes • Humans

3.1 Introduction

The brains of the 5400 or so species of present-day mammals vary greatly in structure and function, especially at the level of the neocortex. Early mammals of 250 million years ago (mya) were cat or rat or even smaller in size (O'Leary et al. 2013; Rowe et al. 2011; Smith et al. 2010), and they had small brains with proportionately less neocortex than in most extant mammals (Kielan-Jaworowska et al. 2004; Rowe et al. 2011). Many of the early mammals were near the sizes of the smallest of extant mammals, meaning that they were near the lower limit in size for mammals (Schmidt-Nielson 1984), and evolved variations in body size were largely restricted

J.H. Kaas
Department of Psychology, Vanderbilt University, 301 Wilson Hall, 111 21st Avenue South, Nashville, TN 37240-7817, USA
e-mail: jon.h.kaas@vanderbilt.edu

S. Watanabe et al. (eds.), *Evolution of the Brain, Cognition, and Emotion in Vertebrates*, Brain Science, DOI 10.1007/978-4-431-56559-8_3

to larger sizes (Cope's rule; Baker et al. 2015; Smith et al. 2010). Larger body sizes are generally associated with larger brains (Jerison 1973), and larger brains present more opportunities for the evolution of impressive variations. Most notably, larger brains within a clade have more neurons (e.g., Herculano-Houzel et al. 2007) and may have more cortical areas (Brodmann 1909; Kaas 2008; Van Essen et al. 2012a, b). However, even mammals with relatively small brains sometimes have specialized systems and areas that are impressively unique (Kaas 2011a). Modifications of the somatosensory system of the duck-billed platypus for mediating electroreception in addition to touch (Krubitzer et al. 1995) and the auditory system for echolocation in microbats (Covey 2005) are good examples. Here, we start with a reconstruction of the brains and especially the neocortex of early mammals, based on evidence from the fossil record and from comparative studies of the brains of extant mammals. Then, we consider the likely organization of the brains of placental mammals that are most closely related to primates, and the inferred organization of the brains of early primates, early anthropoid primates, and finally apes and humans. Thus, we closely follow a single line of mammalian evolution while necessarily neglecting other interesting lines and their outcomes. Yet, a focus on the evolution of modern human brains is quite justified, in part by our general interest in ourselves, and how we got here, and in part because this interest has generated a lot of relevant studies of how primate brains are organized.

3.2 Early Mammals

Early mammals evolved from one line of stem amniotes (those able to lay eggs on dry land), the synapsid clade that gave rise to early "mammal-like reptiles" and then early mammals some 200 or more mya, with only mammals as surviving members. The other surviving branch from stem amniotes, the sauropsid clade, gave rise to present-day reptiles and birds. The emergence of one part of the brain, the neocortex, has been considered a mammalian innovation, but that is not fully correct, as neocortex is homologous with the dorsal cortex of reptiles and wulst of birds (Butler and Hodos 2005). Yet, stem amniotes, some 340 mya, likely had something like the dorsal cortex of reptiles, which seems to have changed little in most reptiles while becoming a thick mass of neurons in birds. In contrast, over the course of the evolution of the synapsid ancestors of early mammals, the small, thin cap of forebrain dorsal cortex, likely consisting a single layer of pyramidal neurons with a scattering of small, mostly inhibitory neurons as in reptiles (Shepherd 2011), was transformed into the thick, multilayered neocortex that has been retained by all extant mammals. Other parts of cortex became the piriform (olfactory) cortex and the hippocampus of mammals.

The primitive neocortex presented a fantastic opportunity for further modification in the decedents of early mammals, as the laminar organization of neocortex allowed for the functional specialization and differentiation of layers, while modular specialization of sets of neurons across depth of cortex also added considerable flexibility (Kaas 2012). Most importantly, neocortex could be subdivided into an increasing number of cortical areas, the so-called organs of the brain (Brodmann 1909), as brains got bigger and had more neurons. The various specializations of these areas and the ability to process information in a series of steps from cortical area to cortical area permitted the evolution of cortical systems with impressive computational powers. The computational flexibility of neocortex suggests why the proportion of brains that consisted of neocortex increased in many lines of mammal evolution, as brains get bigger, and the large brains of humans are 80% neocortex (Hofman 1988; Azevedo et al. 2009; Herculano-Houzel 2012). Thus, much of this review is on the enlargement and areal organization of neocortex in evolution. Developmentally, neocortex emerges late, and extending the generation period of neocortical neurons has been a major mechanism of neocortical enlargement in evolution (Finlay and Darlington 1995).

Our present understanding of the organization of neocortex in early mammals depends on part on the inferences that can be made from the fossil record. The endocasts of skulls indicate that early mammals had small brains with little neocortex, but with relatively large olfactory bulbs and olfactory (piriform) cortex (Kielan-Jaworowska et al. 2004). These early mammals had small eyes and were likely nocturnal and feeding on insects and small vertebrates (e.g., Rowe et al. 2011). The small eyes promoted the nocturnal bottleneck theory (see Hall et al. 2012) that suggested a loss of adaptations for photopic vision. This perhaps resulted in a shift of the focus of visual processing from the optic tectum of reptiles and birds and the superior colliculus of mammals to the visual cortex of mammals (Diamond and Hall 1969). Major adaptations of these early mammals included the evolution of high-frequency hearing (Coleman and Boyer 2012; Allman 1999), the evolution of tactile vibrissae for the detection of nearby objects (Muchlinski 2010), and the domination of olfaction (Streidter 2005).

The small cap of neocortex that was revealed by the most favorable endocasts of the sculls of early mammals was likely subdivided into a small number of cortical areas, but the fossil record does not reveal these subdivisions. However, some of the major fissures of neocortex of mammals with large brains are sometimes revealed, and they may indicate aspects of functional organization (e.g., Radinsky 1976). Inferences about the internal anatomical and functional organizations of the neocortex and brains of early mammals are necessarily based on comparative studies of the brains of extant mammals. In that regard, the most informative studies are likely to be from mammals with small brains, and few obvious behavioral specializations, thus resembling early mammals (Kaas 2016). Common features of brain organization across members of the six major clades of the mammalian radiation likely reflect those that have been retained from early common ancestors (Kaas 2011b).

Comparative studies of the areal organization of neocortex have revealed a number of consistent features (Fig. 3.1). For example, all studied mammals have a primary somatosensory area, S1, including mammals with little neocortex (e.g., Catania et al. 1999). There is good evidence from many such mammals that narrow strips of cortex along the rostral and caudal borders of S1 have patterned connections with S1 and thus are somatosensory (e.g., Ebner and Kaas 2015). In addition,

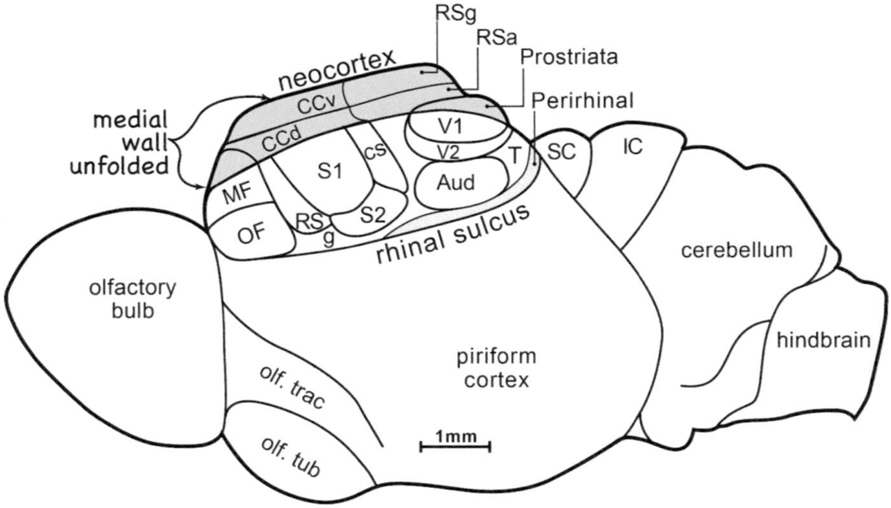

Fig. 3.1 The proposed organization of the neocortex of early mammals. Note the dominance of the large olfactory system, the small cap of neocortex that fails to cover the midbrain, and the lack of motor and premotor areas, as well as a lack of a corpus callosum. Somatosensory areas, with motor functions, included a primary area, S1; a second area, S2; possibly a parietal ventral area, PV; rostral and caudal somatosensory areas, RS and CS; possibly a gustatory area, G; and a small region of posterior parietal cortex. Visual areas included primary and second areas, V1 and V2, prostriata, and a temporal visual area, T. Prefrontal cortex included medial and orbital divisions, MF and OF. Cortex of the medial wall of the cerebral hemisphere included dorsal and ventral divisions of cingulate cortex, CCd and CCv, and granular and agranular divisions of retrosplenial cortex, RSq and RSa. A larger hippocampus (not shown) folded under from medial-caudal entorhinal cortex

the region of the second somatosensory area, S2, has been found to consist of two areas, S2, and parietal ventral area, PV, at least in most mammals (e.g., Remple et al. 2003; Beck et al. 1996). Thus, the neocortex of early mammals likely contained four or five somatosensory areas. While a gustatory region or area, G, has not been studied in many mammals, such an area likely existed in early mammals. All studied mammals also have a region of lateral cortex that responds to sounds and has the architectonic features of primary auditory cortex. However, it is not clear from the limited comparative evidence if early mammals had more than one primary area, as well as other auditory areas, as even small mice have several auditory areas (Kaas 2011c). Primary visual cortex, V1 or area 17, appears to be present in all or most mammals, although it can be reduced to a narrow strip of cortex with few or no visual function in mammals that live underground, such as the "blind" mole rat (Cooper et al. 1993). A second visual area, V2, a small visual area on the medial border of V1, area prostriata, and a visual portion of temporal cortex have all been widely identified across mammals (Rosa and Krubitzer 1999).

Only a few other cortical areas have been widely recognized across members of the major braches of the mammalian radiation. These areas include the architectonically distinct granular and agranular retrosplenial areas, RSa and RSag; dorsal and ventral subdivisions of cingulate cortex, CCd and CCv; and medial and orbital divisions of prefrontal cortex, MF and OF. The perirhinal cortex likely had functional divisions, and the endorhinal cortex was present next to the hippocampus. Overall, the early common ancestors of all extant mammals have roughly 20 cortical areas and possibly more. As the cap of neocortex of early mammals was small, and the proper functioning of any cortical area depends on being large enough to contain all the neurons needed for its function (Kaas 2000), early mammals had few cortical areas. However, some small extant mammals even appear to have lost some areas in comparison to ancestors (Catania et al. 1999).

3.3 Placental Mammals

Early mammals slowly started to diverge soon after they emerged, with the monotreme line separating from the marsupial plus placental line about 165 mya and the marsupials from the placentals some 130–150 mya (Bininda-Emonds et al. 2007; Nilsson et al. 2010). The four placental superorders (Afrotheria, Euarchontoglires, Laurasiatheria, and Xenarthra) separated about 100 mya, and most of the orders of extant mammals were established by 85 mya, with some members surviving the mass extinction event 65 mya. Overall, the major branches of the mammalian tree are old, the rate of change was often slow, and diversification rates were low until 55–50 mya (Bininda-Emonds et al. 2007). This is especially the case for anthropoid primates with New World monkeys, Old World monkeys, apes, and human lineages all emerging in the last 45 mya. This suggests that the brains of many mammalian species have not changed that much from those of early ancestors, while others have changed greatly within the last 40–50 million years. The comparative evidence is largely consistent with this scenario. However, an important advance was the evolution of the placenta, which allowed the embryonic development period for the brain to be greatly extended. This resulted in longer prenatal development times for the brain and reduced the need for the early development of neural mechanisms for postnatal functions. This advantage likely relates to the relatively great success of placental mammals over monotremes and marsupials. Placental mammals also had two major innovations in brain organization. First, only placental mammals have the corpus callosum, which first supplemented the functions of the anterior commissure, and then largely replaced them. Second, present evidence largely supports the conclusion that only placental mammals have a primary motor area, M1, and premotor cortex (e.g., Beck et al. 1996). It is not clear how motor cortex emerged, but the dysgranular cortex on the rostral border of S1 has important motor functions, as does S1 to a lesser extent (Young et al. 2012). Thus, motor cortex may have emerged as an elaboration of dysgranular sensorimotor cortex. The motor functions of neocortex in monotremes and marsupials appear to be mediated by somatosensory cortex (see Ashwell 2013).

While early mammals had roughly 20 or more cortical areas, and a similar number of nuclei in the dorsal thalamus, the architectonic differentiation of cortical areas, thalamic nuclei, and midbrain colliculi was poor. Little changed for most mammals, including early placental mammals. However, cellular and laminar specializations of cortical areas and subcortical nuclei occurred to varying extents independently in many different lines of mammalian evolution. The dorsal lateral geniculate nucleus, as a well-studied example, has poorly differentiated layers, and little variation in neuron sizes and features in most mammals, but markedly different patterns of lamination and neuron anatomy emerged in different lines of mammalian evolution as the importance of vision increased (see Kaas et al. 1972). Other examples of anatomical specializations of thalamic nuclei are illustrated by Jones (2007). These anatomical changes mean that the functions of some structures were substantially changed in some lines of evolution, while the structures themselves were retained from early common ancestors, and thus are homologous as structures.

3.4 The Brain of Early Primates

Primates are part of the Euarchontoglires superorder of placental mammals. Other Superorders of placental mammals include Xenarthra (e.g., sloths, armadillos), Afrotheria (e.g., elephants, tenrecs), and Laurasiatheria (e.g., cats, moles, and bats). Euarchontoglires include Glires (rodents and rabbits) and archontans (gliding lemurs, tree shrews, and primates). The primate line diverged from the other Euarchontoglires lines roughly 100 mya, and the last common ancestors of all present-day primates probably existed as early as 80 or more mya, although fossil evidence of primate evolution goes back only 55 mya or so (e.g., Steiper and Seiffert 2012). Thus, it is not clear when the morphological characteristic of modern primates first emerged. Modern primates have large brains relative to body size, and they have more parietal and temporal cortex, and parts of frontal cortex than other Archontoglires, and depend more on vision and less on olfaction (Martin 1990). The fossil record suggests that many of these brain changes emerged within the last 50 mys (e.g., Silcox et al. 2010). However, the earliest fossil primate already had grasping hands and feet with nails instead of claws and forward-facing eyes. These features are likely adaptations to living in the terminal branches of tropical trees where insects, buds, and fruits are available as food (Bloch and Boyer 2002). Their lifestyle placed an emphasis on the evolution of neural systems for processing visual inputs, especially in the aid of visual predation. Olfaction became less important. As related adaptations, early primates were small and nocturnal. Thus, the small mouse lemurs from the island of Madagascar have been considered as a "living model" of early primates (Gebo 2004). However, the brains of these 40–65 g primates have not been studied enough to provide an overview of their organization. Mouse lemurs are members of one branch of radiation of strepsirrhine primates. Besides the lemurs of Madagascar, other strepsirrhine primates include the loris of Southeast Asia and the galagos of Africa. Fortunately the organization of the forebrain of galagos has been

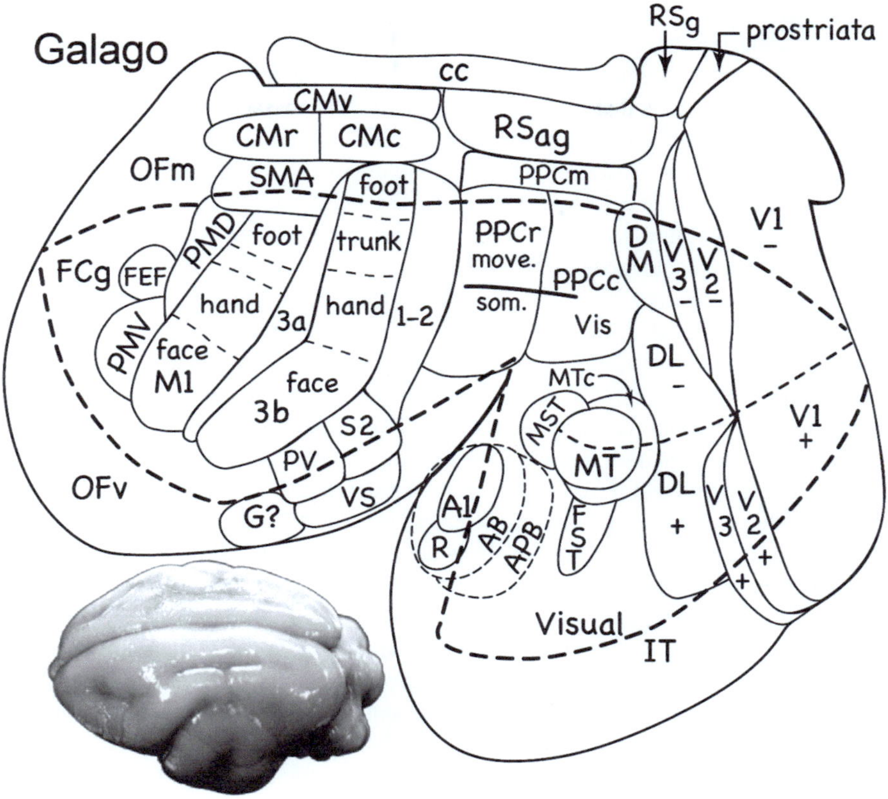

Fig. 3.2 Cortical areas on the flattened cortex of a galago, a strepsirrhine primate. Visual areas V1, V2, and prostriata have been retained from non-primate ancestors. Area V3, the dorsolateral visual area DL (V4); the dorsomedial visual area DM; the middle temporal visual area, MT; the MT crescent, MTc; the middle superior temporal area, MST; and the fundal area of the superior temporal sulcus, FST all appear to be primate innovations. Inferior temporal cortex, IT, likely has 4–5 subdivisions. The auditory core contains primary auditory cortex, A1, and a rostral area, R, as well as an auditory belt (AB) and a parabelt, APB. Somatosensory cortex includes S1 (area 3b), area 3a (see RS in Fig. 3.1), area 1 or 1 plus 2 (see CS in Fig. 3.1), S2 and PV, as well as a new ventral somatosensory area or areas, VS. A gustatory (taste) area (G) is likely; posterior parietal cortex, PPC, is large and divided into a caudal portion dominated by visual inputs, and a rostral portion with motor functions (see text). Motor cortex includes a primary motor area, M1; dorsal and ventral premotor areas, PMV and PMD; a supplementary motor area, SMA; a frontal eye field, FEF; and two or more cingulate motor areas, CMr and CMc. Retrosplenial cortex includes granular (RAg) and agranular (RSag) divisions. Medial and orbital divisions of frontal cortex, OFm and OFv, separate regions of granular prefrontal cortex, PFg, a primate innovation. The corpus callosum, CC, is new with placental mammals

extensively studied, allowing comparisons with the organization of the forebrains of New and Old World monkeys.

The known organization of the neocortex of galagos (Fig. 3.2) shows many of the characteristic features of primates of the closest relatives of primates that have been well studied, rats, mice, squirrels, and tree shrews. For the visual system, areas V1, V2, and prostriata have been retained, but modified from the organization in non-primate ancestors as inferred from present-day rodents and tree shrews. Thus, V1 has primate specializations that include sublaminar segregations in sublayers 4a and 4b of magnocellular and parvocellular classes of thalamic inputs and an orderly distribution of cytochrome oxidase ("blobs" (modules) in layer 3 (Casagrande and Kaas 1994). The grouping of cells that selectively respond to the orientation of visual bars and lines is a modular feature of V1 in primates that is shared with tree shrews but not with rodents (Van Hooser 2007). In galagos, V2 has a weak expression of band-like modules reflecting different processing streams that are more prominent in anthropoid primates (Collins et al. 2001). A third visual area, V3, has been found in galagos (Lyon and Kaas 2002) but not in rodents or tree shrews. Higher-order visual areas include the dorsolateral visual area, DL or V4; the dorsomedial visual area, DM; the middle temporal visual area MT; and the associated areas of the MT cortex (MTc, MST, FST). In addition, the inferior temporal region is visual, and it contains several visual areas that have not yet been well defined. In posterior parietal cortex, a region that greatly expanded in primates, caudal PPC is predominantly visual (Stepniewska et al. 2016). Most of these areas beyond V1 and V2 have no known homologues in non-primate mammals and are likely innovations that emerged in early primates. Subcortically, the superior colliculus of the midbrain has evolved to greatly increase the magnitude of the retinal inputs from the ipsilateral eye while losing the inputs from the temporal retina of the contralateral eye (Kaas 2014). These primates have a superior colliculus that represents the contralateral visual hemifield with inputs from both eyes rather than the whole retina of the contralateral eye as in non-primates. The dorsal lateral geniculate has been altered so that three classes of inputs from the retina are separated into the M (magnocellular), P (parvocellular), and K (koniocellular) layers in a pattern characteristic of primates and different from non-primates (Kaas et al. 1978; Kaas 2014; Casagrande 1994). Overall, visual cortex and subcortical visual structures have been greatly modified in primates and how these changes occurred as primates evolved from non-primate ancestors is not clear.

Much less is known about the organization of the auditory cortex and how it differs from other mammals, but there is evidence for at least two primary areas, A1 and rostral area R, of other primates (see Kaas 2011c), and for two regions of higher-level processing, the auditory belt (AB) and auditory parabelt (APB) cortex. Each of these regions is likely subdivided into functionally specialized areas, as for monkeys (Kaas and Hackett 2000), but this is not yet known. Multiple auditory areas have been identified in other mammals, such as cats and rats (Kaas 2011c). Multiple areas seem to have evolved in different clades, independently, at least in part, so it is not yet certain what areas and features of auditory cortex in primates are primate innovations.

Somatosensory cortex in galagos includes anterior, lateral, and posterior divisions of parietal cortex. Primary somatosensory cortex (S1) corresponds to area 3b of anthropoid monkeys, but it is not as well differentiated architectonically from adjoining cortex as in anthropoid primates (Sur et al. 1980). Yet, area 3b responds well to tactile stimuli, and it represents the contralateral body surface in a foot to face medial to lateral sequence. As in other primates, a narrow strip of dysgranular cortex, area 3a, exists in the rostral border of S1, which has both tactile and proprioceptive functions. Cortex along the caudal border of S1 has an orderly pattern of connections with S1, demonstrating a somatotopic representation parallel to S1, similar to that of area 1 of anthropoid primates. It is not yet clear if this cortex corresponds to area 1 alone or both areas 1 and 2 of anthropoid primates (Wu and Kaas 2003). Overall, this area 1, or area 1 plus area 2, resembles the cortex along the caudal border of S1 in rodents and tree shrews, more than the more distinct, and responsive, representations in areas 1 and 2 of monkeys. It seems likely that a single, poorly differentiated body representation evolved into area 1 and area 2 in early monkeys and that areas 1 and 2 were retained in all or most present-day anthropoids. Areas PV and S2 are just lateral to S1 (Wu and Kaas 2003), much as they are in rats and squirrels. Thus, these fields have been retained from non-primate ancestors. However, there is also evidence for another area or areas next to PV and S2, the ventral somatosensory area (VS) or areas (see Coq et al. 2004). This seems to be a primate innovation. A gustatory area, G, has not been well defined in galagos, but it likely exists in all mammals. Finally, the PPCr region contains a series of about eight small patches of cortex, we call domains, that have been defined by the complex movements produced by electrical stimulation (Stepniewska et al. 2009; Kaas and Stepniewska 2016). The separate domains evoked eye, protective arm, hand to mouth, reaching, running, and other complex movements when stimulated, and the domains are roughly arranged in a lateromedial sequence from head to foot. The domains appear to exist in all or most primates but not to this extent in the close relatives of primates. Thus they likely emerged with the expansion of posterior parietal cortex in early primates or their immediate ancestors.

Galagos also have a number of cortical motor areas that are shared with other primates (Wu et al. 2000). This includes a primary motor area (M1), ventral (PMV) and dorsal (PMD) premotor areas, frontal eye field (FEF), the supplementary motor area (SMA), and rostral, caudal, and ventral cingulate motor areas (CMA). The motor cortex of rodents (Young et al. 2012) and tree shrews (Remple et al. 2006) includes a primary area, M1, and at least one premotor area, and this is the likely pattern of motor cortex for all placental mammals. But additional motor areas appear to have been added with the evolution of early primates, especially of subdivisions of motor areas concerned with the control of hand and forelimb movements. In addition, small regions or domains where electrical stimulation of cortex evokes complex action specific movements exist in M1 and premotor cortex of galagos and likely all primates. These domains are functionally matched to domains of PPCr (Kaas and Stepniewska 2016). Other mammals may have similar domain-like subdivisions of M1 (e.g., Baldwin et al. 2017), but the types and numbers of such domains are different and fewer. The organizations of M1 and premotor cortex changed with the emergence of primates in that the numbers of domains increased, they were interconnected with functionally matched domains in a newly enlarged rostral region of posterior parietal cortex, and more domain were involved in producing movements of the forelimb and hand in such behaviors as grasping, hand to mouth, and protecting the head with an arm.

Other areas of cortex have been less explored in galagos and other prosimians, but these and all other primates have a granular region of prefrontal cortex that is thought to be an innovation of early primates (Preuss and Goldman-Rakic 1991). Areas of primary sensory cortex are characterized by having densely packed small neurons (granule cells) in layer 4 that are individually activated by only few inputs, thereby preserving this information for further processing. Motor areas, in contrast, are characterized by large pyramidal neurons that have large dendritic arbors that allow them to integrate information from many sources (Kaas 2000). The integration of many sources of information has long been considered to be an important feature of prefrontal cortex, and granular prefrontal cortex does have large pyramidal neurons with widespread dendritic arbors, especially in humans (Elston et al. 2006). So it is interesting that this cortex also is specialized for preserving the details of some of the inputs to small neurons of layer 4.

In summary, we see evidence for a remarkable reorganization of the brain in the ancestors of early primates, so that many changes or innovations occurred that were retained in galagos and most likely other prosimian primates, but also in monkeys and possibly all primates. Major changes involved the visual and motor systems, especially at the cortical level, where we see a sign of a greatly expanded role of neocortex, the "corticalization of function" that so characterizes the human brain (Herculano-Houzel et al. 2016). Comparative studies on the relatives of primates, rodents, and tree shrews don't shed much light on when and how these remarkable changes occurred, but it is clear that most or all the these changes occurred after the divergence of Glires (rodents and lagomorphs) from archontans (tree shrews, gliding lemurs, and primates) and even after the divergence of the ancestors of primates from those of tree shrews some 85 mya.

3.5 The Brains of Early Monkeys

The early haplorhines or "dry-nosed" primates emerged as a branch from the strepsirrhine clade of "wet-nosed" primates over 60 mya (e.g., Murphy et al. 2004). An early branch of early haplorhines evolved into present-day tarsiers, while the other surviving anthropoid branches led to the more widespread New World monkeys (platyrrhines) and Old World monkeys (catarrhines), which eventually produced hominins (apes and humans). Early haplorhines were small and diurnal (Ni et al. 2013; Kay et al. 1997; Seiffert et al. 2010). The line leading to present-day tarsiers reverted back to being nocturnal, with appropriate modifications including extremely large eyes for their small head and body. Tarsiers feed in the fine branches and eat insects and small vertebrates as a specialized visual predator that eats no vegetable matter (Fleagle 1999). Not much is known about their brains, as experimental studies have been limited, but histological studies indicate that both the dorsal lateral geniculate nucleus and the pulvinar resemble those of monkeys in organization more than those of strepsirrhines (Collins et al. 2005; Wong et al. 2010). Primary visual cortex is remarkable in its relative size, 20–30% of neocortex, and its degree of histological differentiations into distinct layers and sublayers. However, these visual specializations in a small brain seem to have come at the cost of having less posterior parietal and prefrontal cortex.

Early monkeys diversified in Africa, with some rare individuals somehow rafting from Africa to South America, perhaps more than once, as early as 30–40 mya (Bond et al. 2015; Bloch et al. 2016). Studies of early African and South American primate fossils reveal similarities that have been lost (Bond et al. 2015; Seiffert 2012). However, these early monkeys had relatively larger brains than their strepsirrhine ancestors, with more neocortex and a more extensive temporal lobe. Early monkeys in both Africa and South America likely had dichromatic color vision, having just two classes of cone photoreceptors (as in most mammals), while surviving African anthropoids are trichromatic, having short, middle, and long wave sensitive cones. Trichromatic vision would seem to offer a considerable advantage in identifying colored fruits and buds as food in the dense arrays of leaves in tropical trees (Regan et al. 2001). As early monkeys diversified, many got larger and depended more on fruits, leaves, and buds for food. Most South American monkeys remained dichromats, while the placement of the gene for the long-wavelength gene on the X chromosome and a slight variation in the gene allowed some XX females to code for different cone pigments on each X chromosome and became trichromats, while all XY males remained dichromats (Jacobs 2008). However, one genus of large New World monkeys has both genes for longer-wave sensitivity pigments (medium and long) on each X chromosome, allowing all females and males to have trichromatic vision. Thus, the advantage of trichromatic vision evolved independently in New and Old World monkeys. This suggests that the independent changes in the retina induced comparable changes for the processing of color information in the visual systems of trichromatic New and Old World monkeys.

New World monkeys evolved into roughly 125 species of various body sizes ranging from the 100–250 g of species of marmosets to the 11–12 kg sizes of howler monkeys. One New World monkey, the owl monkey reverted back to nocturnal living, while others remained diurnal. Old World monkeys are found in Africa and Asia and are generally as large or larger than the New World monkeys. There are over 85 species of Old World monkeys in two major subfamilies, the cercopithecines with cheek pouches for storing food and colobines with adaptations for feeding on leaves (Fleagle 1999). As early monkeys were small, the brains of some of the smaller New World monkeys likely reflect more of the organizations of the brains of early monkeys. Larger monkeys have larger brains, with proportionally more neocortex (Chaplin et al. 2013) and more neurons (Herculano-Houzel et al. 2007), and they likely have more cortical areas (Changizi and Shimojo 2005). The presence of larger brains with proportionally more neocortex implies an increase in the "corticalization of function" (Herculano-Houzel et al. 2016).

The most commonly studied brains of New World monkeys include those of squirrel monkeys, owl monkeys, and marmosets. The brains of the larger cebus and spider monkeys have been studied much less. As most studies have focused on aspects of cortical organization, and the areal organization of cortex is subject to major changes in evolution, our emphasis here remains on the cortex. However,

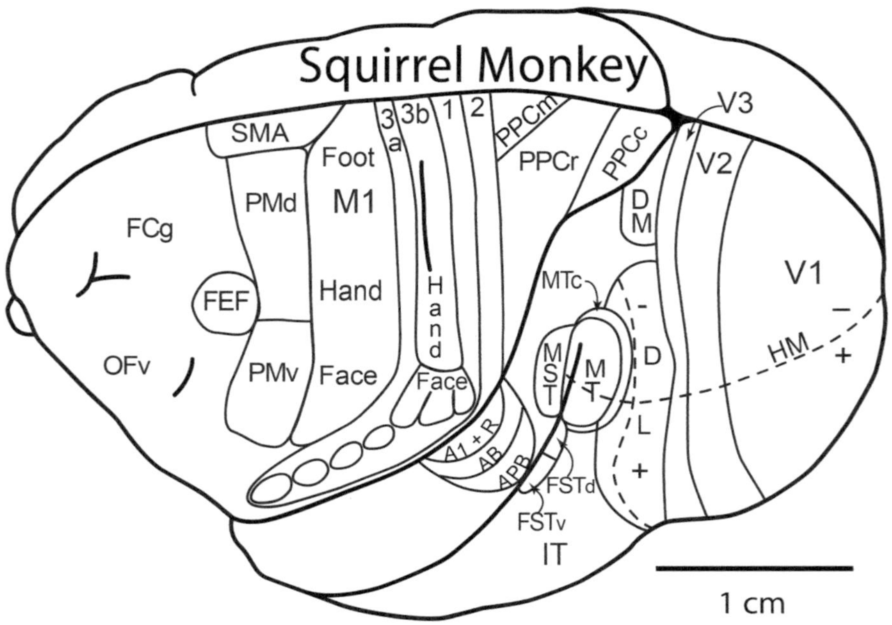

Fig. 3.3 Some of the cortical areas of a squirrel monkey shown on a dorsolateral view of the left cerebral hemisphere. All of the areas of the galago (Fig. 3.2) are present in monkeys, but there is proportionately more neocortex; somatosensory area 2 is distinct from area 1; the ventral somatosensory area is clearly divided into two areas (not shown); visual area FST has dorsal and ventral divisions; and granular frontal cortex, FCg, PPC, and IT cortex have expanded. See Fig. 3.2 for abbreviations and conventions

outside of the primary sensory and motor areas, identifying cortical areas is a difficult process, and many uncertainties and differences of opinion remain. However, the evidence supports the conclusion that even the smaller of the New World monkeys have more cortical areas than strepsirrhine primates, and this means more cortical areas are available for serial processing, and more diversity in cortical specializations is possible.

The proposed areal organization of the neocortex of a squirrel monkey is illustrated in Fig. 3.3. The full number of cortical areas is difficult to estimate as large regions of cortex have been relatively unexplored with experimental procedures, although architectonic features and patterns of connections have been used to define additional areas. The cortex with predominately visual function occupies most of the caudal half of the cortex and includes the visual areas of strepsirrhine primates but likely more areas. The dorsolateral visual area, DL or V4, appears to be subdivided into rostral and caudal halves, with DLr having more connections with the "action" or "dorsal" stream of visual processing, and DLc more connected with the "object identification" or "ventral" processing stream (Kaas and Lyon 2007). The FST region is also divided into dorsal and ventral areas, FSTd with dense connections with MT, and FSTv with dense connections with MTc

(Kaas and Morel 1993). There is clear evidence for V3 (Kaas et al. 2001), although this interpretation has been questioned (e.g., Rosa et al. 2005). The New and Old World monkeys (except owl monkeys) are diurnal, and the amount of cortex devoted to the ventral stream of cortical processing (Ungerleider and Haxby 1994; Goodale and Milner 1992), including a uncertain number of inferior temporal (IT) visual areas, is expanded for better identifying objects including faces for social primates (Ku et al. 2011; Tsao et al. 2008; Hung et al. 2015). The emphasis on visual object identification starts subcortically, with 80% of the retinal ganglion cells projecting to only the parvocellular layers of lateral geniculate nucleus (Weller and Kaas 1989) that activate the cortical ventral stream. Only about 10% of the ganglion cells project to the magnocellular layers, to provide the primary source of visual information for the dorsal "action" stream involving the MT cortex, DM, and much of posterior parietal cortex.

The evolutionary changes in the somatosensory system have been less pronounced. The hand representation in primary somatosensory cortex in New and Old World monkeys is somewhat enlarged, and area 1 is more powerfully driven by touch. Area 2 is also highly responsive to touch, as well as muscle and limb movement. Some of the large spider and cebus New World monkeys have a highly sensitive glabrous pad on the end of their prehensile tail, which serves as an extra limb and has a large representation in somatosensory cortex (Felleman et al. 1983). This is an innovation that is restricted to one line of evolution in New World monkeys.

The larger brains of most Old World monkeys, and perhaps the larger New World monkeys, may have more cortical areas (Changizi and Shimojo 2005; Kaas 2008), but this depends on how they are defined and the types of evidence. Thus, much certainty exists. One recent estimate is that macaque monkeys have 140 areas (Van Essen et al. 2012a), but this may be an underestimate. For comparison, a recent atlas of the small marmoset brain portrays just over 100 cortical areas (Paxinos et al. 2012). However, as this estimate is largely based on architectonic distinctions, it may be an overestimate. In evaluating these estimates, it is important to consider how many neurons would be needed to mediate the proposed functions of an area and what the minimal size of that area might be (Kaas 2000). Some of the proposed cortical areas appear to be too small to be realistic. In addition, modular features of cortex can produce connectional and architectonic differences that are similar to those expressed by areas and thus may be confused with areas.

In summary, the evolved changes in the brains of early to recent monkeys include elaborations of parts of the visual system that enhanced the ability to recognize objects and the faces of individuals in highly social monkeys. This involved an elaboration of the classes of retinal ganglion cells that provided detailed spatial vision and trichromatic color vision and a great expansion of the ventral stream of visual processing. There was also an expansion of parts of the motor and sensory cortex devoted to forepaw use, as more kinds of plant food were used, and food processing with hands and teeth was more needed. As larger species with larger brains emerged, these larger brains provided more capacity, the possibility of more serial processing, and the further specializations of systems, nuclei, and cortical areas.

3.6 The Evolution of Apes and Ape Brains

Apes emerged in East Africa from a line of large bodied catarrhine monkeys some 25–30 mya (Andrews 2015). These monkey-like apes were arboreal, but also partly terrestrial, as suggested by their large size. Their larger sizes and correspondingly larger brains allowed them to compete with monkeys for high-quality food, territory, and stable supplies, such as figs (Jablonski et al. 2000). For a time, the ape radiation was very successful, and they spread to adjoining parts of Europe and Asia. In seasonal climates, the acquired ability to store fat reserves may have become important. Ultimately, the longer maturation times that resulted in slower rates of reproduction, together with the higher energy costs of larger bodies and brains, put apes at a comparative disadvantage with monkeys in times of environmental instability. Apes disappeared from much of their former range. However, the line of African great apes gave rise to a branch leading to present-day gorillas 8 or more mya, the chimpanzee-bonobo line 5–6 mya (producing chimpanzees and bonobos within the last 2 mya), and the line of bipedal hominins that led to modern humans. The last common ancestor of humans and chimpanzees most closely resembled modern chimpanzees in that it was not yet completely bipedal, had a high degree of sexual dimorphism, likely used tools; and probably nested in trees for safety (Andrews 2015). Our ancestors diverged by becoming predominantly bipedal, thus freeing the hands for tool use, food transport, and infant care, with no significant change in brain size. Hominin brains remained in the 400–500 cm^3 range until about 2 mya.

Studies of the brains of the extant great apes provide an opportunity to understand what the brains of our extinct hominin relatives were like. Because of limits on the types of studies that are possible, much of what is known about the brains of apes comes from histological studies of brains obtained after death. As expected some of the sensory and motor areas that have been identified by multiple criteria in monkeys have been identified histologically in chimpanzees. V1 or area 17 in chimpanzees has the distinctive laminar appearance of monkeys, but differs in not having a dense cytochrome oxidase band in the middle of layer 3 (Brodmann's layer 4A), and thus resembles area 17 of human brains in that way (Preuss et al. 1999). This suggests that inputs from the lateral geniculate nucleus have been reduced or eliminated from this sublayer, an interpretation consistent with anatomical results (Tigges and Tigges 1979). Thus, the substrate for processing visual inputs has been modified in great apes with different modifications occurring in humans. What these modifications mean functionally is not yet clear. Area 17 of chimpanzees is larger than in macaque monkeys, as the sheet of neocortex is extremely large, but area 17 is proportionally less of the total of neocortex, suggesting that more cortical areas exist or that other areas have disproportionally enlarged. As in Old World monkeys and humans, area 17 of chimpanzees is divided in layer 4 by broad, distinctive, ocular dominance columns (Tigges and Tigges 1979). Clear evidence for other visual areas, V2, V3, DM, MT, etc., of monkey visual cortex is lacking, but all these areas, and more, one are expected to exist in apes. It is clear that V1 and nearby visual areas, V2 and V3, have high neuron packing densities, as in other primates (Collins et al. 2016).

There is also good architectonic evidence for areas 3a, 3b, 1, and 2 of anterior parietal cortex of chimpanzees (Qi et al. 2008). These areas are arranged in a rostrocaudal sequence from the depth of the central sulcus as they are in monkeys and humans, and they are very similar in architectonic appearance across these primates. Other somatosensory areas such as S2, PV, and VS likely exist but have not been demonstrated. In monkeys, primary auditory cortex consists of a distinctive core of three auditory areas, A1, a rostral area, R, and a rostrotemporal area, RT. A similar core region has been identified by architectonic criteria in chimpanzees, and the identifying features are very similar to those that identify the core in macaque monkeys and human (Hackett et al. 2001). However, it is not yet certain if this core is divided into three primary areas in macaques or if adjoining belt and parabelt auditory regions are functionally organized as in macaques (Kaas and Hackett 2000). Motor cortex organization, at least for primary and secondary motor areas, also seems to be similar in apes and macaques (Woolsey et al. 1960; Bailey et al. 1950; Grunbaum and Sherrington 1901). And a recent neuroimaging study of brain regions activated during grasping revealed a frontoparietal network similar to that of macaques, including presumed homologues of the anterior intraparietal area, AIP, and ventral premotor area of macaques (Hecht et al. 2015). In addition, a study using diffusion tensor imaging (DTI) to reveal the connections of the arcuate fasciculus in primates has suggested that these connections are more similar to those in macaque than in humans (Rilling et al. 2008). However, the chimpanzee brain is not simply a more-than-three-times larger macaque brain. Structural changes in brain organization have occurred, but they are as of yet understood in only a limited way. The proportionally smaller primary sensory and motor areas, for example, suggest that the cortical sheet of chimpanzees is divided into more cortical areas, allowing more steps in cortical processing and greater specialization of cortical areas. Chimpanzees also have marked asymmetries in the gross anatomy of the two cerebral hemispheres, suggesting that different anatomical specializations for hand use and other functions exist as do in humans (Gannon et al. 1998; Hopkins and Cantalupo 2004; Gilissen and Hopkins 2013). As for humans, chimpanzees are predominately right handed (Hopkins et al. 2004).

3.7 The Evolution of the Brains of Modern Humans

The fossil record indicates that the early hominins that diverged from the line that gave size to chimpanzees and bonobos some 6–7 mya were much like early members of the branch that led to chimpanzees and bonobos (Crompton et al. 2008; Robson and Wood 2008). However, these apes adjusted to woodlands that were subject to climate fluctuations and became drier and mixed with grasslands. It became adaptive for these apes to expand their home range for feeding and increase the range of food items in their diet. Walking on two legs allowed these early ancestors to cover more territory, see further and more clearly, and use their forelimbs to carry food to a safe place. Increasingly bipedalism allowed the hand and feet to differently specialize to great advantage. Bipedalism occurred early in hominin evolution, about 6 mya, while the once expected great increase in brain sizes did not occur until 2 mya, and then brain sizes increased from ape size (400 cc) in *Australopithecus* to more than double in early *Homo erectus*, to roughly 1400 cc of modern humans. Over the last 2.5 million years, the hand became increasingly modified from a great apelike beginning to a human hand in ways that would promote tool use (Tocheri et al. 2008). Ancestors from the time of *Homo erectus* spread from Africa to Asia and Europe but retreated or were wiped out with climate change and ice ages. Our species, *Homo sapiens*, emerged about 200,000 years ago in Africa, after a near extinction during a major ice age (Takahata et al. 1995). The evolution of the big human brains provided impressive cognitive and social skills, tool use, and technology and the ability through culture to expand and occupy highly difficult environments. The costs of these big brains include the amount of energy it takes to maintain them, the long developmental period to make them, and maintaining the capacity to store large amounts of (cultural) information. The high-energy cost of the big brains of our ancestors could only be meet by cooking or otherwise processing food, abilities that go back as far as 1.5–2 million years (Wrangham 2009; Herculano-Houzel 2016). The long development time for the bigger brains of our ancestors, resulting in later reproduction in life, was compensated for in a large part by a reduced mortality rate, a longer life span, shorter times between births, and the help of others in child care.

Large brains are important in that they have more neurons, the basic computational units of brains. But for cognition, large numbers of neurons need to be in the neocortex, as elephants have a three-times-larger brain than humans, but fewer neurons in the neocortex (16 billion, humans; 5.6 billion, elephants; Herculano-Houzel 2016). Proportionally more of the neurons in elephants are in the cerebellum, which has important functions, but we largely attribute our cognitive abilities to neocortex. Primates have an advantage over other mammals as primate brains maintain similar levels of neuron packing densities as from smaller to bigger brains, while other mammals have reduced neuronal densities as their brains get bigger (Herculano-Houzel et al. 2007). Additionally, we devote about 80% of the mass of our brains to neocortex (Azevedo et al. 2009; for review see Hofman 2014). Another important factor is where cortical neurons are concentrated. While the results of early studies suggested that neuron densities are the same across all cortices and across mammalian clades, except for a doubling of neurons in primary visual cortex of primates, this is far from the case. As shown most clearly in neuron counts per mm^2 of cortical surface for over 700 squares of cortex in a chimpanzee, neuron densities vary three to fourfold across cortical areas and regions. Neuron counts were very high in primary visual cortex and high in secondary visual areas and in auditory and somatosensory areas but also in frontal granular cortex (Collins et al. 2016). Low neuron packing densities were found in motor and premotor cortex. Similar differences in neuron packing exist in other primates and appear to exist in neocortex of humans (Gabi et al. 2016). Neuron packing densities decrease with increases in average neuron size, and larger neurons sum more inputs. Thus, we can assume that smaller neurons of sensory areas and of frontal granular cortex have the important role of preserving much of the information in their inputs, while the larger neurons of motor areas sum many inputs to best provide motor commands. Another factor that greatly contributes to the impressive abilities of the human brain (Tomasello 2014; Passingham 2008) is the great increase in the numbers of cortical areas and, thus, the steps available for cortical processing (Pinker 2009). These new areas have been added to association cortex, the cortex devoted to higher-order processing. Outside of a few, cortical areas are hard to identify, and the absolute number in the neocortex of human is far from certain. Estimates vary (e.g., Kaas 2006; Van Essen et al. 2012b), but the number is likely over 200, perhaps 4 times more than in early primates. Thus, in both human postnatal development and evolution, the association regions of cortex have expanded most, while the sensory areas expanded little (Hill et al. 2010). Functional imaging studies in human are now starting to present evidence for cortical areas that are unique to humans (e.g., Peeters et al. 2009), in addition to those areas involved in language (e.g., Brauer et al. 2011). Finally, hemispheric specialization has been greatly enhanced in the evolution of modern humans, allowing systems and cortical areas of the two hemispheres to differentially specialize, reduce processing time, and increase the cognitive abilities of humans (Corballis 2007; Van Essen et al. 2012b; Ringo et al. 1994; Brown and Kosslyn 1993).

References

Allman JM (1999) Evolving brains. H. W. Freeman, New York

Andrews P (2015) An ape's view of human evolution. Cambridge University Press, Cambridge

Ashwell K (2013) Neurobiology of monotremes. CSIRO, Collingwood

Azevedo FA, Carvalho LR, Grinberg LT, Farfel JM, Ferretti RE, Leite RE, Jacob FW, Lent R, Herculano-Houzel S (2009) Equal numbers of neuronal and non-neuronal cells make the human brain an isometrically scaled-up primate brain. J Comp Neurol 513:532–541

Bailey D, von Bonin G, McCulloch WS (1950) The isocortex of the chimpanzee. University of Illinois Press, Urbana

Baker J, Meade A, Pagel M, Venditti C (2015) Adaptive evolution toward larger size in mammals. Proc Natl Acad Sci U S A 112:5093–5098

Baldwin MK, Cooke DF, Krubitzer L (2017) Intracortical microstimulation maps of motor, somatosensory, and posterior parietal cortex in tree shrews (Tupaia belangeri) reveal complex movement representations. Cereb Cortex 27(2):1439–1456

Beck PD, Pospichal MW, Kaas JH (1996) Topography, architecture, and connections of somatosensory cortex in opossums: evidence for five somatosensory areas. J Comp Neurol 366:109–133

Bininda-Emonds ORP, Cardillo M, Jones KE, MacPhee Ross DE, Beck RMD, Grenyer R, Price SA, Vos RA, Gittleman JL, Purvis A (2007) The delayed rise of present-day mammals. Nature 446:507–512

Bloch JI, Boyer DM (2002) Grasping primate origins. Science 298:1606–1610

Bloch JI, Woodruff ED, Wood AR, Rincon AF, Harrington AR, Morgan GS, Foster DA, Montes C, Jarmillo CA, Jud NA, Jones DS, MacFadden BJ (2016) First North American fossil monkey and early miocene tropical biotic interchange. Nature 533:243–246

Bond M, Tejedor MF, Campbell K Jr, Chornogubsky L, Novo N, Goin F (2015) Eocene primates of South America and the African origins of New World monkeys. Nature 520:538–541

Brauer J, Anwander A, Friederici AD (2011) Neuroanatomical prerequisites for language functions in the maturing brain. Cereb Cortex 21:459–466
Brodmann K (1909) Vergleichende lokalisationslehre der grosshirnrinde in ihren prinzipien dargestellt auf Grund des Zellenbaues. Barth, Leipzig
Brown H, Kosslyn S (1993) Cerebral lateralization. Curr Opin Neurobiol 3:183–186
Butler AB, Hodos W (2005) Comparative vertebrate neuroanatomy, 2nd edn. Wiley, Hoboken
Casagrande VA, Kaas JH (1994) The afferent, intrinsic, and efferent connections of primary visual cortex in primates. In: Peters A, Rockland K (eds) Cerebral cortex. Plenum Press, New York, pp 201–259
Casagrande VA (1994) A third parallel visual pathway to primate area V1. Trends Neurosci 17:305–310
Catania KC, Lyon DC, Mock OB, Kaas JH (1999) Cortical organization in shrews: evidence from five species. J Comp Neurol 410:55–72
Changizi MA, Shimojo S (2005) Parcellation and area-area connectivity as a function of neocortex size. Brain Behav Evol 66:88–98
Chaplin TA, Yu HH, Soares JG, Gattass R, Rosa MG (2013) A conserved pattern of differential expansion of cortical areas in simian primates. J Neurosci 33:15120–15125
Coleman MN, Boyer DM (2012) Inner ear evolution in primates through the Cenozoic: implications for the evolution of hearing. Anat Rec 295:615–631
Collins CE, Stepniewska I, Kaas JH (2001) Topographic patterns of V2 cortical connections in a prosimian primate (Galago garnetti). J Comp Neurol 431:155–167
Collins CE, Hendrickson A, Kaas JH (2005) Overview of the visual system of Tarsius. Anat Rec A Discov Mol Cell Evol Biol 287:1013–1025
Collins CE, Turner EC, Sawyer EK, Reed JL, Young NA, Flaherty DK, Kaas JH (2016) Cortical cell and neuron density estimates in one chimpanzee hemisphere. Proc Natl Acad Sci U S A 113:740–745
Cooper HM, Herbin M, Nevo E (1993) Visual system of a naturally microphthalmic mammal: the blind mole rat, Spalax ehrenbergi. J Comp Neurol 328:313–350
Coq J-O, Qi H-X, Collins CE, Kaas JH (2004) Anatomical and functional organization of somatosensory areas of the lateral fissure of the New World titi monkey (Callicebus moloch). J Comp Neurol 476:363–387
Corballis MC (2007) The evolution of hemispheric specializations of the human brain. In: Kaas JH, Preuss TM (eds) Evolution of nervous systems, Primates, vol 4. Elsevier, London, pp 379–394
Covey E (2005) Neurobiological specializations in echolocating bats. Anat Rec A Discov Mol Cell Evol Biol 287:1103–1116
Crompton RH, Vereecke EE, Thorpe SK (2008) Locomotion and posture from the common hominoid ancestor to fully modern hominins, with special reference to the last common panin/hominin ancestor. J Anat 212:501–543
Diamond IT, Hall WC (1969) Evolution of neocortex. Science 164:251–262
Ebner FF, Kaas JH (2015) Somatosensory system. In: Paxinos G (ed) The rat nervous system, 4th edn. Elsevier, London, pp 673–699
Felleman DJ, Nelson RJ, Sur M, Kaas JH (1983) Representations of the body surface in areas 3b and 1 of postcentral parietal cortex of Cebus monkeys. Brain Res 268:15–26
Elston GN, Benavides-Piccione R, Elston A, Zietsch B, DeFelipe J, Manger P, Casagrande V, Kaas JH (2006) Specializations of the granular prefrontal cortex of primates: implications for cognitive processing. Anat Rec Part A 288A:26–35
Finlay BL, Darlington RB (1995) Linked regularities in the development and evolution of mammalian brains. Science 268:1578–1584
Fleagle JG (1999) Primate adaptation and evolution, 2nd edn. Academic, San Diego
Gabi M, Neves K, Masseron C, Ribeito P, Ventura-Antunes L, Torres L, Mota B, Kaas JH, Herculano-Houzel S (2016) No relative expansion of the number of prefrontal neurons in primate and human evolution. Proc Natl Acad Sci U S A 113(34):9617–9622

Gannon PJ, Holloway RL, Broadfield DC, Braun AR (1998) Asymmetry of chimpanzee *planum temporale*: human-like pattern of Wernicke's brain language area homolog. Science 279:220–222

Gebo D (2004) A shrew-sized origin for primates. Yearb Phys Anthropol 47:40–62

Gilissen EP, Hopkins WD (2013) Asymmetries of the parietal operculum in chimpanzees (Pan troglodytes) in relation to handedness for tool use. Cereb Cortex 23:411–422

Goodale MA, Milner AD (1992) Separate visual pathways for perception and action. Trends Neurosci 15:20–25

Grunbaum ASF, Sherrington CS (1901) Observations on the physiology of the cerebral cortex of some of the higher apes. Proc R Soc Lond 69:206–209

Hackett TA, Preuss TM, Kaas JH (2001) Architectonic identification of the core region in auditory cortex of macaques, chimpanzees, and humans. J Comp Neurol 441:197–222

Hall MI, Kamilar JM, Kirk EC (2012) Eye shape and the nocturnal bottleneck of mammals. Proc Biol Sci 279:4962–4968

Hecht EE, Gutman DA, Bradley BA, Preuss TM, Stout D (2015) Virtual dissection and comparative connectivity of the superior longitudinal fasciculus in chimpanzees and humans. NeuroImage 108:124–137

Herculano-Houzel S, Collins CE, Wong P, Kaas JH (2007) Cellular scaling rules for primate brains. Proc Natl Acad Sci U S A 104:3562–3567

Herculano-Houzel S (2012) Neuronal scaling rules for primate advantage. Prog Brain Res 195:325–340

Herculano-Houzel S (2016) The human advantage. The MIT Press, Cambridge

Herculano-Houzel S, Kaas JH, de Oliveira-Souza R (2016) Corticalization of motor control in humans is a consequence of brain scaling in primate evolution. J Comp Neurol 524:448–455

Hill J, Inder T, Neil J, Dierker D, Harwell J, Van Essen D (2010) Similar patterns of cortical expansion during human development and evolution. Proc Natl Acad Sci U S A 107: 13135–13140

Hofman MA (1988) Size and shape of the cerebral cortex in mammals. Part II. The cortical volume. Brain Evol 32:17–26

Hofman MA (2014) Evolution of the human brain: when bigger is better. Front Neuroanat 8:15. doi:10.3389/fnana.2014.00015.

Hopkins WD, Cantalupo C (2004) Handedness in chimpanzees (Pan troglodytes) is associated with asymmetries of the primary motor cortex but not with homologous language areas. Behav Neurosci 118:1176–1183

Hopkins WD, Wesley MJ, Izard MK, Hook M, Schapiro SJ (2004) Chimpanzees (Pan troglodytes) are predominantly right-handed: replication in three populations of apes. Behav Neurosci 118:659–663

Hung CC, Yen CC, Ciuchta JL, Papoti D, Bock NA, Leopold DA, Silva AC (2015) Functional mapping of face-selective regions in the extrastriate visual cortex of the marmoset. J Neurosci 35:1160–1172

Jablonski NG, Whitfort MJ, Roberts-Smith N, Qinqi X (2000) The influence of life history and diet on the distribution of catarrhine primates during the Pleistocene in eastern Asia. J Hum Evol 39:131–157

Jacobs GH (2008) Primate color vision: a comparative perspective. Vis Neurosci 25:619–633

Jerison HJ (1973) Evolution of the brain and intelligence. Academic, New York

Jones EG (2007) The thalamus, 2nd edn. Cambridge University Press, Cambridge

Kaas JH (2000) Why is brain size so important: design problems and solutions as neocortex gets bigger or smaller. Brain Mind 1:7–23

Kaas JH (2006) Evolution of the neocortex. Curr Biol 16:R910–R914

Kaas JH (2008) The evolution of the complex sensory and motor systems of the human brain. Brain Res Bull 75:384–390

Kaas JH (2011a) Neocortex in early mammals and its subsequent variations. Ann N Y Acad Sci 1225:28–36

Kaas JH (2011b) Reconstructing the areal organization of the neocortex of the first mammals. Brain Behav Evol 78:7–21
Kaas JH (2011c) The evolution of auditory cortex: the core areas. In: Winer JA, Schreiner CE (eds) The auditory cortex. Springer, New York, pp 407–427
Kaas JH (2012) Evolution of columns, modules, and domains in the neocortex of primates. Proc Natl Acad Sci U S A 109(Suppl 1):10655–10660
Kaas JH (2014) The evolution of the visual system in primates. In: Werner JS, Chalupa LM (eds) The new visual neuroscience. MIT Press, Cambridge, pp 1233–1246
Kaas JH (2016) Approaches to the study of brain evolution. In: Shepherd SV (ed) Handbook of evolutionary neuroscience. Wiley, Chichester
Kaas JH, Guillery RW, Allman JM (1972) Some principles of organization in the dorsal lateral geniculate nucleus. Brain Behav Evol 6:253–299
Kaas JH, Hackett TA (2000) Subdivisions of auditory cortex and processing streams in primates. Proc Natl Acad Sci U S A 97:11793–11799
Kaas JH, Huerta MF, Weber JT, Harting JK (1978) Patterns of retinal terminations and laminar organization of the lateral geniculate nucleus of primates. J Comp Neurol 182:517–553
Kaas J H and Lyon D C (2001) Visual Cortex Organization in Primates: Theories of V3 and adjoining Visual Areas. Progress in Brain Research 134:285–295
Kaas JH, Lyon DC (2007) Pulvinar contributions to the dorsal and ventral streams of visual processing in primates. Brain Res Rev 55:285–296
Kaas JH, Morel A (1993) Connections of visual areas of the upper temporal lobe of owl monkeys: the MT crescent and dorsal and ventral subdivisions of FST. J Neurosci 13:534–546
Kaas JH, Stepniewska I (2016) Evolution of posterior parietal cortex and parietal-frontal networks for specific actions in primates. J Comp Neurol 524:595–608
Kay RF, Ross C, Williams BA (1997) Anthropoid origins. Science 275:797–804
Kielan-Jaworowska Z, Citelli RL, Luo Z-X (2004) Mammals from the age of dinosaurs. NY, Columbia Univ. Press
Krubitzer L, Manger P, Pettigrew J, Calford M (1995) Organization of somatosensory cortex in monotremes: in search of the prototypical plan. J Comp Neurol 351:261–306
Ku SP, Tolias AS, Logothetis NK, Goense J (2011) fMRI of the face-processing network in the ventral temporal lobe of awake and anesthetized macaques. Neuron 70:352–362
Lyon DC, Kaas JH (2002) Connectional evidence for dorsal and ventral V3, and other extrastriate areas in the prosimian primate, Galago garnetti. Brain Behav Evol 59:114–129
Martin RD (1990) Primate origins and evolution. Princeton University Press, Princeton
Muchlinski MN (2010) A comparative analysis of vibrissa count and infraorbital foramen area in primates and other mammals. J Hum Evol 58:447–473
Murphy WJ, Perzner PA, O'Brien SJ (2004) Mammalian phylogenomics comes of age. Trends Genet 20:631–639
Ni X, Gebo DL, Dagosto M, Meng J, Tafforeau P, Flynn JJ, Beard KC (2013) The oldest known primate skeleton and early haplorhine evolution. Nature 498:60–64
Nilsson MA, Churakov G, Sommer M, Van Tran N, Zemann A, Brosius J, Schmitz J (2010) Tracking marsupial evolution using archaic genomic retroposon insertions. PLoS Biol 8(1–9):e1000436
O'Leary MA, Bloch JI, Flynn JJ, Gaudin TJ, Giallombardo A, Giannini NP, Goldberg SL, Kraatz BP, Luo ZX, Meng J, Ni X, Novacek MJ, Perini FA, Randall ZS, Rougier GW, Sargis EJ, Silcox MT, Simmons NB, Spaulding M, Velazco PM, Weksler M, Wible JR, Cirranello AL (2013) The placental mammal ancestor and the post-K-Pg radiation of placentals. Science 339:662–667
Passingham R (2008) What is special about the human brain? Oxford University Press, Oxford
Pinker S (2009) How the mind works. W.W. Norton, New York
Paxinos G, Petrides M, Rosa MG, Tokuno H (2012) The marmoset brain in stereotaxic coordinates. Elsevier, London
Peeters R, Simone L, Nelissen K, Fabbri-Destro M, Vanduffel W, Rizzolatti G, Orban GA (2009) The representation of tool use in humans and monkeys: common and uniquely human features. J Neurosci 29:11523–11539

Preuss TM, Goldman-Rakic PS (1991) Myelo- and cytoarchitecture of the granular frontal cortex and surrounding regions in the strepsirrhine primate Galago and the anthropoid primate Macaca. J Comp Neurol 310:429–474
Preuss TM, Qi H, Kaas JH (1999) Distinctive compartmental organization of human primary visual cortex. Proc Natl Acad Sci U S A 96:11601–11606
Qi HX, Preuss TM, Kaas JH (2008) Somatosensory areas of the cerebral cortex: architectonic characteristics and modular organization. In: Gardner EP, Kaas JH (eds) The senses: a comprehensive reference. Elsevier, London, pp 143–169
Radinsky L (1976) Cerebral clues. Nat Hist 85:54–59
Regan BC, Julliot C, Simmen B, Vienot F, Charles-Dominique P, Mollon JD (2001) Fruits, foliage and the evolution of primate colour vision. Philos Trans R Soc Lond B Biol Sci 356:229–283
Remple MS, Henry EC, Catania KC (2003) The organization of somatosensory cortex in the laboratory rat (*Rattus norvegicus*): evidence for two lateral areas joined at the representation of the teeth. J Comp Neurol 467:105–118
Remple MS, Reed JL, Stepniewska I, Kaas JH (2006) Organization of frontoparietal cortex in the tree shrew (Tupaia belangeri). I. Architecture, microelectrode maps, and corticospinal connections. J Comp Neurol 497:133–154
Rilling JK, Glasser MF, Preuss TM, Ma XY, Zhao TJ, Hu XP, Behrens TEJ (2008) The evolution of the arcuate fasciculus revealed with comparative DTI. Nat Neurosci 11:426–428
Ringo JL, Doty RW, Demeter S, Simard PY (1994) Time is of the essence: a conjecture that hemispheric specialization arose from inter-hemispheric conduction delay. Cereb Cortex 4:331–343
Robson SL, Wood B (2008) Hominin life history: reconstruction and evolution. J Anat 212:394–425
Rosa MG, Krubitzer LA (1999) The evolution of visual cortex: where is V2? Trends Neurosci 22:242–248
Rosa MG, Palmer SM, Gamberini M, Tweedale R, Pinon MC, Bourne JA (2005) Resolving the organization of the new World monkey third visual complex: the dorsal extrastriate cortex of the marmoset (Callithrix jacchus). J Comp Neurol 483:164–191
Rowe TB, Macrini TE, Luo ZX (2011) Fossil evidence on origin of the mammalian brain. Science 332:955–957
Seiffert ER (2012) Early primate evolution in afro-Arabia. Evol Anthropol 21:239–253
Seiffert ER, Perry JMG, Simons EL, Boyer DM (2010) Convergent evolution of anthropoid-like adaptations in eocene adapiform primates. Nature 461:1118–1121
Schmidt-Nielson K (1984) Why is animal size so important? Cambridge University Press, New York
Shepherd GM (2011) The microcircuit concept applied to cortical evolution from three-layer to six-layer cortex. Front Neuroanat 5:1–15
Silcox MT, Benham AE, Bloch JI (2010) Endocasts of microsyops (microsyopidae, primates) and the evolution of the brain in primitive primates. J Hum Evol 58:505–521
Smith FA, Boyer AG, Brown JH, Costa DP, Dayan T, Ernest SK, Evans AR, Fortelius M, Gittleman JL, Hamilton MJ, Harding LE, Lintulaakso K, Lyons SK, McCain C, Okie JG, Saarinen JJ, Sibly RM, Stephens PR, Theodor J, Uhen MD (2010) The evolution of maximum body size of terrestrial mammals. Science 330:1216–1219
Steiper ME, Seiffert ER (2012) Evidence for a convergent slowdown in primate molecular rates and its implications for the timing of early primate evolution. Proc Natl Acad Sci U S A 109:6006–6011
Stepniewska I, Fang PC, Kaas JH (2009) Organization of the posterior parietal cortex in galagos: I. Functional zones identified by microstimulation. J Comp Neurol 517:765–782
Stepniewska I, Cerkevich CM, Kaas JH (2016) Cortical connections of the caudal portion of posterior parietal cortex in prosimian galagos. Cereb Cortex 26:2753–2777
Streidter GF (2005) Principles of brain evolution. Sinauer Associates, Sunderland
Sur M, Nelson RJ, Kaas JH (1980) Representation of the body surface in somatic koniocortex in the prosimian Galago. J Comp Neurol 189:381–402

Takahata N, Satta Y, Klein J (1995) Divergence time and population size in the lineage leading to modern humans. Theor Popul Biol 48:198–221
Tigges J, Tigges M (1979) Ocular dominance columns in the striate cortex of chimpanzee (Pan troglodytes). Brain Res 166:386–390
Tocheri MW, Orr CM, Jacofsky MC, Marzke MW (2008) The evolutionary history of the hominin hand since the last common ancestor of Pan and Homo. J Anat 212:544–562
Tomasello M (2014) A natural history of human thinking. Harvard University Press, Cambridge
Tsao DY, Moeller S, Freiwald WA (2008) Comparing face patch systems in macaques and humans. Proc Natl Acad Sci U S A 105:19514–19519
Ungerleider LG, Haxby JV (1994) 'What' and 'where' in the human brain. Curr Opin Neurobiol 4:157–165
Van Essen DC, Glasser MF, Dierker DL, Harwell J (2012a) Cortical parcellations of the macaque monkey analyzed on surface-based atlases. Cereb Cortex 22:2227–2240
Van Essen DC, Glasser MF, Dierker DL, Harwell J, Coalson T (2012b) Parcellations and hemispheric asymmetries of human cerebral cortex analyzed on surface-based atlases. Cereb Cortex 22:2241–2262
Van Hooser SD (2007) Similarity and diversity in visual cortex: is there a unifying theory of cortical computation? Neuroscientist 13:639–656
Weller RE, Kaas JH (1989) Parameters affecting the loss of ganglion cells of the retina following ablations of striate cortex in primates. Vis Neurosci 3:327–349
Wong P, Collins CE, Kaas JH (2010) Overview of sensory systems of Tarsius. Int J Primatol 31:1002–1031
Woolsey CN, Tasker R, Welt C et al (1960) Organization of pre and postcentral leg areas in chimpanzee and gibbon. Trans Am Neurol Assoc 85:144–146
Wrangham R (2009) Catching fire: how cooking made us human. Basic Books, New York
Wu CW, Bichot NP, Kaas JH (2000) Converging evidence from microstimulation, architecture, and connections for multiple motor areas in the frontal and cingulate cortex of prosimian primates. J Comp Neurol 423:140–177
Wu CW, Kaas JH (2003) Somatosensory cortex of prosimian galagos: physiological recording, cytoarchitecture, and corticocortical connections of anterior parietal cortex and cortex of the lateral sulcus. J Comp Neurol 457:263–292
Young N, Stepniewska I, Kaas JH (2012) Motor cortex. In: Watson C, Paxinos G, Puelles L (eds) The mouse nervous system. Elsevier, London, pp 527–536

Chapter 4
Developmental Sequences Predict Increased Connectivity in Brain Evolution: A Comparative Analysis of Developmental Timing, Gene Expression, Neuron Numbers, and Diffusion MR Tractography

Christine J. Charvet, Chet C. Sherwood, and Emi Takahashi

Abstract A conserved sequence in cell-type specification across mammals suggests that evolutionary changes in developmental timing may give rise to predictable changes in connectivity patterns across species. We here review the regularities in the timing of developmental events across species. We then use them to predict evolutionary changes in the number of cell types in order to identify evolutionary changes in the internal circuitry of the cerebellum as well as the gray and white matter of the isocortex across mammals. We survey what is known about the sequence and timing of cell-type specification in different brain regions and in

---

C.J. Charvet (✉)
Department of Anthropology and Center for the Advanced Study of Human Paleobiology, The George Washington University, Washington, DC, USA

Division of Newborn Medicine, Department of Medicine, Boston Children's Hospital, Harvard Medical School, Boston, MA, USA

Athinoula A. Martinos Center for Biomedical Imaging, Massachusetts General Hospital, Harvard Medical School, Charlestown, MA, USA

Department of Molecular Biology and Genetics, Cornell University, 345 Biotechnology Building, Ithaca, NY 14853, USA
e-mail: charvetcj@gmail.com

C.C. Sherwood
Department of Anthropology and Center for the Advanced Study of Human Paleobiology, The George Washington University, Washington, DC, USA

E. Takahashi
Division of Newborn Medicine, Department of Medicine, Boston Children's Hospital, Harvard Medical School, Boston, MA, USA

Athinoula A. Martinos Center for Biomedical Imaging, Massachusetts General Hospital, Harvard Medical School, Charlestown, MA, USA

S. Watanabe et al. (eds.), *Evolution of the Brain, Cognition, and Emotion in Vertebrates*, Brain Science, DOI 10.1007/978-4-431-56559-8_4

various mammalian species. We find that lengthened developmental schedules predict a disproportionate increase in the number of locally projecting granule cells within the cerebellum and in the number of isocortical neurons projecting within or across cortical areas. Our main conclusion is that, as brains get bigger, neurons increasingly connect within their own major brain region.

Keywords Connections • Cortex • Diffusion MR • Evolution • Layers • Primate

4.1 Introduction

How connectivity patterns evolve has been an enduring question in the study of comparative neurobiology. Variation in brain size has been proposed to entail evolutionary changes in connectivity patterns (Deacon 1990; Striedter 2005). Yet, we still have few lines of empirical evidence to identify how connectivity patterns have evolved. Masterton and his colleagues compared the number of neurons that comprise different pathways in a large sample of mammalian species and showed that corticospinal neuron numbers largely covary with brain size (Nudo and Masterton 1990; Nudo et al. 1995). These data offer a glimpse as to how connectivity patterns evolve.

Connectivity patterns have been well characterized using anatomical tracer methods in a broad range of species (Kawamura 1973a, b, c; Kaas 1989; Striedter 2005; Schmahmann and Pandya 2009). Collectively, this large body of work has shown that projection patterns exhibit stereotypical patterns, which are relatively stable in mammalian evolution. For instance, cortical connectivity patterns exhibit a small-world network, wherein the majority of neurons project locally between cortical areas rather than over long distances (Sporns and Zwi 2004; Bullmore and Sporns 2012). This small-world pattern of connectivity is evident in primates such as macaque monkeys as well as in carnivores such as cats (Bassett and Bullmore 2006). As another example, neurons across the depth of the isocortex exhibit stereotypic patterns of projections. Upper layer neurons (layers II–IV) preferentially project within and across cortical regions, but many lower layer neurons (layers V–VI) project to subcortical structures (Gilbert and Kelly 1975; Barbas 1986; Nudo et al. 1995; Hof et al. 1995). That is, there is conservation in cortical neuron projection patterns in mammalian evolution.

The present review synthesizes findings from the field of evolutionary and developmental biology (evo-devo) to identify how evolutionary changes in developmental timing and conservation in the sequence of cell-type generation yields evolutionary changes in connectivity patterns. Given that different cortical layers consist of different cell types (Hof et al. 1995; DeFelipe et al. 2002; Belgard et al. 2011; Zeng et al. 2012) and neurons in upper (i.e., layers II–IV) versus lower layers (i.e., layers V–VI) exhibit stereotypical patterns of connectivity (Gilbert and Kelly 1975; Nudo et al. 1995; Barbas 1986; Rowell et al. 2010; García-Cabezas and Barbas 2014; Markov et al. 2014; Yamawaki et al. 2014), allometric variation in cell-type numbers yield insights as to how connectivity patterns evolve. We specifically focus on the relative variation in the number of neuron subtypes such as isocortical upper layer (i.e., layers II–IV) and lower layer neurons (i.e., layers V–VI), as well as cerebellar granule and Purkinje neurons. Such an analysis shows that intra-regional projecting neurons within the isocortex and within the cerebellum become disproportionately more numerous as developmental schedules lengthen and brains expand.

4.1.1 Conservation in Developmental Sequences

The evo-devo approach has identified broadly conserved molecular developmental mechanisms and conserved cell-type specification across species (Finlay and Darlington 1995; Puelles and Rubenstein 2003; Puelles and Ferran 2012). Cell birth-dating studies performed in a broad range of mammals (e.g., rodents, primates, marsupials) show that the sequence of cell-type specification is highly conserved across mammals and that different brain regions vary in their duration of neurogenesis (Clancy et al. 2001; Workman et al. 2013). Some brain regions such as the isocortex and the cerebellum undergo neurogenesis for an extended period of time compared with other brain regions such as the thalamus and the medulla (Bayer and Altman 1991; Finlay and Darlington 1995; Rakic 2002; Workman et al. 2013). The protracted neurogenetic schedule of the developing isocortex and cerebellum is particularly evident in species with prolonged developmental schedules such as primates and marsupial mammals but is less evident in faster-developing species such as mice and rats because developmental events become more clearly separated in time in longer-developing species (e.g., primates; Workman et al. 2013).

Although data from cell birth-dating studies are lacking in humans, an inspection of gene expression levels from the Allen Institute for Brain Science can provide some insights into changes in developmental maturation in humans (Hawrylycz et al. 2012; Miller et al. 2014). For instance, the *RBFOX3* gene (i.e., *NeuN*) differentially increases its expression over developmental time in humans in each brain region (Fig. 4.1). The observation that cerebellar *RBFOX3* expression increases well into the postnatal period compared with other brain regions, such as the thalamus, suggests that the prolonged period of neurogenesis timing that has been observed for nonhuman mammals is mirrored in the duration of changes in gene expression in developing humans. Both birth-dating studies of nonhuman primates (Rakic 2002) and variation in gene expression over developmental time in humans (Hawrylycz et al. 2012; Miller et al. 2014) yield similar results in illustrating the extended period of neurogenesis or maturation of some structures (Rakic and Sidman 1970; Rakic 2002).

Early in the development of the isocortex, neurons are generated within the ventricular zone. They undergo mitosis along the ventricular wall. As development progresses, proliferative cells exit the cell cycle and migrate outside of the proliferative zone to become neurons or glia (Bystron et al. 2008). As cells exit the

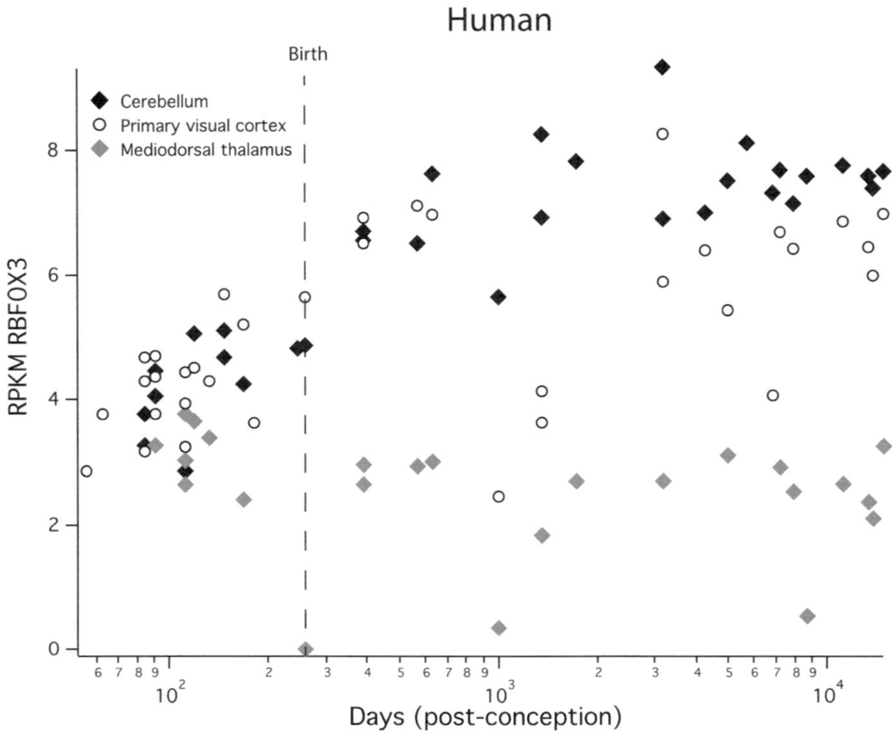

Fig. 4.1 Reads per kilobase of transcript per million (RPKM) of *RBFOX3* (start exon position, 77099243) in the thalamus, primary visual cortex, and cerebellar cortex over developmental time (age in days postconception) in humans. *RBFOX3* expression continues to increase in the isocortex and cerebellum over an extensive period of time, whereas RBFOX3 expression in the thalamus is relatively invariant. These data show that the maturation of the cerebellum extends for longer than the thalamus. These data are from the Allen Institute for Brain Science, brain atlas

proliferative zone, many neurons migrate along radial glia to the cortical plate (Rakic 2003). These radial glia can be seen with high-angular-resolution diffusion MR imaging tractography as shown in a human fetus at 17 gestational weeks (Takahashi et al. 2012; Fig. 4.2). These observations highlight the geometric structure of scaffolds that serve to give rise to the adult isocortex (Wilkinson et al. 1990; Wedeen et al. 2012). Diffusion MR tractography further shows that as neurogenesis wanes, scaffolds also regress and corticocortical tracts become evident (Takahashi et al. 2012).

The sequence of pyramidal cell-type specification is conserved in mammalian evolution. Neurons migrate to the cortical plate in an inside-out fashion such that infragranular layer neurons are born before granular layer neurons, which are in turn born before supragranular layer neurons (Sanderson and Weller 1990; Rakic 1974, 2002). The inside-out sequence of cell birth specification has been observed

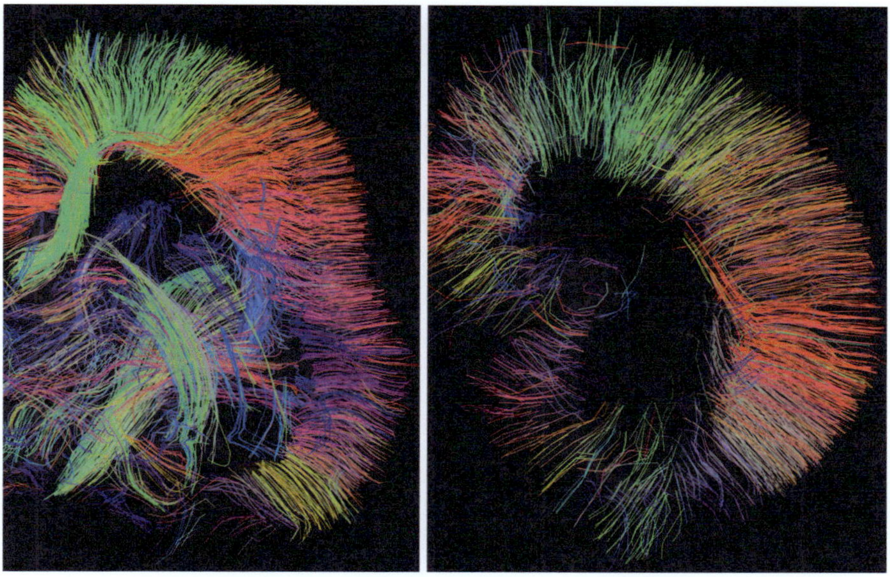

Fig. 4.2 High-angular-resolution MR tractography of a human at 17 gestational weeks. Coronal planes show pathways coursing radially from the proliferative zone along the ventricle to the outer surface of the cortical surface. The color-coding of tractography pathways is based on a standard RGB code, applied to the vector between the end points of each fiber (*red*, left-right; *green*, dorsal-ventral; *blue*, anterior-posterior). Images of these scans are from Takahashi et al. (2012)

in every mammalian isocortex studied so far (e.g., Fig. 4.3, primates, rodents, marsupial mammals; Sanderson and Weller 1990; Polleux et al. 1997; Marotte and Sheng 2000; Workman et al. 2013), but the sequence of neurogenesis that is characteristic of the mammalian isocortex is not observed in the pallium of reptiles and birds (Tsai et al. 1981; Goffinet et al. 1986; Striedter and Keefer 2000; Rowell and Ragsdale 2012). The most parsimonious interpretation of these data is that the inside-out sequence of cortical neurogenesis emerged early in mammalian evolution.

The cerebellum, likewise, exhibits a specific sequence in the birth order of cell types. Early in development, proliferative cells are located toward the ventricular surface and migrate radially outward throughout the developing cerebellum. As development progresses, an additional proliferative pool called the external granular layer forms, which consists of proliferative cells concentrated toward the cerebellar surface (Fujita et al. 1966; Ponti et al. 2006, 2008). As cells exit the external granular layer, these cells migrate inward to occupy the granular cell layer (Rakic and Sidman 1970). This outside-in pattern of neurogenesis observed in the developing cerebellum stands in contrast to the inside-out pattern of neurogenesis observed in the developing isocortex.

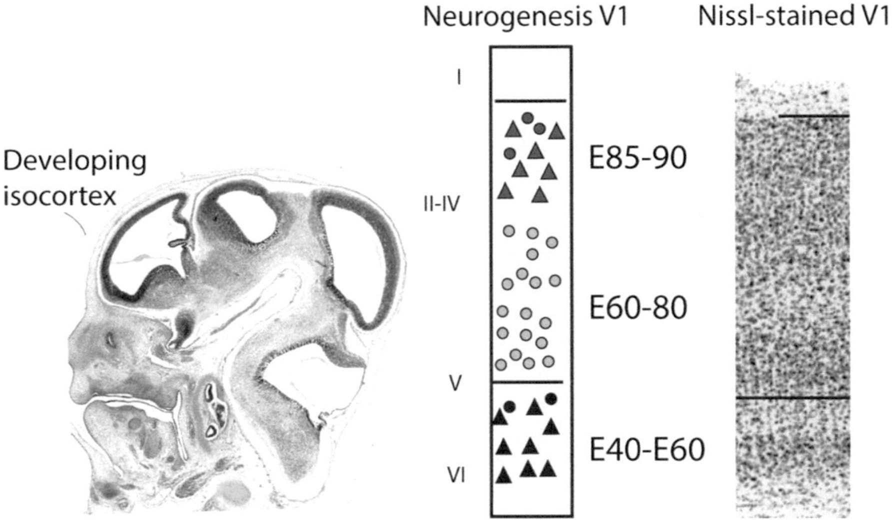

Fig. 4.3 Early in development, the developing isocortex consists of the ventricular zone. Neurons born found at various depths of the isocortex are born at different times. Early in development, the developing isocortex consists of the ventricular zone, which contains proliferative cells as shown here from a sagittal section of an embryonic tarsier. Representation of the birth order of lower and upper layer neurons in the macaque V1 cortex. Neurons found in lower layers are generated between embryonic day (E) 40 and 60. Neurons located in layer IV are generated between E60 and E80. Neurons located in upper layers are generated between E85 and E90. The tarsier (specimen ID# 1012) is part of the Hubrecht collection and was photographed at the Naturkunde Museum. Data on V1 neurogenesis timing are from Rakic (1974, 2002)

By adulthood, late-born granule cells are located within inner layers, whereas early-born Purkinje cells are located toward the outer cellular layer. Cell birth-dating studies in rodents and primates have shown that Purkinje cells are generated over a short interval, but granule cell production is generated for an extended period of time. For instance, Purkinje cell production occurs for roughly 5 days, but granule cell production extends over 140 days in rhesus macaques (Rakic 2002). In primates, as in other mammals, granule cell production extends into the postnatal period (Fig. 4.4; Bayer and Altman 1991; http://braindevelopmentmaps.org; Ponti et al. 2006, 2008). A similar situation is observed in nonmammalian species such as chickens or quail where Purkinje cells are generated over a short interval but granule cell production extends into the post-hatchling period (Gona 1976; Yurkewicz et al. 1981; Uray et al. 1987; Stamatakis et al. 2004). The most parsimonious interpretation of these data is that the sequence of cerebellar cell-type specification evolved early at least in sauropsids.

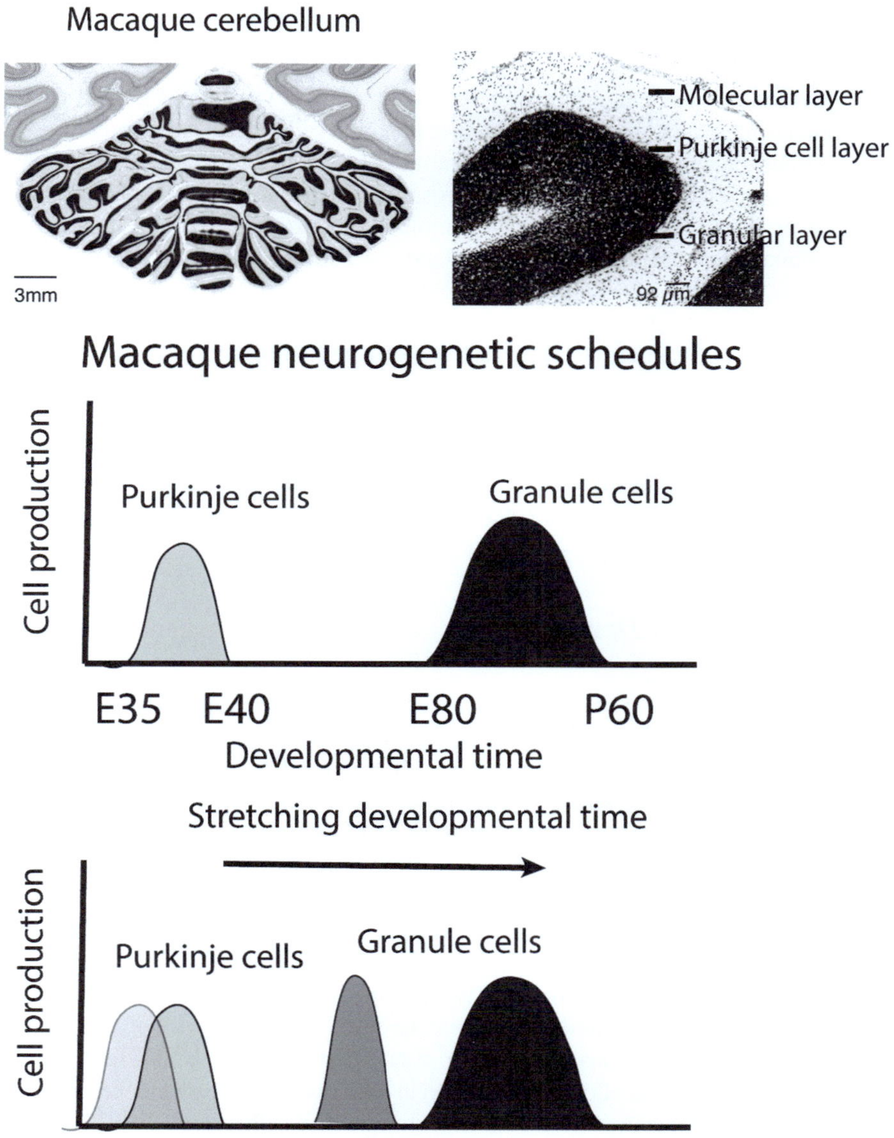

Fig. 4.4 A coronal section through the cerebellum of a macaque. Purkinje cells are born for a few days, but granule cell production extends into the postnatal period in macaques. Evolutionary changes in developmental duration entails that cells that are born late in development become disproportionately increased in numbers compared with neurons that are born early in development. In other words, the duration of granule cell production occurs for proportionately longer in longer-developing species. As a consequence, granule cell production disproportionately increases in bigger-brained species

4.1.2 Variation in Developmental Duration

The sequence in which neurons are born combined with variable lengths of developmental timing entail predictable changes in which cell types become amplified in bigger brains. A model proposed by Barbara Finlay and her collaborators states that as developmental schedules lengthen across species, neurons that are born late in development become disproportionately more numerous than neurons that are born early (Finlay and Darlington 1995; Cahalane et al. 2012, 2014). Consequently, this model predicts that as developmental schedules lengthen, the isocortex and cerebellum should become disproportionately enlarged relative to other regions because isocortical and cerebellar neurogenesis is protracted compared with that of other brain regions (Finlay and Darlington 1995; Reep et al. 2007; Workman et al. 2013). This model explains the allometric variation in the size of brain regions, the number of neurons that are contained within each brain region, as well as the allometric variation in the number of various cell types that comprise a given brain region (Fig. 4.5; Finlay and Darlington 1995).

4.1.3 Isocortical Development

Cell types within a given brain region can be distinguished by their birth order, their position, their gene expression, as well as their patterns of connectivity (Rakic 1974; Gilbert and Kelly 1975; Belgard et al. 2011; Zeng et al. 2012). A comparative analysis of isocortical neuron numbers in primates and rodents shows that upper layer neuron numbers become disproportionately amplified in larger brains relative to lower layer neurons (Fig. 4.6; Finlay et al. 1998; Clancy et al. 2001; Cahalane et al. 2014; Charvet et al. 2015, 2016, 2017a). The protracted production of upper layer neurons accounts for the differential increase in the number of upper layer neurons relative to lower layer neurons in bigger brains.

Because the laminar position of cell types in the isocortex can be distinguished by their specific patterns of connectivity, the disproportionate expansion of upper layer neurons yields specific consequences for connectivity patterns across taxa. As an example, we consider the allometric variation in the number of upper layer neurons and layer V corticospinal neurons (Nudo and Masterdon 1990; Charvet et al. 2015). Many somata of neurons that comprise the corticospinal tract are found within the frontal cortex (e.g., primary motor cortex; Nudo et al. 1995), and they form synapses with neurons within the spinal cord. Interestingly, the precise terminations of corticospinal tract neurons vary between species (Striedter 2005). To quantify corticospinal tract neurons, Nudo et al. (1995) applied horseradish peroxidase to the cervical spinal cord and quantified the labeled somata found in the isocortex. As is evident in Figs. 4.6 and 4.7, upper layer neuron numbers become disproportionately amplified relative to lower

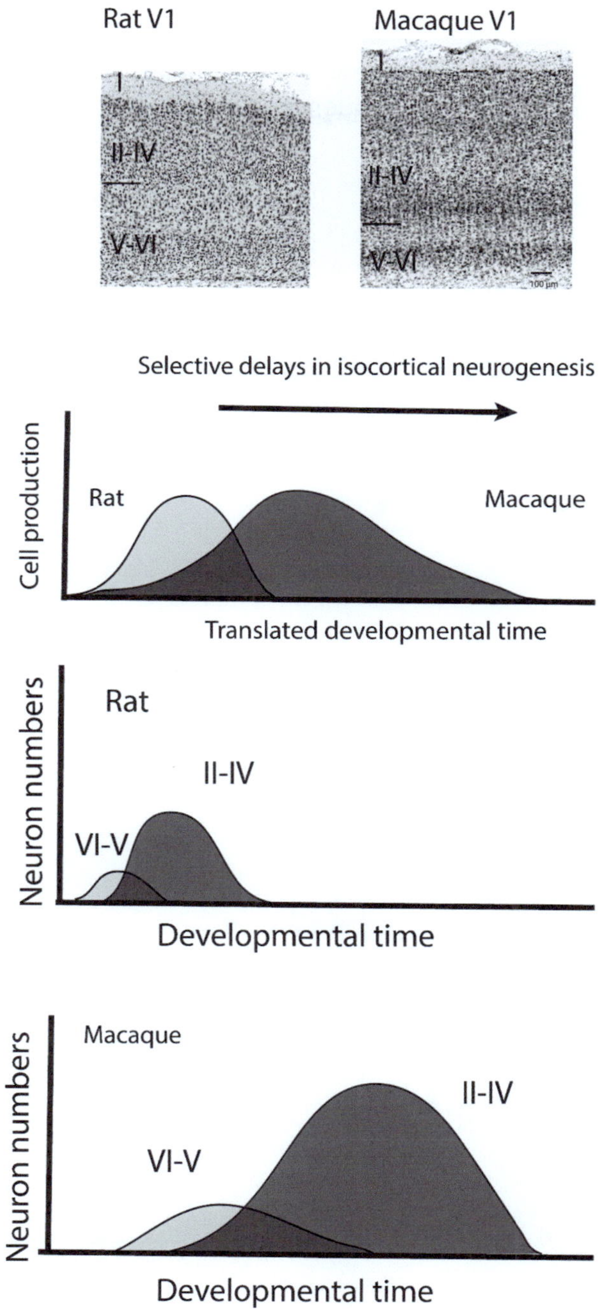

Fig. 4.5 Primates possess expanded upper layers compared with rats. Selective delays in isocortical neurogenesis in primates relative to rodents lead to a disproportionate expansion of cells that are born late in development. Because upper layer neurons are born after lower layer neurons, selective delays in isocortical neurogenesis entail that primates possess disproportionately more upper layer neurons compared with rodents. Images of Nissl-stained sections are screenshots from brainmaps.org

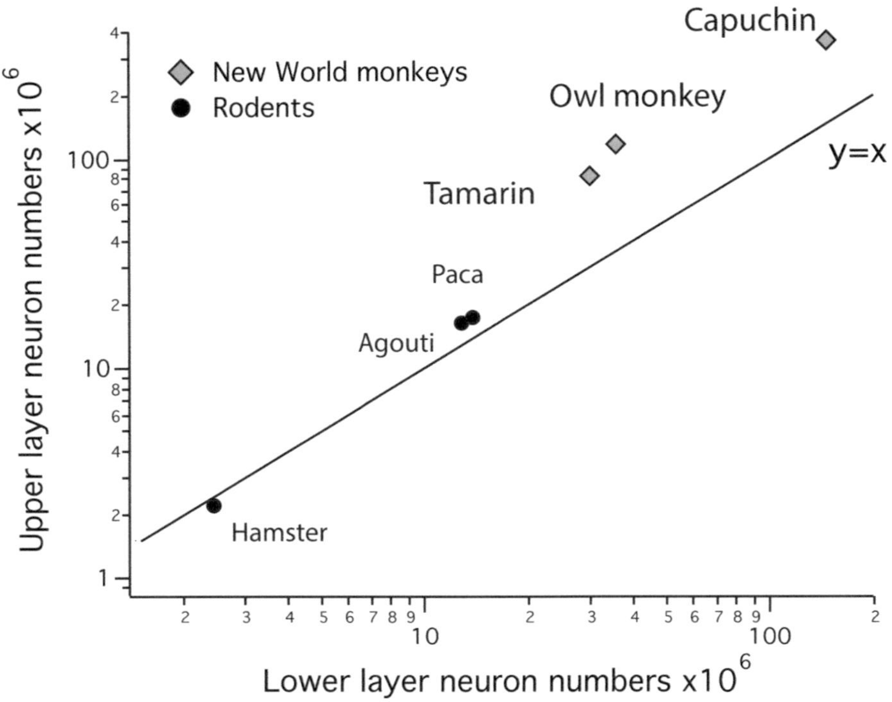

Fig. 4.6 Upper layer neuron numbers are plotted against lower layer neurons in primates and rodents. A linear regression is also plotted to highlight that upper layer neuron numbers expand disproportionately relative to lower layer neurons in primates and rodents

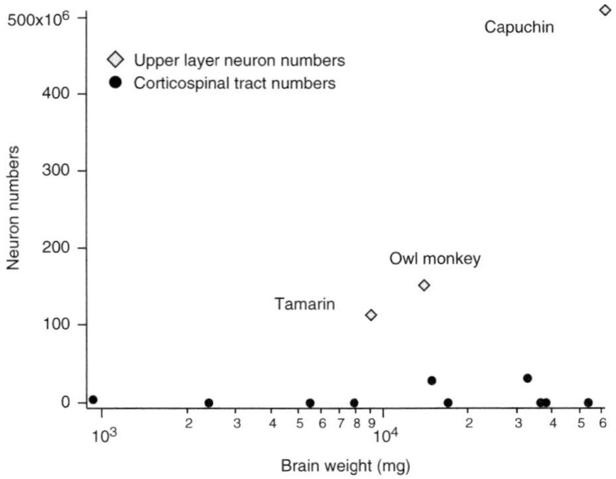

Fig. 4.7 Corticospinal tract neuron numbers (in a lower layer) and upper layer neuron numbers are plotted against brain weight in primate species. Upper layer neuron numbers expand disproportionately relative to corticospinal tract neuron numbers in bigger-brained primates. These findings show that cross-cortically projecting neurons increase with a positive allometry as brains expand. These data are from Nudo and Masterdon 1990 and Charvet et al. (2015)

layers and corticospinal tract neuron numbers as brains expand. The disproportionate expansion of isocortical neuronal numbers projecting within or to other cortical regions relative to isocortical neuron numbers projecting to subcortical structures may thus serve to increasingly modulate incoming and outgoing information in bigger brains.

4.1.4 Cerebellar Development

Similar to what is observed for the isocortex, cerebellar cell types can also be distinguished by their birth order, their position, as well as their patterns of connectivity. Although there are few quantitative studies examining the number of cerebellar cell types across species, a comparison between granule cells and Purkinje cell numbers in humans and rats show that the relative number of granule cells to Purkinje cell numbers is disproportionately increased in humans compared with rats. For instance, the ratio of granule cells to Purkinje cells in rats is less than 500, while in humans, the ratio of granule cells to Purkinje cells is approximately 3500 (Harvey and Napper 1988; Andersen et al. 1992). That is, granule cells are disproportionately more numerous in the bigger-brained humans than in the small-brained rat.

Within the cerebellum, incoming information from pre-cerebellar nuclei located within the pons projects onto granule cells. Granule cells, in turn, project to Purkinje cells (Voogd and Glickstein 1998). Purkinje cells also receive input from inferior olive neurons and project onto cerebellar nuclei as well as onto the neurons of the vestibular complex. Within this circuitry, granule cells become preferentially amplified, project locally, and increasingly synapse with Purkinje cells (Huang et al. 2014). The disproportionate increase in granule cell numbers in bigger-brained species entails increased modulation of incoming and outgoing information.

4.1.5 Selective Changes in Neurogenesis Timing

In evolution, the conservation in the sequence of cell-type specification is superimposed on selective changes in the duration of neurogenesis timing (Workman et al. 2013). Heterochronies (i.e., developmental changes in the timing or rate of events) within some brain regions have been observed in a number of taxa such as primates, carnivores, as well as parrots and songbirds (Charvet et al. 2011; Workman et al. 2013). For instance, distantly related nocturnal species such as owl monkeys and cats exhibit selective delays in retinal neurogenesis relative to diurnal species, which led to a disproportionate amplification of late-born neurons relative to early-born neurons (Finlay 2008; Dyer et al. 2009; Workman et al. 2013). Because rods are born late in development, selective delays in retinal neurogenesis are associated with a disproportionate increase in the number of rods in nocturnal species compared with diurnal species. Evolutionary changes in developmental timing within the peripheral nervous system evolved in very distant lineages such as cats and owl monkeys.

As another example of heterochrony, parrots and songbirds selectively delay telencephalic neurogenesis relative to galliform birds. Evolutionary changes in developmental timing are concomitant with a number of changes in developmental processes. The selective delay in telencephalic neurogenesis in parrots is concomitant with delays in the decline in telencephalic cell cycle rates as well as an amplification of cells undergoing mitosis within the subventricular zone, which is an additional proliferative zone that lies superficial to the ventricular zone in development (Smart et al. 2002; Bystron et al. 2008; Charvet and Striedter 2008; Charvet et al. 2011; Martínez-Cerdeño et al. 2012; Dehay et al. 2015). Unlike ventricular zone cells, proliferative cells in the subventricular zone undergo mitosis at scattered locations throughout the subventricular zone (Smart 1972; Smart et al. 2002; Martínez-Cerdeño et al. 2012). The protracted neurogenetic schedules of parrots and songbirds may have fostered an increased duration of post-hatchling maturation in which juveniles may learn from conspecifics (Charvet and Striedter 2011).

Among mammals, birth-dating studies demonstrate that primates selectively delay isocortical neurogenesis relative to rodents (Workman et al. 2013). In adulthood, the selective delays in isocortical neurogenesis are concomitant with the expansion of the isocortex as well as increased isocortical neuron numbers in primates relative to many other taxa (Workman et al. 2013; Herculano-Houzel 2012). According to the model of "late equals large" cell populations that are born late in development become disproportionately more numerous relative to neurons that are born early in development (Figs. 4.4 and 4.5). Because upper layer neurons are generated late in the developing isocortex, upper layer neuron numbers should become disproportionately expanded in primates relative to other mammals such as rodents.

Although few studies have explicitly compared the number of cortical upper layer and lower layer neurons across mammalian species, data on upper and lower layer neuron numbers and densities have been gathered for primates, rodents, and manatees (Charvet et al. 2015, 2016; Reyes et al. 2015). Rodent brains are generally smaller and contain fewer neurons per brain mass compared with primates, which makes it difficult to specifically address whether primates exhibit disproportionately more upper layer neurons compared with rodents (Figs. 4.5 and 4.6; Herculano-Houzel 2012). A previous study compared upper layer neuron numbers in manatees relative to primates, because manatees have a large brain that is similar to that of some primates, which allows for a comparison between taxa of equivalent brain size and overall cortical neuron numbers. This study found that primates exhibit disproportionately more isocortical upper layer neurons compared with manatees (Charvet et al. 2015, 2016). It would be interesting to quantify upper and lower layer neuron numbers in a broader range of mammals to identify how upper layer neuron numbers in primates deviate from other taxa (Hofman 1985; Charvet et al. 2017a, b). The data so far supports the notion that selective delays in cortical neurogenesis in primates are concomitant with a disproportionate expansion of isocortical neurons and, in particular, upper layer neuron numbers.

4.2 Species Differences in Projection Patterns

The number of isocortical upper layer neurons (i.e., layers II–IV) projecting within or across cortical regions and lower layer neurons (i.e., layers V–VI) projecting to subcortical regions varies predictably with brain size. Yet, there are clear species

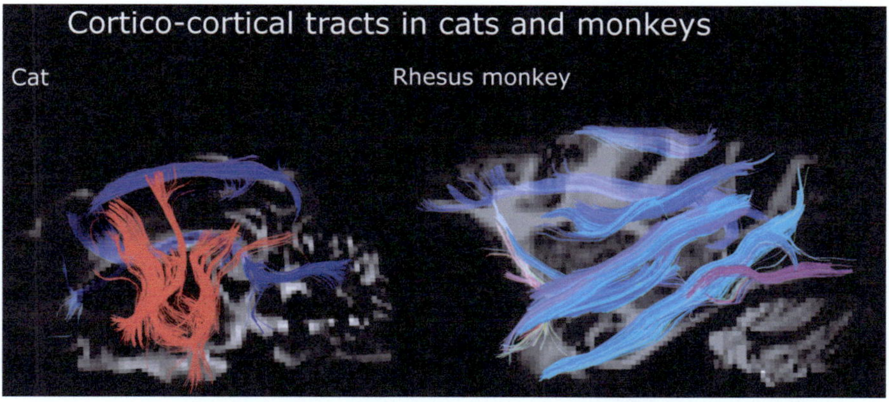

Fig. 4.8 Corticocortical tracts in a cat and a macaque were reconstructed from diffusion spectrum MR scans and in accordance with previous descriptions of tract-tracing studies in cats and macaques. Many tracts course across the anterior to posterior direction in the macaque, but a number of tracts course across the dorsal to ventral direction in cats. Except for the dorsal-ventral pathways in the cat, the color-coding of tractography pathways is based on a standard RGB code, applied to the vector between the end points of each fiber (*red*, left-right; *green*, dorsal-ventral; *blue*, anterior-posterior). The pathways coursing across the dorsal-ventral direction in the cat are shown in a single color (*smoky red*). These scans were published previously (Takahashi et al. 2011, 2012, Charvet et al. 2017a)

differences in the terminations of corticocortical projection patterns. Figure 4.8 shows corticocortical tracts reconstructed with the use of diffusion spectrum imaging (Takahashi et al. 2010, 2011; Wedeen et al. 2012). We referred to tract-tracing or lesion studies that identified corticocortical tracts in cats and macaques to ensure the accuracy of the diffusion MR tractography of these brains (Schmahmann and Pandya 2009; Diamond et al. 1968; Kawamura and Otani 1970; Kawamura 1973a, b, c; Paula-Barbosa et al. 1975; Poldrack and Farah 2015). As is evident in Fig. 4.8, trajectory patterns of corticocortical tracts clearly differ between the two species with macaques exhibiting many tracts aligned along the anterior-posterior axis, whereas cats exhibit a number of tracts coursing in the dorsal-ventral direction. Although cell birth order and position can predict the number of specific cell types projecting within or outside their major subdivisions, the trajectories and therefore terminal locations of corticocortical tracts clearly vary between species.

4.3 Developmental Sources of Change in the Brain and Behavior

Our analysis supports the notion that the number corticocortically projecting neurons in the isocortex and granule cells projecting to Purkinje cells devoted to modulating incoming information becomes disproportionately increased in bigger brains. Evolutionary changes in brain size are concomitant with the amplification of isocortical and cerebellar cell types that are born late in development such as upper layer neurons and granule cells. These observations suggest that allometric variation in cell-type numbers have emerged from wholesale changes in developmental timing rather than selected changes in the developmental mechanisms generating a specific cell type (Finlay and Darlington 1995). The increase in relative numbers of intra-regionally connecting neurons observed in the isocortex and locally projecting granule cells in the cerebellum would provide a powerful matrix to modulate incoming sensory information to mediate various behaviors in bigger-brained species.

Evolutionary changes in behavior might emerge through evolutionary changes in developmental timing, increased neuron numbers, or through a combination of all of these variables. We argue that, among these variables, behavioral changes are likely to have emerged from changes in developmental timing. Developmental schedules are associated to some degree with the duration of postnatal development. An extended duration of developmental timing may promote an extended period of postnatal maturation, a prolonged period of parental care, which may foster learning from conspecifics (Charvet and Finlay 2012). The covariation in developmental timing, brain expansion, and allometric variation in neuron numbers projecting locally within the cerebellum and those isocortical neurons projecting either within or across the cortical areas can together foster increased learning from conspecifics. The brain is intrinsically plastic, and environmental exposure can further sculpt what is learned. The flexible nature of the brain, coupled with variable lengths of developmental duration, may channel what information is modulated in bigger-brained species.

Within this extended developmental schedule and prolonged period from which to learn from conspecifics, what is learned is contingent on what is rewarding (Young and Wang 2004). The mesolimbic system is characterized by dopaminergic projections from the ventral tegmental area to the nucleus accumbens, and this circuit has been implicated in mediating a range of naturally rewarding behaviors such as pair bonding and bird song (Goodson et al. 2009; O'Connell and Hofmann 2011). Evolutionary changes within this circuitry may arise through changes in receptors that serve to modulate reward-related behaviors. For instance, intra- and interspecific variation in the dopamine D4 receptor (DRD4) have been noted and may be associated with a number of changes in behaviors related to reward (Ebstein et al. 1996; Ding et al. 2002; Wang et al. 2004; Vallender 2011, 2012; Yamamoto et al. 2013). These observations are increasingly at odds with the notion that evolutionary changes in brain region size account for changes in select behaviors (Healy and Rowe 2007). Rather, evolutionary changes in reward circuitries and developmental timing may be powerful substrates through which evolutionary changes in behaviors emerge.

4.4 Summary

We have surveyed which cell types become preferentially amplified in bigger brains. We have remained relatively agnostic as to the precise targets of these cell types. Our overview of evolutionary changes in the number of isocortical and cerebellar neuronal populations shows that isocortical neurons increasingly project either within or across cortical areas and that cerebellar neurons increasingly project locally. In other words, neurons projecting within their major brain subdivision become disproportionately amplified in bigger brains.

Acknowledgments This work was supported by the James S. McDonnell Foundation Grant 22002078 (C. C. S.), the Eunice Shriver Kennedy NICHD R01HD078561 and R21HD069001 (E. T.), as well as a DAAD grant (C. J. C). Images of Nissl-stained sections of primary visual cortex of the macaque and rat as well as the cerebellum of the macaque shown in Figs. 4.4 and 4.5 were obtained by taking screenshots from brainmaps.org. We also used gene expression data from the BrainSpan: Atlas of the Developing Human Brain, funded by ARRA Awards 1RC2MH089921-01, 1RC2MH090047-01, and 1RC2MH089929-01. These data are available from http://developinghumanbrain.org

References

Andersen BB, Korbo L, Pakkenberg B (1992) A quantitative study of the human cerebellum with unbiased stereological techniques. J Comp Neurol 326:549–560

Barbas H (1986) Pattern in the laminar origin of corticocortical connections. J Comp Neurol 252:415–422

Bassett DS, Bullmore ED (2006) Small-world brain networks. Neuroscientist 12:512–523

Bayer SA, Altman J (1991) Neocortical development. Raven Press, New York

Belgard TG, Marques AC, Oliver PL et al (2011) A transcriptomic atlas of mouse neocortical layers. Neuron 71:605–616

Bullmore E, Sporns O (2012) The economy of brain network organization. Nat Rev Neurosci 13:336–349

Bystron I, Blakemore C, Rakic P (2008) Development of the human cerebral cortex: Boulder Committee revisited. Nat Rev Neurosci 9:110–122

Cahalane DJ, Charvet CJ, Finlay BL (2012) Systematic, balancing gradients in neuron density and number across the primate isocortex. Front Neuroanat 6:28

Cahalane DJ, Charvet CJ, Finlay BL (2014) Modeling local and cross-species neuron number variations in the cerebral cortex as arising from a common mechanism. Proc Natl Acad Sci U S A 111:17642–17647

Charvet CJ, Finlay BL (2012) Embracing covariation in brain evolution: large brains, extended development, and flexible primate social systems. Prog Brain Res 195:71

Charvet CJ, Hof PR, Raghanti MA, Van Der Kouwe AJ, Sherwood CC, Takahashi E (2017a) Combining diffusion magnetic resonance tractography with stereology highlights increased cross-cortical integration in primates. J Comp Neurol 525:1075–1093

Charvet CJ, Stimpson CD, Kim YD, Raghanti MA, Lewandowski AH, Hof PR, Gomez-Robles A, Krienen FM, Sherwood CC (2017b) Gradients in cytoarchitectural landscapes of the isocortex: Diprotodont marsupials in comparison to eutherian mammals. J Comp Neurol 525:1811–1826

Charvet CJ, Striedter GF (2008) Developmental species differences in brain cell cycle rates between northern bobwhite quail (*Colinus virginianus*) and parakeets (*Melopsittacus undulatus*): implications for mosaic brain evolution. Brain Behav Evol 72:295–306

Charvet CJ, Striedter GF (2011) Developmental modes and developmental mechanisms can channel brain evolution. Front Neuroanat 5:4

Charvet CJ, Striedter GF, Finlay BL (2011) Evo-devo and brain scaling: candidate developmental mechanisms for variation and constancy in vertebrate brain evolution. Brain Behav Evol 78:248–257

Charvet CJ, Cahalane DJ, Finlay BL (2015) Systematic, cross-cortex variation in neuron numbers in rodents and primates. Cereb Cortex 25:147–160
Charvet CJ, Reep RL, Finlay BL (2016) Evolution of cytoarchitectural landscapes in the mammalian isocortex: Sirenians (*Trichechus manatus*) in comparison with other mammals. J Comp Neurol 524:772–782
Clancy B, Darlington RB, Finlay BL (2001) Translating developmental time across mammalian species. Neuroscience 105:7–17
Deacon TW (1990) Rethinking mammalian brain evolution. Am Zool 30:629–705
DeFelipe J, Alonso-Nanclares L, Arellano JI (2002) Microstructure of the neocortex: comparative aspects. J Neurocytol 31:299–316
Dehay C, Kennedy H, Kosik KS (2015) The outer subventricular zone and primate-specific cortical complexification. Neuron 85:683–694
Diamond IT, Jones EG, Powell TP (1968) The association connections of the auditory cortex of the cat. Brain Res 11:560–579
Ding YC, Chi HC, Grady DL et al (2002) Evidence of positive selection acting at the human dopamine receptor D4 gene locus. Proc Natl Acad Sci U S A 99:309–314
Dyer MA, Martins R, da Silva Filho M et al (2009) Developmental sources of conservation and variation in the evolution of the primate eye. Proc Natl Acad Sci U S A 106:8963–8968
Ebstein RP, Novick O, Umansky R et al (1996) Dopamine D4 receptor (D4DR) exon III polymorphism associated with the human personality trait of novelty seeking. Nat Genet 12:78–80
Finlay BL (2008) The developing and evolving retina: using time to organize form. Brain Res 1192:5–16
Finlay BL, Darlington RB (1995) Linked regularities in the development and evolution of mammalian brains. Science 268:1578–1584
Finlay BL, Hersman MN, Darlington RB (1998) Patterns of vertebrate neurogenesis and the paths of vertebrate evolution. Brain Behav Evol 52:232–242
Fujita S, Shimada M, Nakamura T (1966) H3-Thymidine autoradiographic studies on the cell proliferation and differentiation in the external and internal granular layers of the mouse cerebellum. J Comp Neurol 128:191–208
García-Cabezas MÁ, Barbas H (2014) Area 4 has layer IV in adult primates. Eur J Neurosci 39:1824–1834
Gilbert CD, Kelly JP (1975) The projections of cells in different layers of the cat's visual cortex. J Comp Neurol 163:81–105
Goffinet AM, Daumerie CH, Langerwerf B, Pieau C (1986) Neurogenesis in reptilian cortical structures: 3H-thymidine autoradiographic analysis. J Comp Neurol 243:106–116
Gona AG (1976) Autoradiographic studies of cerebellar histogenesis in the bullfrog tadpole during metamorphosis: the external granular layer. J Comp Neurol 165:77–87
Goodson JL, Kabelik D, Kelly AM et al (2009) Midbrain dopamine neurons reflect affiliation phenotypes in finches and are tightly coupled to courtship. Proc Natl Acad Sci U S A 106:8737–8742
Harvey RJ, Napper RMA (1988) Quantitative study of granule and Purkinje cells in the cerebellar cortex of the rat. J Comp Neurol 274:151–157
Hawrylycz MJ, Lein ES, Guillozet-Bongaarts AL et al (2012) An anatomically comprehensive atlas of the adult human brain transcriptome. Nature 489:391–399
Healy SD, Rowe C (2007) A critique of comparative studies of brain size. Proc Biol Sci 274:453–464
Herculano-Houzel S (2012) Neuronal scaling rules for primate brains: the primate advantage. Prog Brain Res 195:325–340
Hof PR, Nimchinsky EA, Morrison JH (1995) Neurochemical phenotype of corticocortical connections in the macaque monkey: quantitative analysis of a subset of neurofilament protein-immunoreactive projection neurons in frontal, parietal, temporal, and cingulate cortices. J Comp Neurol 362:109–133
Hofman MA (1985) Neuronal correlates of corticalization in mammals: a theory. J Theor Biol 112:77–95
Huang C, Gammon SJ, Dieterle M et al (2014) Dramatic increases in number of cerebellar granule-cell-Purkinje-cell synapses across several mammals. Mamm Biol Zeitschrift für Säugetierkunde 79:163–169

Kaas JH (1989) Why does the brain have so many visual areas? J Cogn Neurosci 1:121–135
Kawamura (1973a) Corticocortical fiber connections of the cat cerebrum. III. The occipital region. Brain Res 51:41–60
Kawamura K (1973b) Corticocortical fiber connections of the cat cerebrum. I. The temporal region. Brain Res 51:1–21
Kawamura K (1973c) Corticocortical fiber connections of the cat cerebrum. II. The parietal region. Brain Res 51:23–40
Kawamura K, Otani K (1970) Corticocortical fiber connections in the cat cerebrum: the frontal region. J Comp Neurol 139:423–448
Markov NT, Vezoli J, Chameau P, Falchier A, Quilodran R, Huissoud C, Lamy C, Misery P, Giroud P, Ullman S, Barone P, Dehay C, Knoblauch K, Kennedy H (2014) Anatomy of hierarchy: feedforward and feedback pathways in macaque visual cortex. J Comp Neurol 522:225–259
Marotte LR, Sheng X-M (2000) Neurogenesis and identification of developing layers in the visual cortex of the wallaby (*Macropus eugenii*). J Comp Neurol 416:131–142
Martínez-Cerdeño V, Cunningham CL, Camacho J et al (2012) Comparative analysis of the subventricular zone in rat, ferret and macaque: evidence for an outer subventricular zone in rodents. PLoS One 7:e30178
Miller JA, Ding SL, Sunkin SM et al (2014) Transcriptional landscape of the prenatal human brain. Nature 508:199–206
Nudo RJ, Masterton RB (1990) Descending pathways to the spinal cord, III: sites of origin of the corticospinal tract. J Comp Neurol 296:559–583
Nudo RJ, Sutherland DP, Masterton RB (1995) Variation and evolution of mammalian corticospinal somata with special reference to primates. J Comp Neurol 358:181–205
O'Connell LA, Hofmann HA (2011) Genes, hormones, and circuits: an integrative approach to study the evolution of social behavior. Front Neuroendocrinol 32:320–335
Paula-Barbosa MM, Feyo PB, Sousa-Pinto A (1975) The association connexions of the suprasylvian fringe (SF) and other areas of the cat auditory cortex. Exp Brain Res 23:535–554
Poldrack RA, Farah MJ (2015) Progress and challenges in probing the human brain. Nature 526:371–379
Polleux F, Dehay C, Kennedy H (1997) The timetable of laminar neurogenesis contributes to the specification of cortical areas in mouse isocortex. J Comp Neurol 385:95–116
Ponti G, Peretto P, Bonfanti L (2006) A subpial, transitory germinal zone forms chains of neuronal precursors in the rabbit cerebellum. Dev Biol 294:168–180
Ponti G, Peretto P, Bonfanti L (2008) Genesis of neuronal and glial progenitors in the cerebellar cortex of peripuberal and adult rabbits. PLoS One 3:e2366
Puelles L, Ferran JL (2012) Concept of neural genoarchitecture and its genomic fundament. Front Neuroanat 6:47
Puelles L, Rubenstein JL (2003) Forebrain gene expression domains and the evolving prosomeric model. Trends Neurosci 26:469–476
Rakic P (1974) Neurons in rhesus monkey visual cortex: systematic relation between time of origin and eventual disposition. Science 183:425–427
Rakic P (2002) Pre and post-development of neurogenesis in primates. Clin Neurosci Res 2:29–32
Rakic P (2003) Developmental and evolutionary adaptations of cortical radial glia. Cereb Cortex 13:541–549
Rakic P, Sidman RL (1970) Histogenesis of cortical layers in human cerebellum, particularly the lamina dissecans. J Comp Neurol 139:473–500
Reep RL, Finlay BL, Darlington RB (2007) The limbic system in mammalian brain evolution. Brain Behav Evol 70:57–70
Reyes LD, Stimpson CD, Gupta K, Raghanti MA, Hof PR, Reep RL, Sherwood CC (2015) Neuron types in the presumptive primary somatosensory cortex of the florida manatee (*Trichechus manatus latirostris*). Brain Behav Evol 86:210–231
Rowell JJ, Ragsdale CW (2012) BrdU birth dating can produce errors in cell fate specification in chick brain development. J Histochem Cytochem 60:801–810
Rowell JJ, Mallik AK, Dugas-Ford J, Ragsdale CW (2010) Molecular analysis of neocortical layer structure in the ferret. J Comp Neurol 518:3272–3289

Sanderson KJ, Weller WL (1990) Gradients of neurogenesis in possum neocortex. Brain Res Dev 55:269–274
Schmahmann JD, Pandya D (2009) Fiber pathways of the brain. OUP, Oxford
Smart IH (1972) Proliferative characteristics of the ependymal layer during the early development of the spinal cord in the mouse. J Anat 111:365–380
Smart IH, Dehay C, Giroud P, Berland M, Kennedy H (2002) Unique morphological features of the proliferative zones and postmitotic compartments of the neural epithelium giving rise to striate and extrastriate cortex in the monkey. Cereb Cortex 12:37–53
Sporns O, Zwi JD (2004) The small world of the cerebral cortex. Neuroinformatics 2:145–162
Stamatakis A, Barbas H, Dermon CR (2004) Late granule cell genesis in quail cerebellum. J Comp Neurol 474:173–189
Striedter GF, Keefer BP (2000) Cell migration and aggregation in the developing telencephalon: pulse-labeling chick embryos with bromodeoxyuridine. J Neurosci 20:8021–8030
Striedter GF (2005) Principles of brain evolution. Sinauer Associates, Sunderland
Takahashi E, Dai G, Wang R (2010) Development of cerebral fiber pathways in cats revealed by diffusion spectrum imaging. NeuroImage 49:1231–1240
Takahashi E, Dai G, Rosen GD et al (2011) Developing neocortex organization and connectivity in cats revealed by direct correlation of diffusion tractography and histology. Cereb Cortex 21:200–211
Takahashi E, Folkerth RD, Galaburda AM, Grant PE (2012) Emerging cerebral connectivity in the human fetal brain: an MR tractography study. Cereb Cortex 22:455–464
Tsai HM, Garber BB, Larramendi LMH (1981) 3H-Thymidine autoradiographic analysis of telencephalic histogenesis in the chick embryo: I. Neuronal birthdates of telencephalic compartments in situ. J Comp Neurol 198:275–292
Uray NJ, Gona AG, Hauser KF (1987) Autoradiographic studies of cerebellar histogenesis in the premetamorphic bullfrog tadpole: I. Generation of the external granular layer. J Comp Neurol 266:234–246
Vallender EJ (2011) Comparative genetic approaches to the evolution of human brain and behavior. Am J Hum Biol 23:53–64
Vallender EJ (2012) Genetic correlates of the evolving primate brain. Prog Brain Res 195:27–44
Voogd J, Glickstein M (1998) The anatomy of the cerebellum. Trends Cogn Sci 2:307–313
Wang E, Ding YC et al (2004) The genetic architecture of selection at the human dopamine receptor D4 (DRD4) gene locus. Am J Hum Genet 74:931–944
Wedeen VJ, Rosene DL, Wang R et al (2012) The geometric structure of the brain fiber pathways. Science 335:1628–1634
Wilkinson M, Hume R, Strange R, Bell JE (1990) Glial and neuronal differentiation in the human fetal brain 9–23 weeks of gestation. Neuropathol Appl Neurobiol 16:193–204
Workman AD, Charvet CJ, Clancy B et al (2013) Modeling transformations of neurodevelopmental sequences across mammalian species. J Neurosci 33:7368–7383
Yamamoto K, Mirabeau O, Bureau C et al (2013) Evolution of dopamine receptor genes of the D1 class in vertebrates. Mol Biol Evol 30:833–843
Yamawaki N, Borges K, Suter BA, Harris KD, Shepherd GM (2014) A genuine layer 4 in motor cortex with prototypical synaptic circuit connectivity. eLife 3:e05422
Young LJ, Wang Z (2004) The neurobiology of pair bonding. Nat Neurosci 7:1048–1054
Yurkewicz L, Lauder JM, Marchi M, Giacobini E (1981) 3H-Thymidine long survival autoradiography as a method for dating the time of neuronal origin in the chick embryo: the locus coeruleus and cerebellar Purkinje cells. J Comp Neurol 203:257–267
Zeng H, Shen EH, Hohmann JG et al (2012) Large-scale cellular-resolution gene profiling in human neocortex reveals species-specific molecular signatures. Cell 149:483–496

Part II
Evolution of Cognition

Chapter 5
The Evolution of Cognitive Brains in Non-mammals

Andrew N. Iwaniuk

Abstract The evolution of cognitive abilities is inextricably linked to the evolution of the brain. Although the bulk of the research on brain-cognition evolution has focused on primates and other mammals, increasing evidence from non-mammals also supports a link between cognitive ability and the brain. This is especially true for relationships between cognition and relative brain and brain region sizes. Cross-species comparisons of innovativeness, tool use, and other aspects of cognition indicate that the sizes of the brain and telencephalic regions are important for cognition in birds and fishes. A recent selection experiment has even demonstrated that larger-brained fish have cognitive advantages over smaller-brained fish. However, other studies have yielded mixed evidence. For example, it is unclear whether the social brain hypothesis is applicable to non-mammals or if hippocampus volume accurately reflects spatial abilities in fishes or birds. Some of these uncertainties are due to relatively poor data sets on cognition and brain anatomy in non-mammals. Alternatively, connectivity and neuron numbers might be more appropriate proxies of information processing capacity of a species and hence its cognition. Relating these other aspects of brain anatomy to cognition will deepen our understanding of neural and cognitive evolution and even generate common design principles that support cognitive processes across species.

Keywords Birds • Fish • Hippocampus • Brain size • Spatial cognition Social behavior • Tool use • Innovative behavior

5.1 Introduction

Cognitive abilities vary across species, and this variation has long been associated with anatomical differences in the brain. The anatomical traits associated with "enhanced" or greater cognitive abilities include brain size, brain region size,

---

A.N. Iwaniuk
Department of Neuroscience, Canadian Centre for Behavioural Neuroscience, University of Lethbridge, Lethbridge, AB, Canada, T1K 3M4
e-mail: aniwaniuk@gmail.com

S. Watanabe et al. (eds.), *Evolution of the Brain, Cognition, and Emotion in Vertebrates*, Brain Science, DOI 10.1007/978-4-431-56559-8_5

interconnectedness of brain regions, neuron size, neuron number, and synapse density (Jerison 1973; Striedter 2005; Azevedo et al. 2009; van den Heuvel et al. 2016). The majority of this research into the neural basis of cognitive abilities has been strongly biased toward mammals. The bias arose because non-mammals were viewed as having poor cognitive skills and "simpler" brains compared with mammals. Over the past 20–30 years, however, these perceptions have proven incorrect. Birds are capable of deception, episodic-like memory, tool manufacture, and a range of other "advanced" cognitive abilities (Emery 2006; Seed and Byrne 2010). In addition, it is now clear that there are many similarities between avian and mammalian brains, despite the differences in gross morphology (Jarvis et al. 2005). Similar conclusions have been reached in analyses of squamates, amphibians, and fishes; all of these species are far more capable of problem-solving and other behaviors indicative of "advanced" cognition, and similarities in brain structure tend to outweigh differences (Wilkinson and Huber 2012; Demski 2013; Northcutt 2013; Burghardt 2013; Bshary et al. 2014). Thus, both cognitive abilities and the neural structure that underlie cognition are highly conserved across vertebrates.

Although there are many different ways of examining the relationship between cognition and brain function, a frequently used approach is to use the size of the brain, or individual brain regions, in relation to performance in a specific behavioral task or species/individual variation in behavior. This relationship between brain or brain region size and behavior is based upon Jerison's principle of proper mass, which he defined as "The mass of neural tissue controlling a particular function is appropriate to the amount of information processing involved in performing the function" (p. 8) (Jerison 1973). In other words, if a behavior requires a lot of processing power, there is a concomitant increase in the size of the brain region that controls or modulates that behavior. A classic demonstration of this principle is the association between the size of a sensory region in the brain in relation to the sensitivity and/or acuity of that sensory modality. For example, the expansion of the representation of the star nose in the star-nosed mole's (*Condylura cristata*) somatosensory cortex is clearly associated with the remarkable sensitivity of this appendage and its importance in foraging behavior (Catania 2011). Jerison (1973), however, also applied this principle to cognition and information processing in general. More specifically, he suggested that the expansion of the vertebrate brain over evolutionary time is associated with improved perceptual and cognitive capacities. In contrast to comparisons of sensory systems, relating cognitive abilities to the size of the brain or brain regions has been fraught with contention. Several authors have suggested that such comparisons are inherently flawed, and any purported relationship between cognitive ability and brain or brain region size is either spurious or is not biologically relevant (Healy and Rowe 2007; Brodin and Bolhuis 2008; Roth et al. 2010). Other authors have emphasized that despite some pitfalls and shortcomings, the size of the brain and its constituent regions matter, and this size variation reflects some component of cognition (Lefebvre and Sol 2008; Kotrschal et al. 2013; Lefebvre 2013). Indeed, as I will outline below, there is a large amount of evidence that supports a link between the size of the brain and specific brain regions and the evolution of cognition.

5.2 Relative Brain Size and Cognition

Brain size has a long association with cognition. Darwin (1871) commented that the size of the human brain, relative to body size, is closely connected with "higher mental powers" and even extended this comparison to "cerebral ganglia" and behavioral differences between hymenopterans and beetles. Since then, a host of other researchers have sought to find a relationship between cognitive abilities and brain size in animals spanning vertebrates and invertebrates. Many of these studies focus on relative brain size: the size of the brain in relation to the body. On average, non-mammals have relatively smaller brains than mammals, but as shown in Fig. 5.1, there is a lot of overlap among different vertebrate groups. In fact, some birds have brains that are relatively larger than mammals of the same body mass, and these bird species are the same ones that are considered to be "highly intelligent": corvids and parrots (Iwaniuk et al. 2005; Emery 2006). Both corvids and parrots outperform other bird species on cognitive tests and engage in what appear to be cognitively demanding behaviors, such as tool use, deception, and complex social relationships (Emery 2006). At a gross level then, it would appear that relative brain size and cognitive ability are correlated in birds, but does this reflect a general trend? Is relative brain size related to quantitative components of cognitive ability in non-mammals?

5.2.1 *Innovative Behavior and the Brain*

The answers to both of these questions are yes, based upon a series of comparative studies by Lefebvre and colleagues on birds (Lefebvre et al. 2004; Lefebvre and Sol 2008; Lefebvre 2013). They scoured the literature for reports of novel-feeding

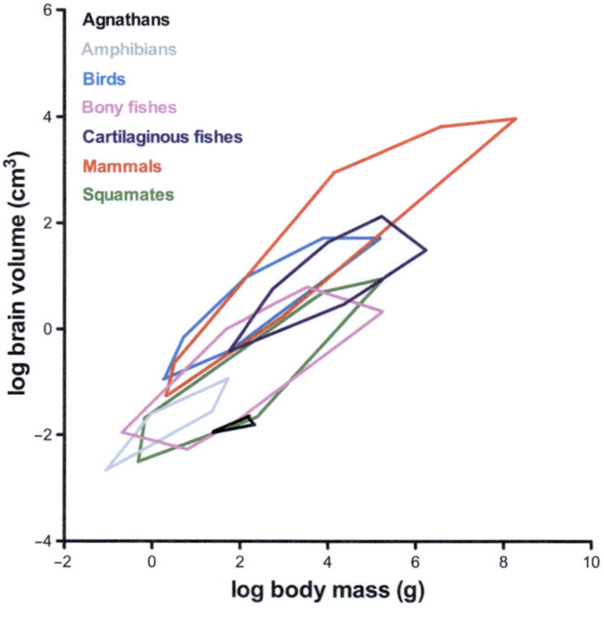

Fig. 5.1 A plot of brain volume and body mass (both log-transformed) with minimum convex polygons surrounding the data for each of the major clades of vertebrates derived from a data set of over 6000 vertebrate species (Yopak and Iwaniuk unpubl data)

behaviors in birds, such as a sparrow hawk (*Accipiter nisus*) drowning its prey (Broadley 1985) or turnstones (*Arenaria interpres*) raiding the gastric cavities of sea anemones (Donoghue et al. 1986). These novel behaviors are thought to be indicative of behavioral flexibility in a general sense. Behavioral flexibility is an important indicator of cognition because it reflects a greater reliance on learning and the need to store and use more information about the environment than simply relying upon preprogrammed, innate behavior (Dunbar 1992; Lefebvre et al. 1997). After tabulating these feeding innovation reports, the number of reports for each species is scaled relative to research effort, defined as the number of papers published on that species. This creates an innovation index across species that was then compared with relative brain size, relative brain region sizes, and a variety of behavioral and life history traits. Across several sets of analyses, species that were more innovative had relatively larger brains and telencephala than species that were less innovative (Lefebvre et al. 1997, 2004; Nicolakakis and Lefebvre 2000). The data was then further subdivided for more focused analyses into what kinds of innovations are related to relative brain size. Overington et al. (2009) categorized innovations according to what specific behaviors were involved. For example, some feeding innovations involved taking advantage of a novel food source, whereas others represent the development of a new foraging technique or the use of a tool. When the innovations were broken down into these finer categories, species that exhibited a more diverse range of innovations had relatively larger brains (Fig. 5.2). Further, technical innovations (e.g., use of tools, novel foraging techniques) explained more variation in relative brain size than other types of innovations (e.g., novel food item or feeding in an atypical habitat). The relatively larger brains of the more innovative species are driven largely by expansion of the forebrain, especially

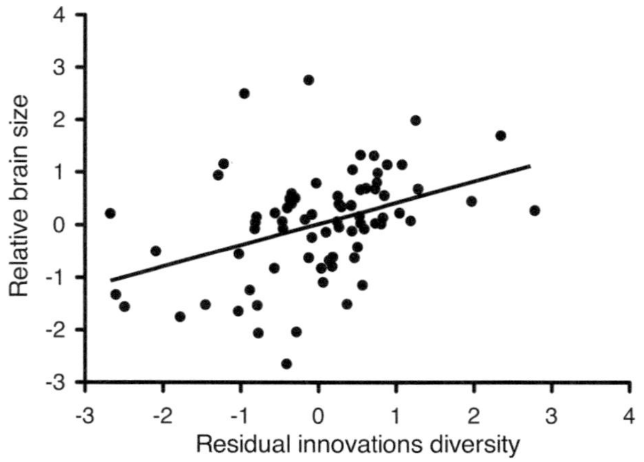

Fig. 5.2 Relative brain size is significantly correlated with the diversity of feeding innovations expressed across bird families (adapted from Overington et al. 2009). Here, relative brain size is expressed as residuals from a regression of brain mass on body mass. Feeding innovations diversity reflects the types of feeding innovations expressed by species within a family, correct for research effort

the pallial components (Sayol et al. 2016). Indeed, species that are more innovative have relatively larger forebrains, and in particular a larger mesopallium, than less innovative species (Nicolakakis and Lefebvre 2000; Timmermans et al. 2000). Given that information integration and decision-making occur within forebrain regions, this strongly suggests that the evolution of cognitive abilities is related to relative brain and forebrain sizes.

The correlation between behavioral flexibility and relative brain size is one aspect of a theory of increasing interest in recent years: the cognitive-buffer hypothesis. This hypothesis posits that the behavioral flexibility enabled by having a relatively large brain allows individuals to buffer the effects of environmental challenges (Sol 2009). Thus, the larger-brained, more innovative species are better able to cope with new ecological opportunities, changes in the environment, and avoidance of novel predators. In support of the cognitive-buffer hypothesis, larger-brained bird, reptile, and amphibian species are more likely to be successful when introduced into a novel environment than smaller-brained species (Sol et al. 2005; Amiel et al. 2011). The successful avian invaders are also more innovative, thus indicating that invasion success and innovativeness are together related to relative brain size. The extent to which these evolutionary trends apply to the expansion of forebrain regions has yet to be fully tested, but based on the initial description of parrots and corvids, it is likely that this reflects the expansion of the mesopallium and nidopallium.

5.2.2 Making a Smarter Fish

Comparative analyses across species have yielded some strong correlations between relative brain size and some aspects of cognition but are often criticized because they are correlational (Healy and Rowe 2007). A mechanistic link between cognition and brain (or forebrain) size requires either analyses of putative molecular mechanisms or experiments within species. Recent advances in avian genomics demonstrate convergent patterns of gene expression between songbirds and humans that likely underlie language and vocal learning (Pfenning et al. 2014), but to date there are no comparable analyses examining more general aspects of cognition. Strong evidence for a link between cognitive and relative brain size is, however, provided by a recent experiment on Trinidadian guppies (*Poecilia reticulata*).

To determine whether selection for relative brain size affected behavior, Kotrschal et al. (2013) performed a line-breeding experiment in which they selected guppies to bred based on brain size. Initially, three groups of breeding pairs were allowed to reproduce and the brains of the parents weighed. The offspring of those parents with the largest and smallest brains were then placed in large-brained and small-brained pairs and the procedure repeated. The resulting generation of brain size-selected guppies were then subjected to a numerical learning test. The fish were placed into a testing tank and trained to discriminate between cards that had two or four objects printed on them. The large-brained females significantly outperformed the small-brained females, indicating that relatively large brains conferred a cognitive advantage in this

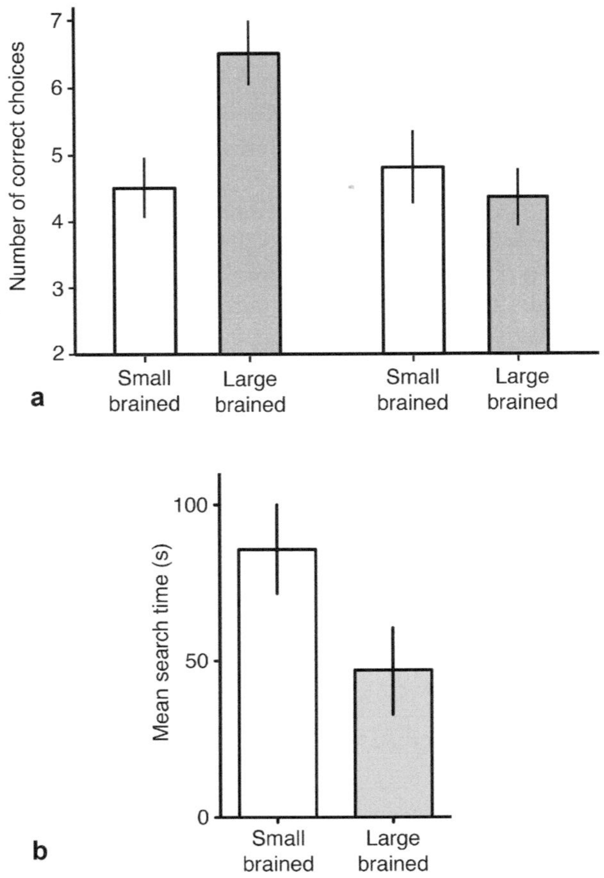

Fig. 5.3 Results of behavioral testing of selected lines of small-brained and large-brained guppies (*Poecilia reticulata*). (**a**) Large-brained females significantly outperform small-brained females in a numerical learning task, but no difference is detected in males (Kotrschal et al. 2013). The bars indicate the mean and standard error for the number of times (out of eight tests) that an individual chooses the correct option. The figure is courtesy of A. Kotrschal. (**b**) In a learning task in which males have to navigate a maze in order to find a receptive female, large-brained males are nearly twice as fast as small-brained males. Shown here are the mean search times for large- and small-brained males (± standard error) derived from a generalized linear model (Kotrschal et al. 2015a)

task (Fig. 5.3). The larger-brained guppies also outperform smaller-brained guppies in other behavioral tasks. For example, males were trained to navigate a maze and be "rewarded" with a receptive female at the end (Kotrschal et al. 2015a). After 2 weeks of training, the large-brained males were nearly twice as fast as small-brained males in finding the female (Fig. 5.3), suggesting that the larger-brained males learned faster or had better spatial cognition than their smaller-brained counterparts. When exposed to a predator, large-brained female guppies also experienced higher survivorship (15% higher) than small-brained females (Kotrschal et al. 2015b). Although no survivorship difference was detected in the males, this could be attributed to a correlation between brain size and color whereby large-brained males are more colorful than small-brained males (Kotrschal et al. 2015c). Nevertheless, their results suggest that a larger brain confers an advantage in predator avoidance in females. Whether the large-brained guppies have more neurons or differential enlargement of specific brain regions (e.g., pallial components of the telencephalon) is unknown at this time, but these experimental results strongly suggest that there are cognitive advantages of having a large brain.

5.3 Tool Use in Birds

Tool use is a behavior that many consider to be cognitively demanding. It is not only dependent on fine motor control but also requires a great deal of somatosensory-visual integration and learning to use the tool effectively and involves physical reasoning and, in many species, planning (Seed and Byrne 2010). Together, this means that tool-using species likely require a lot of information processing in sensory, motor, and integrative regions of the telencephalon in order to use tools (Lefebvre 2013). The cognitive requirements of tool use and the progressive evolution of tool use and manufacture are often associated with an increase in relative brain size in hominids, and a similar pattern is present in birds. Using a similar data set to that of innovations (see above), Lefebvre et al. (2002) tested whether tool use is associated with relative brain size in birds. True tool users, those species that manipulate items that are detached from the substrate and held in the beak or foot (e.g., hammers, probes, Fig. 5.4), have

Fig. 5.4 Species that use tools have a more folded cerebellar cortex than species that do not use tool. (**a**) is a photo of a Goffin cockatoo (*Cacatua goffiniana*) using a tool it manufactured in order to access a nut (Auersperg et al. 2012). Photo courtesy of A. Auersperg. (**b**) depicts midsagittal sections of two species with a similar brain size but different degrees of folding in the cerebellum and different cognitive abilities. At the top is the cerebellum of a sulfur-crested cockatoo (*Cacatua galerita*), which has one of the most highly folded cerebella of any birds. At the bottom is the cerebellum of the Australian pelican (*Pelecanus conspicillatus*), a large bird that does not use tools or engage in the range of other innovative behaviors typical of parrots and cockatoos. Although the pelican cerebellum also has a lot of folds, it is much less folded than that of the cockatoos. Both drawings are adapted from Iwaniuk et al. (2006). Scale bars = 1 mm

relatively larger brains than species using "proto-tools" (e.g., anvils or other tools that are not detached from the substrate) or not using tools at all. This difference in relative brain size between true tool users and other species appears to be drive by the nidopallium. Of several telencephalic brain regions examined, only the relative size of the nidopallium was significantly correlated with true tool use. Again, given the multifunctional organization of the nidopallium (Shanahan et al. 2013), it is likely involved in both the motor planning and cognitive components of tool use.

Corroborative evidence for both the enlargement of the brain and telencephalic brain regions comes from the New Caledonian crow (*Corvus moneduloides*). The New Caledonian crow has been the subject of intense study by several labs because this species manufactures tools and uses them regularly to forage in the wild. In comparison with other crows and ravens, the New Caledonian crow appears to have a relatively larger brain (Cnotka et al. 2008). A subsequent study of brain region sizes found that the mesopallium was significantly enlarged in New Caledonian crows compared with the carrion crow (*Corvus corone*), Eurasian jay (*Garrulus glandarius*), and house sparrow (*Passer domesticus*) (Mehlhorn et al. 2010). The mesopallium is a highly integrative region (Reiner et al. 2005; Shanahan et al. 2013) and is likely involved in tool using behavior. The nidopallium was also moderately enlarged in the New Caledonian crow, but not significantly enlarged compared with the same three species. Thus, at least two lines of evidence indicate that the enlargement of the brain and integrative regions of the telencephalon are associated with the use and manufacture of tools in birds.

Brain and brain region sizes are not, however, the only neuroanatomical features that are associated with tool use. The cerebellum plays a key role in motor planning as well as cognitive components of tool use in primates (Cantalupo and Hopkins 2010), and comparative evidence suggests that there is also an association between cerebellar anatomy and tool use in birds. Rather than a correlation between size of the cerebellum and tool use, however, it is the degree of folding or surface area of the cerebellar cortex that is greater in tool-using birds (Iwaniuk et al. 2009) (Fig. 5.4). The greater amount of cerebellar folding translates to more Purkinje cells, which would provide the cerebellum with greater information processing power and a larger amount of output to extra-cerebellar targets elsewhere in the brain. This would include feedback to telencephalic regions responsible for motor planning and learning, both of which are essential for effective tool use. Thus, tool use in birds relies not only on telencephalic regions but also on the cerebellum, an evolutionary pattern that parallels the evolution of tools and brains in hominids (Weaver 2005; Lefebvre 2013).

5.4 Social Cognition and Brains in Non-mammals

The social brain hypothesis (Dunbar 1998) is one of the major theories in evolutionary neurobiology and evolutionary psychology. The basic premise is that the complexities of social behavior, such as remembering previous social interactions and subsequently using that information, have driven an increase in information processing capacity of the brain, especially in the neocortex in mammals. Thus, species that live in large social groups with a diverse array of social interactions have relatively larger neocortices than species living only in pairs or asocial species. The social brain hypothesis was originally developed to explain the coevolution of large brains and social complexity in primates (Dunbar 1998) but has since been applied to a broader range of mammals as well as birds and fish. Although some studies suggest that the social brain hypothesis is broadly applicable across vertebrates, the evidence is mixed and likely depends on the types of social interactions that occur within specific social systems.

In fish, two studies have addressed the evolution of the brain in relation to mating systems in African cichlids. Cichlids from Lake Tanganyika are an important group in evolutionary biology because they represent one of the most rapid adaptive radiations of any vertebrate group, and this adaptive radiation generated considerable behavioral diversity among species, especially with respect to habitat, diet, and mating system (Kocher 2004). The close phylogenetic relationships among species combined with this behavioral diversity make them an ideal group within which to test for relationships between behavior and the brain. An initial study across seven cichlid species found that monogamous species have telencephalic volumes that are significantly larger than that of polygamous species (Pollen et al. 2007). Polygamous species do not form pair bonds, so this could be taken as evidence in support of the social brain hypothesis, but monogamy and biparental care occur together, so it is unclear whether this difference truly has to do with social relationships between adults or some relationship with parental care. Parental care is associated with significant changes in relative brain size within and across species (Gonzalez-Voyer et al. 2009; Samuk et al. 2014) in fishes, so this could account for the difference between monogamy-biparental care and polygamy-uniparental care. In addition, the species with the largest telencephalic volume lives in a complex habitat and is monogamous, suggesting that habitat complexity and social system could be related. Due to correlations across behaviors and a small number of species, it is therefore unclear whether social bonding is linked to the expansion of the telencephalon or not.

Much larger comparisons (30+ species) of cichlids tested for the effects of diet, habitat, parental care, and mating system on both brain size and the size of several brain regions (Gonzalez-Voyer et al. 2009; Gonzalez-Voyer and Kolm 2010). Mating system was not associated with brain or telencephalon in this larger data set, but was correlated with dimorphism in telencephalon size. In polygamous species, telencephalon size is monomorphic, whereas in monogamous species, the telencephalon is dimorphic with males having larger volumes than females. Again, parental care and mating system coevolve in Tanganyikan cichlids, so it is difficult to disentangle the effects of these two behaviors, especially when uniparental care is associated with an increase in relative brain size in female cichlids and not males (Gonzalez-Voyer and Kolm 2010). The lack of clear support for the social brain hypothesis in African cichlids could reflect differences in the amount or types of social information used in cichlid social interactions compared with primates or other mammals (Shultz and Dunbar 2007). It should be noted that African cichlids are the only clade of fishes in which mating system has been compared with brain size and composition. Whether the lack of a clear association between mating system and the brain is a trait unique to African cichlids or not is unknown, and further testing is needed in other clades that vary in sociality or mating system, such as seahorses and pipefishes (Syngnathidae) and butterfly fishes (Chaetodontidae). Evidence from birds, however, suggests that the social brain hypothesis might not be as universal as suggested by some authors (Dunbar and Shultz 2007; Shultz and Dunbar 2007).

Like mammals, birds exhibit a diverse range of social and mating systems from essentially asocial species to complex, multi-male, multi-female fission-fusion social systems. Social systems characterized as being "fission-fusion" vary in the size, dispersion, and membership of the social group such that individuals experience variable degrees and types of social interactions over time (Aureli et al. 2008). It is often considered to be the most cognitively demanding form of social organization because of the dynamic nature of their social interactions. Because birds have evolved social systems with high degrees of fission-fusion as well as varying degrees of monogamy and asociality (Emery et al. 2007), they are an ideal nonmammalian group within which to test the social brain hypothesis. Corvids and parrots have relatively large brains and complex social behaviors that include fission-fusion social systems and complex forms of communication among conspecifics (Emery 2006; Emery et al. 2007; Hobson et al. 2014). Despite these "primate-like" social interactions, there is limited evidence that variations in the size of social group or type of mating system are associated with an increase in relative brain size in birds. Emery et al. (2007) in their review on social bonding in birds analyzed relative brain size across hundreds of species and found that relative brain size varied with social and mating system in birds (Fig. 5.5). In terms of social system, bird species that live in small groups (5–30 individuals) have relatively larger brains than those in extremely large groups (70+) or solitary species. Species in living in large groups are communal breeders in which affiliative social interactions are primarily limited to pairs and their offspring (e.g., gulls, gannets, penguins), so they would lack the long-term social relationships that can occur in smaller groups. However, species that live in small groups tend to be either in fission-fusion systems or are cooperative breeders, and this is supported by their analysis of mating system; species in long-term pair bonds or cooperative breeding systems have relatively larger brains than species in other mating systems (Fig. 5.5). Although the authors acknowledge that these are proxies for social complexity, the results parallel findings in non-primate mammals (Shultz and Dunbar 2007) and support the social brain hypothesis.

Although the analysis of Emery et al. (2007) is convincing, analyses within avian orders have yielded markedly different results. Across corvids and related songbird families (superfamily Corvoidea), there is no significant difference in relative brain size between cooperative and non-cooperative species and no significant correlation between relative brain size and cooperative breeding group size (Iwaniuk and Arnold 2004). This is somewhat surprising given that highly social corvids learn dyadic relationships faster and with fewer errors than less social corvids (Bond et al.

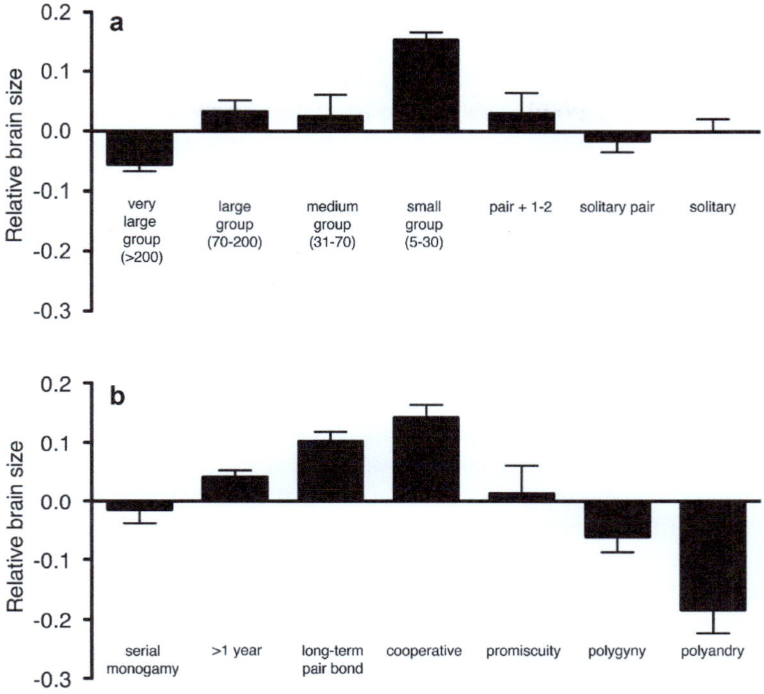

Fig. 5.5 Relative brain size varies significantly across social and mating systems in birds. Both plots are adapted from Emery et al. (2007) in which relative brain size is expressed as residuals from a regression of brain volume against brain mass. (**a**) Mean relative brain size (± standard deviation) is shown for seven different types of social systems that vary from species that live very large groups (over 200 individuals) to species that are solitary. Relative brain size is largest in those species that live in small groups of 5–30 individuals. (**b**) Mean relative brain size (± standard deviation) is shown for seven different types of mating systems. Relative brain size is largest in cooperatively breeding species and smallest in polyandrous species

2003). Perhaps corvids evolved relatively large brains in relation to innovativeness and behavioral flexibility (see above). Under this evolutionary scenario, corvids would have the brain size needed to process complex social information but do not necessarily have to live a highly social lifestyle. Alternatively, cooperative breeding might not represent the kinds of social complexity that occurs in other types of social or mating systems (Thornton and McAuliffe 2015; Thornton et al. 2016). By definition, cooperative breeding involves the cooperation of related conspecifics in the raising and protection of offspring (Iwaniuk and Arnold 2004). Most cooperative breeding groups therefore contain a breeding pair, several offspring from previous breeding seasons and the brood from the current reproductive cycle. In other words, it is an extended family living together. As such, there is very little flux in the composition of the social group, and manipulative or Machiavellian behavior is selected against because it would harm an individual's inclusive fitness. In contrast, the complexities of social living as a primate involve dominance hierarchies and changes in group composition, which together would place significant demands on integrative regions of the brain, especially frontal cortex (Dunbar 1998; Kudo and Dunbar 2001; Shultz and Dunbar 2007). Keeping track of family members in a cooperatively breeding bird is unlikely to be as cognitively demanding because an individual grows up within the breeding group with relatively few changes. This explains not only the lack of association between cooperative breeding and relative brain size in birds but also how cooperative breeding can evolve in species with relatively small brains, such as moorhens (*Gallinula* spp.) and cuckoos (e.g., *Crotophaga ani*, *Guira guira*).

A final comparison focused on mating systems and relative brain size in shorebirds (Charadriiformes). Gulls, terns, puffins, sandpipers, plovers, and their relatives are somewhat unique among birds in that they exhibit a very wide range of mating systems, including polyandry. In shorebirds, polyandrous species have relatively smaller brains than monogamous species (García-Peña et al. 2013), but like the African cichlids, mating system and parental care have coevolved in shorebirds, so it is unclear if mating system or parental care is the salient variable. Thus, despite the findings of Emery et al. (2007), analyses of both shorebirds and the Corvoidea do not support the social brain hypothesis.

The lack of consistent empirical support for the social brain hypothesis in non-mammals parallels findings across mammalian species. Although primates have a robust relationship between social system and relative brain (and neocortex) size, the same is not true of other mammals (Holekamp 2007; Shultz and Dunbar 2007). Indeed, non-primate mammals living in multi-male societies have relatively small brains than those living in simple pairs, whereas primates in multi-male societies have relatively larger brains than those living in pairs or solitarily (Shultz and Dunbar 2007). Of course, not all variations in relative brain size necessarily reflect social cognition (see above and Holekamp 2007), but some authors have suggested that the lack of correlation between brain size and sociality in non-primate mammals is due to a "lack of cognitive demand." That is, bats, ungulates, and carnivores that live in large groups have highly unstable group composition compared with primates, which would not place the same demands on processing social information. Although the same argument could also explain the inconsistent support for the social brain hypothesis in birds, there is ample evidence that many bird species exhibit primate-like social interactions. Rooks (*Corvus frugilegus*) live in highly complex societies in which individuals form social affiliations with offspring, mates, and unrelated conspecifics, and they appear to actively manage these relationships and cooperative and coordinate group activities (Emery et al. 2007). Similarly, many parrot species live in fission-fusion social systems in which social interactions occur among many individuals, and the specific nature of those social interactions change throughout the day and from one day to the next (Rowley 1990; Hobson et al. 2014; Toft and Wright 2015). This includes the formation of juvenile flocks, which are roving bands of young parrots that forage and interact socially with one another (Rowley 1990; Toft and Wright 2015). Parrots can even refer to one another by "name" using specific contact calls for other individuals within a flock (Wanker et al. 2005). All of this is strongly indicative of complex social interactions in parrots, but details are scant. Even though Marler (1996) highlighted a profound lack of data on social interactions in birds decades ago, there have been few attempts to truly document social behavior in corvids or parrots in ways that are comparable to that of primates (see Hobson et al. 2014 for an exception). To effectively test the social brain hypothesis in birds, more detailed information is required on social interactions and brain anatomy. Network analyses similar to those shown in Emery et al. (2007) and Hobson et al. (2014) across several species are needed to establish network size and complexity. In addition, more quantitative neuroanatomical studies are needed of corvids, parrots, and other species to test if sociality is correlated with the size of the pallium and subregions within pallium (e.g., caudolateral nidopallium).

5.5 Spatial Cognition and the Hippocampus

Thus far, I have provided many examples of relationships between cognition, or proxies of cognition, and the relative size of the brain and regions of the telencephalon. Clearly, the telencephalon is critical for most aspects of cognition, but therein lies the problem with comparing the telencephalon with cognitive abilities. Large brain regions, like the nidopallium and mesopallium in birds, perform a wide range of functions (Shanahan et al. 2013), which can make it difficult to directly associate the anatomy of these regions with specific cognitive domains or general cognitive ability. In contrast, the hippocampus (and hippocampal homologs) offers a more specific means of examining the role between anatomy and cognitive ability. The hippocampus plays a key role in spatial memory; lesioning or disruption of hippocampal activity generally results in poor learning ability in spatial tasks (Andersen et al. 2007). The function of the hippocampus appears to be highly conserved across vertebrates such that similar lesion experiments performed in birds, lizards, turtles, and bony fishes all demonstrate deficits in spatial memory (Sherry 2006; Demski 2013; Striedter 2015). This has led to considerable research into the evolution of hippocampal homologs in relation to behavior across vertebrates, especially birds (see below).

Despite the importance of the hippocampus in spatial memory, intraspecific studies of hippocampus size in non-mammals have yielded mixed results with respect to spatial cognition and behavior. Within some species, home range size covaries with hippocampus size such that sexes or populations that occupy larger home ranges have correspondingly larger hippocampal volumes (Roth et al. 2005; LaDage et al. 2009). However, for many of these species, home range size is used as a proxy for spatial cognition, and testing of spatial abilities is not conducted. When spatial cognition is tested, the relationship between hippocampus size and actual spatial abilities is unclear.

In one of the few examples in fishes, (Costa et al. 2011) tested for sex differences in the ventral dorsolateral telencephalon (DLv, the hippocampal homolog in fishes) in two species of blenny that live in rock pools. In these two species, males defend small territories in which they build a nest, and females visit multiple males during the breeding season. Thus, females of both species occupied home ranges that were 4–10× larger than that of males. These same females also had larger DLv volumes, but the sex difference was much smaller in magnitude (1.1–1.2× larger). Although the authors concluded that this likely reflected sex differences in spatial cognition, subsequent testing of homing behavior in one of the species revealed no sex differences (Thyssen et al. 2013).

Behavioral and neuroanatomical studies of cowbirds also reveal some discrepancies between performance in spatial tasks and sexual dimorphism in hippocampus size. The brown-headed cowbird (*Molothrus ater*) is an obligate brood parasite, meaning that it can only reproduce through laying eggs in the nests of other species (Johnsgard 1999). Only females engage in nest searching and visit dozens of nests each breeding season (Johnsgard 1999), which presumably results in a sex difference in processing spatial information. In support of a sex difference in cognition, female cowbirds have relatively larger hippocampal volumes than males (Sherry et al. 1993). This sexual dimorphism is not unique to brown-headed cowbirds and was also found in the shiny cowbird (*Molothrus bonariensis*) in which only the females search for nests (Reboreda et al. 2006). Recent learning and memory tests in both species, however, yielded mixed results in terms of sex differences. In the first test of learning and memory in shiny cowbirds Astie et al. (1998), females outperformed males in a food-finding task based on appearance cues, but not on location. More recent studies in brown-headed cowbirds found that females made fewer errors and took more direct paths than males in a spatial task (Guigueno et al. 2014), but males outperformed females in a spatial touchscreen task (Guigueno et al. 2015). Spatial ability can be task dependent, so these mixed results could arise from differences in the testing paradigm across studies. In addition, nothing is known about what features female cowbirds are attending to when searching for and presumably remembering the locations and status of host nests. Nevertheless, the conclusion reached is that the relationship between hippocampus size and spatial cognition is not straightforward. This lack of a consistent relationship between spatially dependent behaviors and hippocampal anatomy is enforced by studies of food-caching songbirds.

Many animals cache or hide food, but several songbird families rely heavily on scattered food caches (or scatter hoarding) as part of their overall foraging strategy. These are the corvids (jays, crows, magpies), parids (chickadees, tits, and titmice), and nuthatches. Very little is known about nuthatch spatial cognition or hippocampus (but see Petersen 1995), but studies of corvids and parids have proven to be important for understanding spatial cognition and hippocampal function in birds (Sherry 2006). In the initial comparative studies of both groups, food-caching species had significantly larger hippocampal volumes than non-caching species (Krebs et al. 1989; Sherry et al. 1989; Healy and Krebs 1992, 1996; Basil et al. 1996). Further analyses then suggested that the type or amount of food caching was correlated with the relative size of the hippocampus (Healy and Krebs 1992; Lucas et al. 2004). These comparative studies were corroborated by behavioral testing; food-caching species performed better than non-caching species in spatial memory tasks, and species that cached more tended to outperform species that cached fewer items

(Sherry 2006; Kamil and Gould 2008). Intraspecific studies of chickadees (*Poecile atricapilla* and *Poecile sclateri*) across altitudinal and latitudinal gradients also support a relationship between spatial memory and hippocampal anatomy. Chickadees living in harsher environments (i.e., higher altitudes or latitudes) have better spatial memory, larger hippocampus, more hippocampal neurons, and a higher neurogenesis rate than chickadees living in more benign habitats (Pravosudov and Roth 2013). However, several authors questioned the validity of the comparative studies (Brodin and Bolhuis 2008), and reanalyses of all of the available hippocampal data yielded contradictory results: one study supported a correlation between food caching and hippocampus size, and the other did not (Brodin and Lundborg 2003; Lucas et al. 2004). An analysis of hippocampus size across caching and non-caching woodpeckers also found no differences across species (Volman et al. 1997). This has led some to suggest that hippocampus size is not an appropriate neuroanatomical measurement and that neuron numbers and/or rate of neurogenesis are more meaningful correlates of spatial memory in birds (Roth et al. 2010). In a common garden experiment in which black-capped chickadees from northern regions were housed in the same aviaries as chickadees from southern regions, there is no difference in hippocampal volume, but the differences in spatial memory, neuron numbers, and neurogenesis persist (Roth et al. 2011). Thus, neuron number and neurogenesis are more reliably associated with spatial ability in chickadees. Similarly, a new analysis of corvids found no evidence to support a difference in hippocampal volume between scatter hoarding and other species, but the number of hippocampal neurons did vary (Gould et al. 2013). More specifically, the Clark's nutcracker (*Nucifraga columbiana*) has the highest number of neurons and has the best spatial memory of any of the corvids tested (Kamil and Balda 1985; Kamil and Gould 2008), suggesting that the number of neurons might be a more important variable than hippocampal volume. Hippocampus size might not be the best anatomical measurement, but the fact that neuron numbers are associated with spatial memory variation still indicates that hippocampal anatomy does play a significant role in individual and species differences in spatial cognition.

An additional group in which there seems to be a cognitive advantage to having a relatively large hippocampus is hummingbirds (Fig. 5.6). Hummingbirds visit hundreds of flowers per day and can remember not only the spatial location and distribution of flowers but also the nectar quality and content of flowers and refilling rates of individual flowers (Healy and Hurly 2013). To accompany their memories for both spatial and temporal properties of flowers, hummingbirds have proportionally the largest hippocampus of any bird examined to date (Ward et al. 2012; González-Gómez et al. 2014) (Fig. 5.6). As with other studies in birds, there is a disconnect between hippocampus size and spatial memory within species. Male green-backed firecrowns (*Sephanoides sephanoides*) outperform females in a spatial memory task but have significantly larger hippocampal volumes (González-Gómez et al. 2014). Further study is needed to determine whether hummingbirds have superior spatial cognition compared to other birds, but given their strong reliance on spatial memory for successful foraging, much could be gleaned by further study of their hippocampal anatomy, especially neuron numbers and neurogenesis.

Fig. 5.6 Hummingbirds have remarkable memories and the largest hippocampal volumes, relative to brain size, of any bird. Shown here are (**a**) a male rufous hummingbird (*Selasphorus rufus*) feeding from a flower and (**b**) boxplots indicating the relative size of the hippocampus. Relative size here is the volume of the hippocampus divided by the volume of the telencephalon and then log-transformed (as in Ward et al. 2012). The hummingbird photo is courtesy of B. Goller

5.6 Constraints on Understanding Neural and Cognitive Evolution

Throughout this review, I have provided evidence that the size of the brain and brain regions is an important correlate of cognitive abilities in non-mammals. Relative brain size is clearly associated with some aspects of cognition, such as innovativeness and tool use, and the selective breeding experiment in guppies demonstrates that there is a cognitive advantage to having a relatively larger brain. Similarly, there is evidence that the size of individual brain regions is associated with several aspects of cognition, including innovativeness, tool use, spatial abilities, and (possibly) sociality. Apart from the experimental studies of the guppies (Kotrschal et al. 2013) and intraspecific studies of chickadees (Pravosudov and Roth 2013), the bulk of the evidence in favor of brain and brain region sizes covarying with cognition is constrained by several problems.

First, the evidence provided by comparative studies is largely correlative and not causal (Healy and Rowe 2007; Brodin and Bolhuis 2008). It is therefore not possible to determine whether relatively larger brains are required for certain cognitive tasks or changes in behavior drive the expansion of the brain and telencephalic brain regions. Evolutionary path analyses can be used to infer the timing of evolutionary changes in cognition, behavior, and the brain (Boerner and Krüger 2008; Sol et al. 2010), but these methods have been used infrequently in evolutionary neurobiology thus far.

Second, there is the ubiquitous and long-standing issue of how best to measure cognition across species and whether certain measurements are true proxies of general cognition or specific cognitive domains. This is not an easy question to address because species-specific sensory system design and behaviors can often prevent using the same tests equally across all species. The innovation and tool use indices developed by Lefebvre and colleagues provide at least a partial solution to this problem

(Lefebvre et al. 1997; Lefebvre and Sol 2008; Lefebvre 2013) but are constrained by the number of reports published and range of species that they cover.

Third, and related to the previous point, the ability to execute any comparative study of brain and/or cognition is dependent on data availability. For many species, cognitive testing data is available for a very small number of species, which is why so many comparative studies use proxies, such as social system, food-caching propensity, and other behaviors that can be readily observed in the wild or gleaned from the literature. Even so, there are no broad studies of amphibian, squamate, turtle, or crocodilian brains that have employed proxies of cognition (see Sect. 5.6.1). Quantitative brain data is also missing for most nonmammalian vertebrates. Unlike mammalian taxa, such as bats, primates, and "insectivores," which were the subject of intense study in the last century, the same attention has not been paid to collecting quantitative brain data in nonmammalian vertebrates. In fishes, significant effort by a couple of authors has improved our understanding of neuroanatomical differences in cartilaginous fishes (Yopak 2012), but most studies of bony fishes have focused on a small subset of taxa (Kotrschal et al. 1998). The number of species and range of brain regions measured in birds has increased progressively over the past 20 years, but data is still deficient for many orders (e.g., hornbills, grebes, suboscine songbirds) and most families within orders (e.g., songbirds, shorebirds) (Iwaniuk 2010). Without more data, it is simply not possible to test for other evolutionary patterns, include potential confounding variables, or generate general principles of brain composition and design that are related to cognition. This is especially true for amphibians, turtles, squamates, and crocodilians because they occupy key points in vertebrate evolution. Without data on these key taxa, reconstructing the evolutionary history of cognition or brain composition is impossible.

5.6.1 *Brain and Cognition in Amphibians, Turtles, Squamates, and Crocodilians*

Although there seems to be an abundance of studies on birds and fishes, relatively little is known about cognitive or neuroanatomical variation among amphibians, squamates, and crocodilians. In terms of brain organization, homologs of mammalian and avian brain regions can generally be identified, and much of their hodology is reasonably well understood (Butler and Hodos 2005). However, there is a gross lack of quantitative neuroanatomical data within these taxa and that constrains our ability to determine.

The cognitive abilities of crocodilians have rarely been investigated, probably owing to the logistics of animal husbandry and handling. In terms of basic biology, they can be highly social and exhibit parental care and have a telencephalic organization similar to birds (Butler and Hodos 2005; Northcutt 2013). Whether these behaviors covary with brain size or brain region sizes across species is unknown. Data on brain size is only available for 4/24 extant species of crocodilians and volumetrics of brain regions for 1 species (Northcutt 2013). Thus, even comparing relative brain or telencephalon size of crocodilians with other taxa, irrespective of cognitive testing, cannot be done at present.

Data availability is slightly better for amphibians. Amphibians are often used in studies of sensory systems, but cognitive/behavioral testing is limited. Some amphibians can learn spatial tasks and are capable of avoidance learning and discrimination, but there is no evidence of complex social interactions or individual recognition (Burghardt 2013). There is also very little information on brain size and composition variation among amphibians (20/655 species of newts and salamanders, 24/4800 species of frogs and toads). The paucity of data on frogs and toads also means that species diversity in brain size and composition is entirely unknown. Similarly, a lack of cognitive testing and brain data for all amphibians means that comparisons of brain and cognition that have been conducted in birds and fishes simply cannot be done in amphibians.

Squamates and turtles have been the study of a far greater range of cognitive studies. Turtles and lizards are capable of sensory discrimination, reversal learning, habituation, problem-solving, and various forms of spatial learning (Wilkinson and Huber 2012; Burghardt 2013). There are even reports of sensitivity to gaze direction and social learning in some species (Wilkinson and Huber 2012). Snakes also appear to be capable of various types of learning but are studied far less intensively (Burghardt 2013). Brain size data is available for a reasonable range of squamates (nearly 200 species), and data on brain region sizes has been improving in recent years (e.g., Powell and Leal 2014). The same cannot be said for turtles with brain size data for only 6/300 species, and brain regions have only been measured in a single species (Northcutt 2013). Despite improvements in quantitative measures of squamate brains, there have been only a few studies in which brain region sizes have been compared with behavior (Roth et al. 2005; LaDage et al. 2009; Powell and Leal 2012, 2014; Robinson et al. 2015) and none with cognitive testing.

If one of the many goals of comparative cognition and neurobiology is to address fundamental questions related to cognitive and neural evolution, these knowledge gaps need to be filled. As outlined above, data on cognitive abilities of crocodilians and a broader range of amphibians, squamates, and turtles is required to better understand cognitive differences across these groups. In addition, quantitative data on brain size and brain region sizes are needed for all these groups to determine if patterns observed in fishes, birds, and/or mammals can be replicated in other vertebrate taxa or if other patterns emerge. For example, the available data suggests that relative brain size is greatly reduced in snakes compared with lizards, but data for more species is needed to enable effective statistical tests.

5.6.2 *Is Size Everything?*

The sizes of the brain and brain regions are often the easiest measurements to obtain, but reflect only one component of brain anatomy that we can compare with cognition. Connectivity among brain regions is likely a crucial component of cognition. The number of synapses, strength of connections between brain regions, and the amount of interconnectedness throughout the brain all contribute to how the brain makes decisions (Bota et al. 2015; van den Heuvel et al. 2016). Quantifying connectivity is not, however, a trivial task, and the field of connectomics is still in its early stages. Relating overall or regional connectivity to cognition across species is therefore not possible yet. However, as computational methods become more refined and they are applied to neuroanatomical data across species, it will eventually be possible to relate connectomes with behavior (van den Heuvel et al. 2016).

Over the past 10 years, there has also been renewed interest in neuron numbers. This is largely due to the development of the isotropic fractionator technique, which enables neuron and non-neuronal cell numbers to be quantified in a fraction of the time needed for a stereological approach (Herculano-Houzel et al. 2015a). Data is now available for 39 mammals, 28 birds, and the Nile crocodile (*Crocodylus niloticus*) (Herculano-Houzel et al. 2015b; Ngwenya et al. 2016; Olkowicz et al. 2016). Some authors have suggested that neuron numbers are more meaningful than the size of brains or brain regions because the neuron is the basic computational unit of the brain (Azevedo et al. 2009). In fact, much of Jerison's (1973) principle of proper mass, and his book as a whole, is based on the idea that larger brains have more neurons. Already there is evidence, neuron numbers might reflect spatial memory more accurately than hippocampal volume (Roth et al. 2011; Gould et al. 2013). As data is collected for more species, it will eventually become possible to relate neuron numbers to several measures of cognition, which will finally address why variation in brain and brain region sizes varies and why it matters.

References

Amiel JJ, Tingley R, Shine R (2011) Smart moves: effects of relative brain size on establishment success of invasive amphibians and reptiles. PLoS One 6:e18277. doi:10.1371/journal.pone.0018277.g002

Andersen P, Morris R, Amaral D et al (eds) (2007) The hippocampus book. Oxford University Press, New York

Astie AA, Kacelnik A, Reboreda JC (1998) Sexual differences in memory in shiny cowbirds. Anim Cogn 1:77–82

Auersperg AMI, Szabo B, Bayern von AMP, Kacelnik A (2012) Spontaneous innovation in tool manufacture and use in a Goffin's cockatoo. Curr Biol 22:R903–R904. doi:10.1016/j.cub.2012.09.002

Aureli F, Schaffner CM, Boesch C et al (2008) Fission-fusion dynamics. Curr Anthropol 49:627–654

Azevedo FAC, Carvalho LRB, Grinberg LT et al (2009) Equal numbers of neuronal and nonneuronal cells make the human brain an isometrically scaled-up primate brain. J Comp Neurol 513:532–541. doi:10.1002/cne.21974

Basil JA, Kamil AC, Balda RP, Fite KV (1996) Differences in hippocampal volume among food storing corvids. Brain Behav Evol 47:156–164

Boerner M, Krüger O (2008) Why do parasitic cuckoos have small brains? Insights from evolutionary sequence analyses. Evolution 62:3157–3169. doi:10.1111/j.1558-5646.2008.00490.x

Bond AB, Kamil AC, Balda RP (2003) Social complexity and transitive inference in corvids. Anim Behav 65:479–487. doi:10.1006/anbe.2003.2101
Bota M, Sporns O, Swanson LW (2015) Architecture of the cerebral cortical association connectome underlying cognition. Proc Natl Acad Sci U S A 112:E2093–E2101. doi:10.1073/pnas.1504394112
Broadley BJ (1985) Unusual method of killing prey by sparrowhawk. Scott Birds 13:268
Brodin A, Bolhuis JJ (2008) Memory and brain in food-storing birds: space oddities or adaptive specializations? Ethology 114:633–645. doi:10.1111/j.1439-0310.2008.01508.x
Brodin A, Lundborg K (2003) Is hippocampal volume affected by specialization for food hoarding in birds? Proc R Soc B Biol Sci 270:1555–1563. doi:10.1098/rspb.2003.2413
Bshary R, Gingins S, Vail AL (2014) Social cognition in fishes. Trends Cogn Sci 18:465–471. doi:10.1016/j.tics.2014.04.005
Burghardt GM (2013) Environmental enrichment and cognitive complexity in reptiles and amphibians: concepts, review, and implications for captive populations. Appl Anim Behav Sci 147:286–298. doi:10.1016/j.applanim.2013.04.013
Butler AB, Hodos W (2005) Comparative vertebrate neuroanatomy: evolution and adaptation. Wiley, New York
Cantalupo C, Hopkins W (2010) The cerebellum and its contribution to complex tasks in higher primates: a comparative perspective. Cortex 46:821–830. doi:10.1016/j.cortex.2009.10.004
Catania KC (2011) The sense of touch in the star-nosed mole: from mechanoreceptors to the brain. Philos Trans R Soc Lond Ser B 366:3016–3025. doi:10.1093/cercor/8.3.227
Cnotka J, Güntürkün O, Rehkämper G et al (2008) Extraordinary large brains in tool-using New Caledonian crows (*Corvus moneduloides*). Neurosci Lett 433:241–245. doi:10.1016/j.neulet.2008.01.026
Costa SS, Andrade R, Carneiro LA et al (2011) Sex differences in the dorsolateral telencephalon correlate with home range size in blenniid fish. Brain Behav Evol 77:55–64. doi:10.1159/000323668
Darwin C (1871) The descent of man, and selection in relation to sex. John Murray, London
Demski LS (2013) The pallium and mind/behavior relationships in teleost fishes. Brain Behav Evol 82:31–44. doi:10.1159/000351994
Donoghue AM, Quicke DL, Brace RC (1986) Turnstones apparently preying on sea anemones. Br Birds 79:91
Dunbar RIM (1992) Neocortex size as a constant on group size in primates. J Hum Evol 20:469–493
Dunbar RIM (1998) The social brain hypothesis. Evol Anthropol 6:178–190
Dunbar RIM, Shultz S (2007) Evolution in the social brain. Science 317:1344–1347
Emery NJ (2006) Cognitive ornithology: the evolution of avian intelligence. Philos Trans R Soc Lond Ser B 361:23–43. doi:10.1038/358152a0
Emery NJ, Seed AM, Bayern von AMP, Clayton NS (2007) Cognitive adaptations of social bonding in birds. Philos Trans R Soc Lond Ser B 362:489–505. doi:10.1006/hbeh.2001.1691
García-Peña GE, Sol D, Iwaniuk AN, Székely T (2013) Sexual selection on brain size in shorebirds (Charadriiformes). J Evol Biol 26:878–888. doi:10.1111/jeb.12104
Gonzalez-Voyer A, Kolm N (2010) Sex, ecology and the brain: evolutionary correlates of brain structure volumes in Tanganyikan cichlids. PLoS One 5:e14355. doi:10.1371/journal.pone.0014355.s001
Gonzalez-Voyer A, Winberg S, Kolm N (2009) Social fishes and single mothers: brain evolution in African cichlids. Proc R Soc B Biol Sci 276:161–167. doi:10.1159/000100037
González-Gómez PL, Madrid-Lopez N, Salazar JE et al (2014) Cognitive ecology in hummingbirds: the role of sexual dimorphism and its anatomical correlates on memory. PLoS One 9:e90165. doi:10.1371/journal.pone.0090165.t002
Gould KL, Gilbertson KE, Hrvol AJ et al (2013) Differences in relative hippocampus volume and number of hippocampus neurons among five corvid species. Brain Behav Evol 81:56–70. doi:10.1159/000345560

Guigueno MF, MacDougall-Shackleton SA, Sherry DF (2015) Sex differences in spatial memory in brown-headed cowbirds: males outperform females on a touchscreen task. PLoS One 10:e0128302. doi:10.1371/journal.pone.0128302.s007
Guigueno MF, Snow DA, MacDougall-Shackleton SA, Sherry DF (2014) Female cowbirds have more accurate spatial memory than males. Biol Lett 10:20140026–20140026. doi:10.1037/0735-7044.122.3.527
Healy SD, Hurly TA (2013) What hummingbirds can tell us about cognition in the wild. Compar Cogn Behav Rev 8:13–28. doi:10.3819/ccbr.2013.80002
Healy SD, Krebs JR (1992) Food storing and the hippocampus in corvids: amount and volume are correlated. Proc R Soc Lond Ser B 248:241–245
Healy SD, Krebs JR (1996) Food storing and the hippocampus in Paridae. Brain Behav Evol 47:195–199
Healy SD, Rowe C (2007) A critique of comparative studies of brain size. Proc Biol Sci 274:453–464. doi:10.1016/S0047-2484(02)00162-8
Herculano-Houzel S, Bartheld v CS, Miller DJ, Kaas JH (2015a) How to count cells: the advantages and disadvantages of the isotropic fractionator compared with stereology. Cell Tissue Res 360:29–42. doi:10.1007/s00441-015-2127-6
Herculano-Houzel S, Catania K, Manger PR, Kaas JH (2015b) Mammalian brains are made of these: a dataset of the numbers and densities of neuronal and nonneuronal cells in the brain of Glires, Primates, Scandentia, Eulipotyphlans, Afrotherians and Artiodactyls, and their relationship with body mass. Brain Behav Evol 86:145–163. doi:10.1159/000437413
Hobson EA, Avery ML, Wright TF (2014) The socioecology of Monk Parakeets: insights into parrot social complexity. Auk 131:756–775. doi:10.1642/AUK-14-14.1
Holekamp KE (2007) Questioning the social intelligence hypothesis. Trends Cogn Sci 11:65–69. doi:10.1016/j.tics.2006.11.003
Iwaniuk AN (2010) Comparative brain collections are an indispensable resource for evolutionary neurobiology. Brain Behav Evol 76:87–88. doi:10.1159/000320214
Iwaniuk AN, Arnold KE (2004) Is cooperative breeding associated with bigger brains? A comparative test in the Corvida (Passeriformes). Ethology 110:203–220
Iwaniuk AN, Dean KM, Nelson JE (2005) Interspecific allometry of the brain and brain regions in parrots (Psittaciformes): comparisons with other birds and primates. Brain Behav Evol 65:40–59. doi:10.1159/000081110
Iwaniuk AN, Hurd PL, Wylie DRW (2006) Comparative morphology of the avian cerebellum: I. Degree of foliation. Brain Behav Evol 68:45–62. doi:10.1159/000093530
Iwaniuk AN, Lefebvre L, Wylie DR (2009) The comparative approach and brain–behaviour relationships: a tool for understanding tool use. Can J Exp Psychol 63:150–159. doi:10.1037/a0015678
Jarvis ED, Güntürkün O, Bruce LL et al (2005) Avian brains and a new understanding of vertebrate brain evolution. Nat Rev Neurosci 6:151–159
Jerison HJ (1973) Evolution of the brain and intelligence. Academic, New York
Johnsgard PA (1999) The avian brood parasites: deception at the nest. Oxford University Press, New York
Kamil AC, Balda RP (1985) Cache recovery and spatial memory in Clark's nutcrackers (*Nucifraga columbiana*). J Exp Psychol Anim Behav Process 11:95–111
Kamil AC, Gould KL (2008) Memory in food caching animals. In: Menzel R, Byrne JH (eds) Learning and memory: a comprehensive reference. Learning theory and behaviour, vol 1. Elsevier, Amsterdam, pp 419–439
Kocher TD (2004) Adaptive evolution and explosive speciation: the cichlid fish model. Nat Rev Genet 5:288–298. doi:10.1038/nrg1316
Kotrschal A, Buechel SD, Zala SM et al (2015b) Brain size affects female but not male survival under predation threat. Ecol Lett 18:646–652. doi:10.1111/ele.12441
Kotrschal A, Corral-Lopez A, Amcoff M, Kolm N (2015a) A larger brain confers a benefit in a spatial mate search learning task in male guppies. Behav Ecol 26:527–532. doi:10.1093/beheco/aru227

Kotrschal A, Corral-Lopez A, Zajitschek S et al (2015c) Positive genetic correlation between brain size and sexual traits in male guppies artificially selected for brain size. J Evol Biol 28:841–850. doi:10.1111/jeb.12608
Kotrschal A, Rogell B, Bundsen A et al (2013) Artificial selection on relative brain size in the guppy reveals costs and benefits of evolving a larger brain. Curr Biol 23:168–171. doi:10.1016/j.cub.2012.11.058
Kotrschal K, van Staaden MJ, Huber R (1998) Fish brains: evolution and environmental relationships. Rev Fish Biol Fish 8:373–408
Krebs JR, Sherry DF, Healy SD et al (1989) Hippocampal specialization of food-storing birds. Proc Natl Acad Sci U S A 86:1388–1392
Kudo H, Dunbar RIM (2001) Neocortex size and social network size in primates. Anim Behav 62:711–722. doi:10.1006/anbe.2001.1808
LaDage LD, Riggs BJ, Sinervo B, Pravosudov VV (2009) Dorsal cortex volume in male side-blotched lizards, *Uta stansburiana*, is associated with different space use strategies. Anim Behav 78:91–96. doi:10.1016/j.anbehav.2009.03.020
Lefebvre L (2013) Brains, innovations, tools and cultural transmission in birds, non-human primates, and fossil hominins. Front Hum Neurosci 7:245. doi:10.3389/fnhum.2013.00245/abstract
Lefebvre L, Nicolakakis N, Boire D (2002) Tools and brains in birds. Behaviour 139:939–973
Lefebvre L, Reader SM, Sol D (2004) Brains, innovations and evolution in birds and primates. Brain Behav Evol 63:233–246. doi:10.1159/000076784
Lefebvre L, Sol D (2008) Brains, lifestyles and cognition: are there general trends? Brain Behav Evol 72:135–144. doi:10.1159/000151473
Lefebvre L, Whittle P, Lascaris E, Finkelstein A (1997) Feeding innovations and forebrain size in birds. Anim Behav 53:549–560. doi:10.1006/anbe.1996.0330
Lucas JR, Brodin A, de Kort SR, Clayton NS (2004) Does hippocampal size correlate with the degree of caching specialization? Proc R Soc B Biol Sci 271:2423–2429. doi:10.1098/rspb.2004.2912
Marler P (1996) Social cognition: are primates smarter than birds? In: Nolan V, Ketterson ED (eds) Current ornithology. Plenum Press, New York, pp 1–32
Mehlhorn J, Hunt GR, Gray RD et al (2010) Tool-making New Caledonian crows have large associative brain areas. Brain Behav Evol 75:63–70. doi:10.1159/000295151
Ngwenya A, Patzke N, Manger PR, Herculano-Houzel S (2016) Continued growth of the central nervous system without mandatory addition of neurons in the Nile crocodile *(Crocodylus niloticus)*. Brain Behav Evol 87:19–38. doi:10.1159/000443201
Nicolakakis N, Lefebvre L (2000) Forebrain size and innovation rate in European birds: feeding, nesting and confounding variables. Behaviour 137:1415–1429
Northcutt RG (2013) Variation in reptilian brains and cognition. Brain Behav Evol 82:45–54. doi:10.1159/000351996
Olkowicz S, Kocourek M, Lucan RK et al (2016) Birds have primate-like numbers of neurons in the forebrain. Proc Natl Acad Sci U S A 113:7255–7260. doi:10.1073/pnas.1517131113
Overington SE, Morand-Ferron J, Boogert NJ, Lefebvre L (2009) Technical innovations drive the relationship between innovativeness and residual brain size in birds. Anim Behav 78:1001–1010. doi:10.1016/j.anbehav.2009.06.033
Petersen K (1995) Seasonal patterns in food-storing behaviour, memory and hippocampal size in white-breasted nuthatches (*Sitta carolinensis*). M.Sc. thesis, University of Western Ontario, Canada
Pfenning AR, Hara E, Whitney O et al (2014) Convergent transcriptional specializations in the brains of humans and song-learning birds. Science 346:1256846–1256846. doi:10.1126/science.1256846
Pollen AA, Dobberfuhl AP, Scace J et al (2007) Environmental complexity and social organization sculpt the brain in Lake Tanganyikan cichlid fish. Brain Behav Evol 70:21–39. doi:10.1159/000101067

Powell BJ, Leal M (2014) Brain organization and habitat complexity in *Anolis* lizards. Brain Behav Evol 84:8–18. doi:10.1159/000362197
Powell BJ, Leal M (2012) Brain evolution across the Puerto Rican anole radiation. Brain Behav Evol 80:170–180. doi:10.1159/000341161
Pravosudov VV, Roth TC II (2013) Cognitive ecology of food hoarding: the evolution of spatial memory and the hippocampus. Annu Rev Ecol Evol Syst 44:173–193. doi:10.1146/annurev-ecolsys-110512-135904
Reboreda JC, Clayton NS, Kacelnik A (2006) Species and sex differences in hippocampus size in parasitic and non-parasitic cowbirds. Neuroreport 7:505–508
Reiner A, Yamamoto K, Karten HJ (2005) Organization and evolution of the avian forebrain. Anat Rec 287A:1080–1102. doi:10.1002/ar.a.20253
Robinson CD, Patton MS, Andre BM, Johnson MA (2015) Convergent evolution of brain morphology and communication modalities in lizards. Curr Zool 61:281–291
Roth ED, Lutterschmidt WI, Wilson DA (2005) Relative medial and dorsal cortex volume in relation to sex differences in spatial ecology of a snake population. Brain Behav Evol 67:103–110. doi:10.1159/000089183
Roth TC, Brodin A, Smulders TV et al (2010) Is bigger always better? A critical appraisal of the use of volumetric analysis in the study of the hippocampus. Philos Trans R Soc Lond B Biol Sci 365:915–931. doi:10.1159/000112985
Roth TC, LaDage LD, Freas CA, Pravosudov VV (2011) Variation in memory and the hippocampus across populations from different climates: a common garden approach. Proc R Soc B Biol Sci 279:402–410. doi:10.1002/dneu.20736
Rowley I (1990) Behavioural ecology of the galah *Eolophus roseicapillus* in the wheatbelt of Western Australia. Surrey Beatty & Sons, Chipping Norton
Samuk K, Iritani D, Schluter D (2014) Reversed brain size sexual dimorphism accompanies loss of parental care in white sticklebacks. Ecol Evol 4:3236–3243. doi:10.1002/ece3.1175
Sayol F, Lefebvre L, Sol D (2016) Relative brain size and its relation with the associative pallium in birds. Brain Behav Evol 87:69–77. doi:10.1159/000444670
Seed A, Byrne R (2010) Animal tool-use. Curr Biol 20:R1032–R1039. doi:10.1016/j.cub.2010.09.042
Shanahan M, Bingman VP, Shimizu T et al (2013) Large-scale network organization in the avian forebrain: a connectivity matrix and theoretical analysis. Front Comput Neurosci 7:89. doi:10.3389/fncom.2013.00089/abstract
Sherry DF (2006) Neuroecology. Annu Rev Psychol 57:167–197. doi:10.1146/annurev.psych.56.091103.070324
Sherry DF, Forbes MRL, Khurgel M, Ivy GO (1993) Females have a larger hippocampus than males in the brood-parasitic brown-headed cowbird. Proc Natl Acad Sci U S A 90:7839–7843
Sherry DF, Vaccarino AL, Buckenham K, Herz RS (1989) The hippocampal complex of food-storing birds. Brain Behav Evol 34:308–317
Shultz S, Dunbar RIM (2007) The evolution of the social brain: anthropoid primates contrast with other vertebrates. Proc R Soc B Biol Sci 274:2429–2436. doi:10.1146/annurev.es.11.110180.001213
Sol D (2009) Revisiting the cognitive buffer hypothesis for the evolution of large brains. Biol Lett 5:130–133. doi:10.1098/rspb.2006.3765
Sol D, Duncan RP, Cassey P, Lefebvre L (2005) Big brains, enhanced cognition, and response of birds to novel environments. Proc Natl Acad Sci U S A 102:5460–5465
Sol D, Garcia N, Iwaniuk A et al (2010) Evolutionary divergence in brain size between migratory and resident birds. PLoS One 5:e9617. doi:10.1371/journal.pone.0009617
Striedter GF (2005) Principles of brain evolution. Sinauer Associates, Sunderland
Striedter GF (2015) Evolution of the hippocampus in reptiles and birds. J Comp Neurol 524:496–517. doi:10.1002/cne.23803
Thornton A, McAuliffe K (2015) Cognitive consequences of cooperative breeding? A critical appraisal. J Zool 295:12–22. doi:10.1111/jzo.12198

Thornton A, McAuliffe K, Dall SRX et al (2016) Fundamental problems with the cooperative breeding hypothesis. A reply to Burkart & van Schaik. J Zool 299(2):84–88. doi:10.1111/jzo.12351

Thyssen L, Triay-Portella R, Santana del Pino A, Castro JJ (2013) Homing behaviour of rock pool blenny *Parablennius parvicornis* (Pisces: Blenniidae). J Nat Hist 48:1169–1179. doi:10.1080/00222933.2013.859314

Timmermans S, Lefebvre L, Boire D, Basu P (2000) Relative size of the hyperstriatum ventrale is the best predictor of innovation rate in birds. Brain Behav Evol 56:196–203

Toft CA, Wright TF (2015) Parrots of the wild: a natural history of the world's most captivating birds. University of California Press, Oakland

van den Heuvel MP, Bullmore ET, Sporns O (2016) Comparative connectomics. Trends Cogn Sci 20:345–361. doi:10.1016/j.tics.2016.03.001

Volman SF, Grubb TCJ, Schuett KC (1997) Relative hippocampal volume in relation to food-storing behavior in four species of woodpeckers. Brain Behav Evol 49:110–120

Wanker R, Sugama Y, Prinage S (2005) Vocal labelling of family members in spectacled parrotlets, *Forpus conspicillatus*. Anim Behav 70:111–118. doi:10.1016/j.anbehav.2004.09.022

Ward BJ, Day LB, Wilkening SR et al (2012) Hummingbirds have a greatly enlarged hippocampal formation. Biol Lett 8:657–659. doi:10.1002/cne.21098

Weaver AH (2005) Reciprocal evolution of the cerebellum and neocortex in fossil humans. Proc Natl Acad Sci U S A 102:3576–3580

Wilkinson A, Huber L (2012) Cold-blooded cognition: reptilian cognitive abilities. In: Shackelford TK, Vonk J (eds) The Oxford handbook of comparative evolutionary psychology. Oxford University Press, Oxford, pp 129–141

Yopak KE (2012) Neuroecology of cartilaginous fishes: the functional implications of brain scaling. J Fish Biol 80:1968–2023. doi:10.1111/j.1095-8649.2012.03254.x

Chapter 6
Evolution of Cognitive Brains: Mammals

Gerhard Roth and Ursula Dicke

Abstract In mammals, detailed information about higher cognitive abilities or "intelligence" is restricted to representatives of rodents, artiodactyls, carnivores, cetaceans, elephants, and primates. Tool use as well as "technical" problem-solving is present in most species of these taxa. In string-pulling experiments, apes, monkeys, dogs, and elephants were successful but with no sign of insight into mechanisms. Mirror use is demonstrated in apes, monkeys, and pigs, while mirror self-recognition is found only in the great apes, magpies, and possibly dolphins and elephants. Gaze following is documented in primates, dogs, and wolves. Metacognition was demonstrated in apes, macaques, dolphins, and rats. Finally, signs of a theory of mind are found in chimpanzees and rhesus monkeys and questionable in dogs and wolves. Neither absolute nor relative brain size (uncorrected or corrected for body size) are good predictors for higher cognitive abilities. The number of cortical neurons appears to be a better predictor of intelligence but does not solve the paradox of elephants and cetaceans, which have at least several billion cortical neurons like the great apes, while being less intelligent. The best fit is obtained, when parameters that directly determine neuronal information processing capacity, i.e., cortical interneuronal distance and axonal conduction velocity, are also taken into account. Here, primates excel, followed by carnivores, while the large-brained elephants and cetaceans perform poorly.

Keywords Intelligence • Cognitive abilities • Absolute brain size • Relative brain size • Number of cortical neurons • Information processing capacity

---

G. Roth (✉) • U. Dicke
University of Bremen, Brain Research Institute, D-28334 Bremen, Germany
e-mail: gerhard.roth@uni-bremen.de

S. Watanabe et al. (eds.), *Evolution of the Brain, Cognition, and Emotion in Vertebrates*, Brain Science, DOI 10.1007/978-4-431-56559-8_6

6.1 Introduction

Mammals have always been considered to be smart. However, while in primates the presence of higher cognitive abilities like tool fabrication, imitation, metacognition and, at least, in the great apes, mirror self-recognition and aspects of a theory of mind are undisputed (cf. Byrne 1995; Roth 2013), the presence of such abilities in "insectivores" (i.e., Afrosoricida, Eulipotyphla), ungulates (i.e., Perissodactyla, Artiodactyla), elephants, and cetaceans (whales, dolphins) is either undocumented or disputed. There is an ongoing debate about higher cognitive abilities in elephants and cetaceans. While some authors attribute astonishingly high mental abilities to these large-brained animals, particularly to the cetaceans (cf. Marino 2004, Marino et al. 2007, 2008; Herman 2012), others come to the conclusion that elephant and cetacean intelligence is modest compared to primates and even carnivores (cf. Byrne et al. 2009; Manger 2013; Guentuerkuen 2014).

There have been and still are many attempts to correlate degrees of intelligence found in mammalian species to properties of their brains, the most popular being absolute brain size, relative brain size (uncorrected or corrected for body size) as well as absolute or relative size of the cerebral cortex (cf. Jerison 1973; Lefebvre 2012). However, none of these approaches have yielded convincing results. More recently, it has been argued that the number of cortical neurons is a more reliable predictor of intelligence (cf. Roth and Dicke 2005; Herculano-Houzel 2012; Roth 2013), but it turns out that even this factor does not fully explain why some monkeys with about one billion cortical neurons in many respects appear to be as intelligent as great apes with five to eight times more cortical neurons and why elephants and cetaceans with numbers of cortical neurons equal to or even higher than those found in great apes exhibit only modest degrees of intelligence—at least in the eyes of many experts (cf. Manger 2013).

In earlier publications, we have argued that in addition to the number of cortical neurons, other parameters relevant for information processing capacities like the speed of interneuronal signal transfer as well as cortical network properties must be taken into consideration (Roth and Dicke 2005, 2012; Roth 2013). In the following article, we will review recent literature on mammalian intelligence and their relationship to brain properties in the light of this hypothesis.

6.2 Phylogeny and Taxonomy of Mammals

Ancestors of mammals appeared about 224 mya in the Triassic. Early mammals conducted an inconspicuous life until the end of the Mesozoic. Around 170 mya, the modern type of mammals, the multituberculates, evolved which had small bodies and conducted a nocturnal and/or arboreal life. The split between the Prototheria (monotremes) and the Theria (marsupials and placental mammals) is believed to have occurred 150 mya, and the split among the Theria between

Metatheria and Eutheria 125 mya or earlier. The great time of mammals began near the end of the Cretaceous, around 70 mya, and particularly with the extinction of the dinosaurs 65 mya.

The first infraclass of mammals comprises the Prototheria with only one order, the monotremes (Monotremata) with the platypus (*Ornithorhynchus*) and echidna (Echidna, 12 species). The second infraclass, the Theria, includes the Metatheria or Marsupialia, comprising seven extant superorders with a total of 334 species. The larger metatherian group, Australidelphia (five superorders with 234 species, among them the cangaroos), live in Australia and New Guinea, whereas a smaller group, the Ameridelphia (about 100 species, among them the opossum *Didelphis*) live in North, Central, and South America. The Eutheria or Placentalia originated about 100 mya and for a long time conducted a modest life as insect eaters parallel to the marsupials.

Placental mammals are commonly divided into four superorders. The first group, Afrotheria, includes among others the orders Afrosoricida (tenrecs and golden moles), Macroscelidea (elephant shrews), Tubulidentata (aardvarks), Hyracoidea (hyraxes and allies), Proboscidea (elephants, Elephantidae, as single family with three species), and Sirenia (dugongs and manatees). The second group Xenarthra comprises the group Cingulata (armadillos) and Pilosa (sloths and anteater). The third group, Euarchontoglires, includes the orders Scandentia (treeshrews), Dermoptera (colugos), Primates (lemurs, bushbabies, lorises, monkeys, and apes including humans; more than 400 species), Rodentia (rodents; about 2300 species), and Lagomorpha (pikas, rabbits, hares; about 80 species). Finally, the fourth group, Laurasiatheria comprises the orders Eulipotyphla (hedgehogs, moles, shrews; about 388 species), Chiroptera (bats; about 1100 species), Pholidota (pangolins or scaly anteaters), Carnivora, (dogs, cats, bears, seals etc.; about 270 species), Perissodactyla (odd-toed ungulates like horses, zebras, rhinos and tapirs; about 19 species), Artiodactyla (even-toed ungulates like cattle, pigs, sheep, deer, camels, antelopes; about 260 species), and Cetacea (whales, dolphins, porpoises; about 80 species) (largely after Tarver et al. 2016).

6.3 What Is Animal Intelligence?

In humans, intelligence is commonly defined as the sum of "higher" mental capacities such as abstract thinking, understanding, communication, reasoning, learning and memory formation, action planning, and problem-solving. Usually, human intelligence is measured by intelligence tests and expressed in intelligence quotient (IQ) values related to different contents (e.g., visual–spatial, verbal, numerical). Evidently, such a definition and measurement of intelligence cannot be applied directly to nonhuman animals, because any test depending on verbalization is inapplicable. According to the majority of behaviorists and animal psychologists (cf. Roth 2013), "intelligence" can be understood as mental or behavioral flexibility or the ability of an organism to solve problems occurring in its natural and social environment culminating in the appearance of novel solutions not part of the animal's normal repertoire. A number of authors distinguish between *ecological intelligence* consisting in the ability to master challenges of an environment; *social intelligence* targeting social group size, complexity of social relationships, and means of social communication (gaze following, theory of mind, knowledge attribution etc.); and *general intelligence* consisting in efficient information processing and mental functions like abstract thinking, insight, metacognition, mirror self-recognition, etc. Byrne and Bates (2011) have added *physical intelligence* which includes, among others, tool use and tool fabrication and "technical" problem-solving, e.g., carnivores opening a puzzle box, orangutans using water to fill a tube containing a peanut, and crow using stones to weight a container to get access to food, while others consider these cognitive abilities as part of ecological intelligence. Among authors, there is a great overlap in attributing cognitive functions to these categories of animal intelligence.

6.4 Cognitive Abilities and Intelligence in Mammals

In the following, we will concentrate on frequently used paradigms for measuring intelligence in mammals (as rated in birds), i.e., tool use and tool fabrication, technical problem-solving including understanding of principles, mirror use, gaze following, imitation and observational learning, mirror self-recognition, metacognition, "theory of mind," and mental time travels. Data on primates and birds will be presented only briefly for comparative reasons, because they are presented in more detail in other articles of this volume.

6.4.1 Tool Use and Tool Fabrication

Tool use has been extensively studied in primates and birds, and some of them, e.g., chimpanzees or corvid birds, have been shown to regularly fabricate tools in the wild or in captivity (for an overview see Emery and Clayton 2009; Roth and Dicke 2012; Roth 2013). The first studied case of tool use in non-primate mammals was made in the Californian sea otter (*Enhydra lutris*). Otters were observed to open mussels and sometimes crabs and urchins by pounding them against a stone lying on their breast, and the same stone was used (Hall and Schaller 1964).

In a similar way, wild banded mongooses (*Mungos mungo*) regularly use anvils like rocks or the stems of trees, but also the sidewalls of gullies, to open hard-shelled food objects such as bird eggs or snail shells. It was demonstrated in this context that imitation as well as practicing plays an important role (Müller 2010). Dingoes (*Canis lupus dingo*) have been observed to use a table for getting food (Smith et al. 2012). North American badgers (*Taxidea taxus*), while hunting ground squirrels, often plug openings of ground-squirrel tunnels. They usually take soil from the area around the tunnel opening, but sometimes objects moved from greater distances are used for plugging (Michener 2004). Degus (*Octodon degus*, a rodent) can be trained to retrieve otherwise out-of-reach rewards by using a rake-like tool holding it with their forelimbs (Okanoya et al. 2008). According to the authors, the degus managed to distinguish between useful and nonuseful tools after extensive training.

Elephants use sticks for scratching their body and removing ticks and bushes for fly-switching, which they modify until they are long and effective enough, which is considered by some authors as an evidence for tool fabrication. However, a number of experiments (Hobhouse 1915; Rensch and Altevogt 1955; Hart et al. 2008; Nissani 2004) with captive elephants revealed that the animals did not learn to retrieve food by using a stick. Other experiments, in which elephants had been trained to remove a lid from a bucket in order to retrieve a food reward, revealed great difficulties of the animals with adopting their behavior to slightly different experimental conditions (Nissani 2006). In a similar experiment by Irie-Sugomoto and colleagues (2008), one elephant learned to pull a baited tray in order to retrieve the food. Minuzo et al. (2015) demonstrated that elephants acquire inaccessible food by blowing the food, until it comes in accessible places. In summary, elephants mostly reveal only simple forms of tool use with no or little understanding of the underlying mechanism (Byrne et al. 2009).

Among cetaceans, dolphins may kill scorpion fish in order to use their stingy body to poke after a moray eel hidden in a crevice (Brown and Norris 1956). Dolphins, beluga whales, and humpback whales blow bubble rings, and dolphins seem to do this for amusement or for catching fish (McCowan et al. 2000). One matrilinear group of bottlenose dolphins was observed to use marine sponges while foraging (“sponging”) (Krützen et al. 2005; Patterson and Mann 2011). However, the significance of “sponging” for foraging remains unclear (Manger 2013).

6.4.2 *Puzzle Box Problem-Solving*

Problem-solving abilities often have been tested by using puzzle boxes, in which animals had to open a baited box. Such a puzzle box has been used in a recent study by Benson-Amram et al. (2015) in 39 species from nine families of carnivores. The authors measured the time spent by the animal to open the puzzle box as well as changes in work time over successive trials. In addition, body mass, manual dexterity, and absolute and relative brain volume were determined. All subjects except mongooses succeeded in opening the box by manipulating it, and work time significantly decreased, as the number of trials increased. The authors take this as evidence that successful individuals improved their performance with experience on the basis of trial-and-error learning, while there was no sign of insight. The highest success rate was found in the families Ursidae (bears; 69.2% of trial), followed by Procyonidae (raccoons and allies; 53.8%), and Mustelidae (otter, badger, weasels etc., 47.1%), while the members of the family Herpestidae (mongooses) failed. The authors found no correlation between success rate on one hand and social complexity or manual dexterity on the other hand, while there was a significant correlation with relative brain size.

Similar experiments using "puzzle feeders" have been executed with lemurs in the wild (Kendal et al. 2010) as well as in the laboratory. The gray mouse lemur mastered to open boxes in different ways including the use of reversed images and aye-ayes demonstrated basic understanding of features of tools by solving a can-pulling task (Fichtel and Kappeler 2010). All monkey and ape species tested so far likewise succeeded in such tasks, but again evidence for insight into the opening mechanism was either absent or equivocal.

6.4.3 String-Pulling Experiments

Numerous experiments on intelligence in mammals and birds used the string-pulling setup. Recently, Riemer et al. (2014) conducted string-pulling experiments with Border Collies. Previous experiments had revealed that dogs have a tendency to choose the string that is nearest to them instead of the baited string ("proximity bias"). In these experiments, dogs performed above chance when the baited and unbaited strings' ends were equidistant, irrespective of being straight or curved, but seemed to be unable to overcome their proximity bias in a parallel or diagonal string task, when proximity of the unconnected string's end to the reward was misleading. The authors conclude that the dogs can learn to pay attention to connectivity of the strings, when proximity is not a confounding factor.

Nissani (2004) reported string-pulling as well as sucking-blowing experiments with Asian elephants. All tested elephants mastered the string-pulling experiment, but they acquired the behavior gradually by trial and error, and most of them seemed unable to transfer their skills to similar tasks. In another series of experiments, elephants learned to remove food from a narrow tube either by sucking or blowing. In a competitive situation (with one elephant on either side of the tube), they always sucked the food in order to get it first. However, they were unable to transfer their experience to a competitive situation, in which they had to remove the food from a wider tube by either pushing or pulling it with their trunk.

Recently, Mayer et al. (2014) conducted string-pulling experiments in different taxa of primates, i.e., capuchin monkeys, bonobos, chimpanzees, and children, with two versions of a broken-string problem. In the standard condition, subjects had to choose between an intact and a broken string as means to a reward. In the critical condition, the functional parts of the strings were covered up and replaced by perceptually similar but nonfunctional cues. Apes, monkeys, and young children at or above an age of 3.5 years succeeded in the standard, but only children at or above an age of 5.5 years mastered the test with covered strings. Interestingly, there was no difference between monkeys and apes. Again, there was evidence for the important role of experience, but not for insight in monkeys, apes, and children younger than 5.5 years.

Corvid birds as well as primates regularly master these tasks quite often. In the experiments with New Caledonian crows by Taylor and colleagues (cf. Taylor et al.

2009), difficulties arose for the otherwise "intelligent" crows, when visual control was restricted or absent. The authors conclude that spontaneous string pulling in New Caledonian crows may not be based on insight but on operant conditioning mediated by a perceptual-motor feedback cycle.

6.4.4 Use of Mirrors

A larger number of mammals are able to use mirrors in order to identify otherwise unobservable objects or body parts or look around a corner. Rhesus monkeys, which failed the standard version of the mirror self-recognition (MSR) test (see below), learn to use mirrors in order to study otherwise hidden parts of their bodies (genitals, head implants). Similarly, Old World mangabeys (*Cercocebus torquatus*) learn to grope for a peanut on the backside of a board, when their hand is guided by a mirror (McKiggan, quoted in Byrne 1995). Pigs learn to turn to an invisible food bowl within 5 h, when they could see it in a mirror (Broom et al. 2009).

6.4.5 Gaze Following

Gaze following has been demonstrated in lemurs, macaques, capuchin, spider monkeys, and marmosets but without signs of understanding visual perspective (Roth and Dicke 2012). Great apes are able to track gaze to hidden targets and look back to the human experimenter, when they do not find a target (Bräuer et al. 2005; Tomasello et al. 2007).

For decades, the only case of gaze following in non-primate animals was demonstrated in dogs. As shown by Miklósi et al. (2003), dogs can understand human pointing and inform humans about hidden objects, look at the faces of humans and follow their gaze, while this ability appeared to be absent in wolves. The authors reported that dogs "look back" to the human face when confronted with unsolvable problems but that wolves do not. Later, however, Range and Virányi (2011) demonstrated that hand-raised wolves develop both the ability of gaze following into distant space and to look behind barriers. The former developed earlier than the latter. This suggests that gaze following and "looking back" in dogs and wolves is strongly dependent on training by humans.

More recently, Téglás et al. (2012) demonstrated that following gaze or directional gestures by their owner requires preceding human communicative signals, e.g., direct gaze or addressing. Wallis et al. (2015) showed that dogs can follow human gaze into distant space. However, with increasing experience and age, dogs prefer fixating the human face instead of distant targets. The authors speculate that with increasing training, the human face becomes more attractive.

6.4.6 Imitation

Imitation has long been considered an inferior kind of learning and was typically called “aping” or “monkeying” in the sense of a meaningless copying of a certain behavior. Only in recent years did it become clear that imitation is a higher cognitive ability. However, to date, there is no universally accepted definition of imitation, and some kinds of behavior previously viewed as imitation are now interpreted differently. One of these imitation-like behaviors is response facilitation or emulation, found in a wide range of animals, which means that seeing an action “primes” the individual to do the same, and the individual, by trial and error, finds the same or a very similar solution to the problem (cf. Byrne 1995; Bates and Byrne 2010).

In primates, imitation and emulation play an important role in social learning. For example, young baboons (*Papio*) quickly learn which kinds of fruit are edible, after one group member has tasted a fruit. Vervet monkeys (*Chlorocebus*), again Old World monkeys, learn this task more slowly, although they live in the same environment as baboons. The greatest imitation abilities appear to be present in dolphins. Bottlenose dolphins spontaneously imitate other dolphins and humans and do this on demand (“elicited imitation”), and they seem to be superior in this respect to all other nonhuman mammals (Kuczaj II and Yeater 2006). This ability apparently plays an important role in social life as well as in coordinated behaviors (e.g., synchronous swimming).

Dogs are highly social animals, and one could expect that imitation of conspecifics or of their owner is well developed. However, there is little evidence for that. In contrast, a recent study by Range and Virányi (2014) demonstrated that wolves are much better than dogs at imitating problem-solving behavior of a conspecific, i.e., the opening of a baited puzzle box. While the wolves readily opened the box after a demonstration by a conspecific, the dogs failed to solve the problem. The authors try to explain these differences by pointing to the fact that the dependency of wolves on close cooperation with conspecifics, including breeding but also territory defense and hunting, created selection pressures on motivational and cognitive processes enhancing their propensity to pay close attention to conspecifics’ actions. In contrast, during domestication, dogs’ dependency on conspecifics has been relaxed, leading to reduced motivational and cognitive abilities to interact with conspecifics.

Learning by observation was demonstrated in mice in an experiment by Carlier and Jamon (2006). Here, female mice performed reliably and immediately a sequence of actions, i.e., pushing a piece of food into a tube attached to the side of a puzzle box and recovering it by opening a drawer in front of the box after having observed a conspecific “demonstrator.” None of the naive mice was able to solve the task. Similar effects of observational learning or “insight” was demonstrated in rats by Blaisdell et al. (2006). According to the authors, the rats were able to make causal inferences by passively observing task-solving conspecifics, which could not be explained by Pavlovian learning.

6.4.7 Mirror Self-Recognition

The ability of mirror self-recognition (MSR) is often taken as evidence for "higher" mental states eventually leading to self-consciousness. However, the presence of this ability in nonhuman animals is hotly debated. While the MSR test yields clear results in children at and after an age of 18 months, all tests in nonhuman animals have turned out to be complicated. MSR was first demonstrated by Gallup (1970) in chimpanzees and later by various authors including Gallup in orangutans and bonobos and finally and with great difficulty in one gorilla. Regularly, the tests were successful only in less than half of animals tested and not always in those that passed the test.

Reiss and Marino (2001) succeeded in demonstrating that captive-born bottleneck dolphins (*Tursiops truncatus*) are capable of mirror self-recognition. Hints of MSR were observed in the killer whale, the false killer whale, and the California sea lion (Delfour and Marten 2001). After a number of failures, Plotnik and colleagues demonstrated mirror self-recognition at least in one out of three Asian elephants, *Elephas maximus* (Plotnik et al. 2006). But here again the successful elephant lost interest quickly. Given the very poor visual acuity, the results of these experiments have been questioned (cf. Manger 2013; Guentuerkuen 2014). Interestingly, some years ago the common magpie (*Pica pica*), a corvid, was found to pass the MSR test (Prior et al. 2008).

6.4.8 Metacognition

Metacognition is the ability to know what a subject knows and what it does not know. The principle of such experiments is that suitable animal are confronted with tasks, in which they have to discriminate between two tones of different pitch or length or two pictures showing grains of different size. The differences between the two tones or pictures are reduced stepwise such that they become increasingly difficult to distinguish. Correct answers are rewarded; for incorrect answers, there is a "time out." However, in addition to the decision between the two stimuli, there is the possibility to carry out an *uncertainty response (UR)*, if animals have great difficulty with choosing the correct answer, and this allows them to immediately perform the next trial (Smith 2009). Foote and Crystal (2007) demonstrated that rats passed an acoustic duration discrimination test, in which animals had to distinguish between short (2–3.6 s) and long (4.42–8 s) noises. The rats increasingly declined the test, when the difference in duration became intermediate, i.e., 3.6–4.42 s, and thus very difficult to distinguish. In experiments with macaques, chimpanzees, and dolphins, URs occurred exactly in moments, when human participants likewise had problems with distinguishing the pattern, and started disappearing, when it became increasingly easy for the human observer to distinguish the patterns. Usually, the URs were preceded by hesitation. Remarkably, capuchin monkeys believed to be highly intelligent, failed in these experiments.

6.4.9 Theory of Mind

Under the topic "theory of mind (ToM)" experts study to which degrees animals are able to understand the intentions and knowledge of others and accordingly are capable of predicting their behavior. Premack and Woodruff (1978) were the first to ask this question with respect to chimpanzees, and despite numerous studies on a variety of species, this topic is still debated. To date, some experts believe that at least some animals understand some mental states, while others find the evidence unconvincing (for an overview see van der Vaart and Hemelrijk 2014; Roth 2013). It appears reasonable to assume that both chimpanzees and rhesus monkeys understand the goals and perceptions of others as well as what others know but not what others falsely believe (Lyons and Santos 2006; Premack 2007; Byrne and Bates 2011; Seed and Tomasello 2010). The same appears to hold for jays and ravens (Emery and Clayton 2009; Byrne and Bates 2011).

There has been an extensive discussion about the presence of ToM in dogs and wolves (for an overview see Horowitz 2011). To date, there is no unequivocal evidence that hand-raised wolves and domestic dogs possess a ToM, although they pass on some trials of a putative theory-of-mind test and fail on others. While some authors argue in favor of a more relaxed or "rudimentary" definition of ToM (cf. Horowitz 2011), others like Udell et al. (2011) argue that the term "theory of mind" has outgrown its usefulness in comparative cognition studies.

6.4.10 Mental Time Travel

Mental time travel (MTT) is the subject's ability to travel backward and forward mentally from the present moment to remember certain past events and to anticipate future activities. The ability to remember *what* happened *when* and *where* (WWW memory) is widespread among birds and mammals (cf. Roberts 2012), for example, in the context of food caching or past experience of places, where food was obtained. There is also evidence in birds as well as mammals (rats, primates) for prospective MTT, because they readily learn to use both time of day and elapsed time intervals as cues for important events such as food delivery. Rats, for example, learned to suppress the immediate intake of saccharine in favor of a later intake of more palatable and nutritive sucrose up to 30 min. Similarly, chimpanzees and orangutans could wait for 3–8 min for a delayed larger reward, and monkeys did so for 15 min (McKenzie et al. 2004). Thus, at least some birds and mammals are able to learn to anticipate the consequences of different choices, e.g., among kinds of food or among useful and nonuseful tools, and some of them (e.g., squirrel monkeys) are even able to anticipate their own future drive states (thirst or hunger) (cf. Roberts 2012).

6.5 Conclusions Regarding Ranks of Intelligence Among Mammals

Insight into higher cognitive abilities in mammals is compromised by the fact that the majority of mammalian taxa have not or only rarely been studied in this respect, e.g., Prototheria, Metatheria, Afrosoricida, Tubulidentata, Xenarthra, Pholidota, Eulipotyphla, Chiroptera, and Perissodactyla. This restricts our knowledge to representatives of rodents, artiodactyls, carnivores, cetaceans, elephants, and above all primates.

Tool use and tool fabrication are extensively present in primates and to a more limited degree in degus, otters, mongooses, badgers, dogs, dolphins, and elephants. Problem-solving by means of a puzzle box is found in all primates tested including lemurs and, among carnivores, in bears, raccoons, otter, badgers and weasel and presumably in dogs and wolves. In string-pulling experiments, apes, monkeys, dogs, and elephants were successful but with no sign of insight into mechanisms. Mirror use was found in apes, monkeys, and pigs but might be more widespread. Gaze following was demonstrated in all primates tested as well as in dogs and wolves. Extensive imitation and learning by observation was found in apes and monkeys as well as in dolphins and to a limited degree in elephants, wolves, dogs, mice, and rats. Mirror self-recognition was found in chimpanzees, orangutans, bonobos, gorillas, magpies, and possibly dolphins and elephants. Metacognition was demonstrated in apes, macaques, dolphins, and rats. Finally, stronger or weaker signs of a theory of mind were found in chimpanzees and rhesus monkeys and questionably in dogs and wolves. Mental time travels appear to be widespread across mammals and were demonstrated in rats, monkeys, and apes.

6.6 The Correlation of Brain Properties and Intelligence in Mammals

Many attempts have been made to correlate intelligence with brain properties, the most influential work being Harry Jerison's book *Evolution of the Brain and Intelligence* (1973). A much discussed brain trait is absolute size, followed by relative brain size, i.e., percent of body size or the relative size of alleged "seats" of intelligence like the cerebral cortex in mammals. Given the fact that much of brain size is determined by body size and therefore represents "confounding factor" (Jerison 1973), a number of authors have tried to determine the degree of "encephalization," i.e., brain size corrected for body size. The best known of such attempts is Jerison's "encephalization quotient (EQ)" (for a critical overview, see Lefebvre 2012). Other authors argued in favor of more "functional" brain properties like the number of neurons in the entire brain or in the pallium or cortex, their packing density, pattern of connectivity, and other parameters relevant for information processing capacity (IPC) (Roth and Dicke 2012; Roth 2013).

In mammals, there is an enormous variation in body size (volume or weight). The smallest mammal is the Etruscan shrew (*Suncus etruscus*) with a body weight of 1.8 g, and the largest mammal and animal is the blue whale (*Balaenoptera musculus*) with a length of 30 m and a body weight up to 180 tons. The largest living terrestrial animal is the African elephant (*Loxodonta africana*) with a body weight up to 7.5 tons. Thus, among mammals there is a range in body size or weight of eight orders of magnitude. The volumes or weights of brains likewise vary enormously. Among mammals, the smallest brain is found in the bat (*Tylonycteris pachypus*) which weighs 74 mg, and the largest brains of all animals are found in the sperm whale (*Physeter macrocephalus*) and "killer whale" (*Orcinus orca*), with up to 9 kg. African elephant brains weigh up to 6 kg. This is again an enormous range, here roughly five orders of magnitude (cf. Table 6.1).

As to the major mammalian taxa, "insectivores" (i.e., the unrelated Afrosoricida and Eulipotyphla) range from about 10 to 600 g in body size and 0.1 to 6 g in brain size. In the order Carnivora, body size varies between 1.5 and 365 kg and brain size between 17 and 460 g; in the order Artiodactyla body size varies between 1.5 kg and 4.5 tons, with brains up to 762 g; Perissodactyla taxa vary in body size between 150 kg and 3.5 tons and have brains up to 540 g. Among cetaceans, body sizes vary between 40 kg and 180 tons and brain sizes between 470 g and 8.2 kg. Primates, with the exception of prosimians and tarsiers, generally exhibit larger brains compared to body size than the other orders. New World monkeys have brains of 7–118 g and Old World monkeys of 36–222 g, with the largest brains found in baboons. Among apes, gibbons have brain sizes of 88–105 g, which lie within the range of Old World monkeys, while the large apes, i.e., orangutans (*Pongo pygmaeus*), gorillas (*Gorilla gorilla*), and chimpanzees (*Pan troglodytes*), have brain weights between 330 and 570 g (males) (cf. Table 6.1).

6.6.1 Absolute Brain Size

Comparing the data on cognition in mammals as presented above, with absolute brain size, we yield no clear-cut correlation. On the one hand, "insectivores" with very small brains on average reveal no signs of higher cognitive abilities, while the order Carnivora, large-brained species like bears and sea lions appear to be more intelligent than small-brained species, which would fit. Within primates, there appears to be a loose correlation between brain weight and intelligence given that small-brained prosimians appear to be less intelligent than Old and New World monkeys with brain weights of 50–220 g, and the great apes with brain weights of 330–570 g are on average believed to be more intelligent than the monkeys, let alone humans with a brain weight of about 1.350 g. However, some monkeys like the capuchin appear to be almost equal in at least some aspects of intelligence with gorillas or chimpanzees, despite a much smaller brain (about 80 g), and the capuchin is believed to be more intelligent than carnivores or ungulates with brains around 500 g. Furthermore, in carnivores and "ungulates" (artiodactyls and perissodactyls), the range of brain size

Table 6.1 Brain weight, encephalization quotient, and number of cortical neurons in selected mammalian taxa

Taxa	Brain weight(in g)[a]	Encephalization quotient[b,c]	Number of cortical neurons (in millions)
Sperm whale	8183	0.45	
African elephant	4200–7500	1.3–2.4	11,000[d]; 5600[e]
Killer whale	4779–5059	2.57–5.55	
False killer whale	3650	4.03[f]	10,500[d]
Blue whale	3636	0.21	
Man	1250–1450	7.4–7.8	15,000[d]
Bottlenose dolphin	1350–1,88	3.61–5.3	5800[d]
Walrus	1130	1.2	
Camel	520–762	0.6–1.2	1700[g]
Ox	490	0.5–0.8	
Horse	510–600	0.9	1200[h]
Gorilla	570	1.5–1.8	4300[d]; 9000[i]
Chimpanzee	430	2.2–2.5	6200[d]
Tiger	279	0.78	
Lion	260	0.6	
Sheep	140	0.8	
Rhesus monkey	88	2.1	840[d;] 1710[j]
Long-tailed monkey	36		840[d]
Gibbon	88–105	1.9–2.7	
White-fronted capuchin	57	4.8	720[d]; 1140[j]
Dog	64–135	0.7–1.6	160[d]
Fox	43–53	0.9–1.6	
Cat	25–37	1.0	300[d]
Squirrel monkey	23	2.3	450[d]; 1340[j]
Rabbit	11	0.4	
Marmoset	7	1.7	
Opossum	7.6	0.2	27[d]
Squirrel	7	1.1	
Hedgehog	3.3	0.3	24[d]
Rat	2	0.4	21[d]; 31[k]
Mouse	0.3	0.5	5[d]; 13,6[k]

[a]Data from Haug (1987), Jerison (1973), and Russel (1979)
[b]Indicates the deviation of the brain size of a species from brain size expected on the basis of a "standard" species of the same taxon, in this case of the cat
[c]Data after Jerison (1973) and Russel (1979)
[d]Roth and Dicke (2012)
[e]Herculano-Houzel et al. (2014)
[f]Manger (2013)
[g]Kazu et al. (2014)
[h]Cozzi et al. (2014)
[i]Herculano-Houzel (2012)
[j]Herculano-Houzel et al. (2007)
[k]Herculano-Houzel et al. (2006)

is nearly identical, while the former are believed to be considerably more intelligent than the latter. Finally, elephants and cetaceans, with brains 5–30 times larger than those of monkeys certainly are less intelligent. Thus, the maxim "bigger is better" appears to hold for comparisons *within* orders like carnivores or artiodactyls at best, but not for comparisons *across* orders of mammals. Furthermore, among cetaceans, dolphins, in general, appear to be more intelligent than the large-brained whales, while many of them have much smaller brains.

6.6.2 Relative Brain Size and EQ

Given these inconsistencies, experts have studied the impact of *relative* brain size, either uncorrected or corrected for body size. As in all vertebrates, brain size generally increases with body size, but in most cases the relationship between brain size and body size (BBR) is negatively allometrical, i.e., at a phylogenetic increase in body size, the increase in brain size "lags behind" and, thus, becomes relatively smaller (for details cf. Jerison 1973). As a consequence of negative brain allometry, with increasing body weight, relative brain weight decreases from more than 10% in very small mammals like shrews to less than 0.005% in the blue whale (van Dongen 1998). The human brain again ranks relatively high with roughly 2% of body weight, but in close proximity to apes and dolphins.

Comparing relative rather than absolute brain size with the ranks of intelligence in mammals, as presented above, yields even worse results, because the relatively largest brains are found in the "insectivores" with up to 12%, while humans have "only" 2% and many cetaceans and the elephants less than 1%. Thus, there is more of an anticorrelation than a correlation due to the fact that small animals tend to have relatively larger brains. For those reasons, Jerison (1973) tried to correct relative brain sizes for body size by his encephalization quotient EQ. This quotient indicates the extent to which relative brain size of a given species deviates statistically from the expected or average relative brain size of the larger taxon (e.g., order) under consideration.

As shown in Table 6.1, the lowest EQs are found in very large cetaceans like the blue whale (0.21). The sperm whale, hares, mice, and rats have an EQ well below average, followed by mice, sheep, and horse. The cat has an average relative brain, while dogs, camels, and elephants have EQs slightly above average. Among primates, Old World monkeys have slightly higher EQs on average (1.7–2.7) than New World monkeys (1.7–2.3), with the exception of the white-fronted capuchin. Gorillas and chimpanzees have astonishingly low EQs, while the highest EQs are found in dolphins and finally humans. While the high EQ of humans is of no surprise, the relatively high EQ (up to 5.3) of dolphins is unexpected compared to the relatively low EQs of chimpanzees and gorilla given the undeniably higher intelligence of the great apes. Thus, Jerison's EQ does not remove major inconsistencies in BBR.

6.6.3 *Size of Cortex*

An alternative is to look for a correlation between intelligence and absolute or relative cortex size given the reasonable assumption that the mammalian cortex is the site of their intelligence. However, this does not yield better results. With increasing body and brain size, mammalian cortices increase in surface area as well as in volume. The smallest mammals, for example, shrews, have a cortical surface (both hemispheres together) of 0.8 cm^2 or less, and in the rat we find 6 cm^2, in the cat 83 cm^2, in humans about 2400 cm^2, in the elephant 6300 cm^2, and in large-brained cetaceans a maximum of 7400 cm^2. Thus, from shrews to large whales we find a nearly 10,000-fold increase in cortical surface area.

However, the increase in cortical surface area contrasts with a modest increase in cortical thickness, i.e., from 0.4 mm in very small shrews and mice to 2–4 mm in humans and the great apes. Most large-brained cetaceans have surprisingly thin cortices between 1.2 and 1.8 mm (cf. Kern et al. 2011), and even the elephant, again with a very large brain, has an average cortical thickness of 1.9 mm (Haug 1987). In most mammals, the cortex grows somewhat faster than the rest of the brain, whereas in whales as well as in the elephant cortical volume, while increasing in absolute volume, decreases in relative volume. Yet, both elephants and cetaceans possess the largest cortices among mammals, which again is not a good predictor of intelligence.

6.6.4 *Number of Cortical Neurons*

Some authors argue that, instead of absolute or relative brain or cortical size, a much better predictor of mammalian intelligence is the number of cortical neurons as well as the effectiveness of their wiring and processing speed (cf. Roth and Dicke 2005; Herculano-Houzel 2012; Roth 2013). Brains and cortices of the same volume may contain very different numbers of neurons depending on their neuron packing density (NPD). Cortical NPD of mammalian species is highest in small eulipotyphlans and small rodents but is nearly equally high in small primates, which however are much larger in brain size than the former. In primates cortical NPD ranges from 75,000 neurons/mm^3 in the mouse lemur (*Microcebus* sp.) and the marmoset (*Callithrix jacchus*) to 25–30,000 neurons/mm^3 in gorillas and humans. By contrast, with 6000–7000 neurons mm^3, the cortices of whales and elephants have the lowest NPD among mammals (Haug 1987). Herculano-Houzel et al. (2015) report NPD values that are about half of those reported by Haug, but their measurements included both gray and white matter, while Haug's data are based only on gray matter. Despite this technical difference, the NPD ranking order reported by Herculano-Houzel et al. (2015) is the same as in Haug (1987). Thus, primates including humans stand out by having much higher cortical NPD than non-primate mammals of the same brain and cortex size. In addition, while in nonmammalian taxa cortical NPD strongly decreased with increasing brain and cortex volume with an exponent of −0.5, it decreases only slightly in primates with an exponent of −0.17 (Herculano-Houzel et al. 2015).

The number of cortical neurons of a taxon can be determined either indirectly by calculating it on the basis of the cortex volume and the cortical NPD, as we have done on the basis of the data by Haug (1987), or directly by cytometric techniques like the "isotropic fractionator" method used by Herculano-Houzel and colleagues (cf. Herculano-Houzel 2012) or a "stereological" method used by Pakkenberg and colleagues (cf. Eriksen and Pakkenberg 2007). Due to the differences resulting from the different methods, in a considerable number of cases, the data obtained by the different authors strongly vary.

Mice, rats, and hedgehogs have cortical neurons between 12 and 45 million (Roth and Dicke 2012; Herculano-Houzel et al. 2015). Among "ungulates" studied, we find a range between about 300 million (pig) and 1670 million (giraffe) (Kazu et al. 2014). There are almost no cell counts for carnivores except for the dog (160 million), the cat (300 million), and the raccoon (453 million) (Roth and Dicke 2012). Very large carnivores like the polar bear or the sea lion with brain weights around 500 g are expected to have at least 1000 million cortical neurons. Due to higher NPD, carnivores have more cortical neurons than "ungulates" of the same brain size (Herculano-Houzel 2012; Kazu et al. 2014).

Owing to their small neurons and high NPD, primates have many more cortical neurons than other mammals of the same brain and cortex size. However, here we find large differences in the cell counts. According to Roth and Dicke (2012), based on the NPD data by Haug, the small New World squirrel monkey (*Saimiri sciureus*) has 450 million, the larger New World white-fronted capuchin 720 million. Herculano-Houzel et al. (2007), however, report 1340 million for the former and 1140 million for the latter. For the rhesus monkey (*Macaca mulatta*), Roth and Dicke (2012) calculated 840 million, and Herculano-Houzel et al. report (2007) 1710 million cortical neurons, while in the larger long-tailed macaque (*Macaca fascicularis*) they found "only" 800 million cortical neurons, which would be close to our data concerning the rhesus monkey.

Herculano-Houzel and colleagues (2012) and Roth and Dicke (2012) state that chimpanzees have about 6000 million, while there is a discrepancy concerning the gorilla (4300 million by Roth and Dicke 2012 and up to 9000 million by Herculano-Houzel 2012). As to the human cortex, Roth and Dicke (2012) arrived at 14,000 and Herculano-Houzel (2009) at 16,000 million neurons, while Pakkenberg and Gundersen (1997) counted 19,000 million neurons in women and 23,000 million in men. Herculano-Houzel (2009) as well as Roth and Dicke (2012) consider this value as being too high.

Cell counts in elephants and cetaceans likewise vary widely among authors, again mostly because of different methods applied. While Roth and Dicke (2005) calculated about 11,000 million cortical neurons in the African elephant, Herculano-Houzel et al. (2014) found "only" 5600 million. Both authors, however, agree that elephants have fewer cortical neurons than humans despite their much larger brain. There is likewise debate about the number of cortical neurons in cetaceans. Roth and Dicke had arrived at about 11,000 million in a 7-kg cetacean brain, and in the Minke whale (*Balaenoptera acutorostrata*), with a brain weight of about 6 kg, Eriksen and Pakkenberg (2007) counted 12,000 million cortical neurons. However, in a recent paper published by the Pakkenberg group (Mortensen et al. 2014), the authors reported 37,200 million cortical neurons in the long-finned pilot whale (dolphin) (*Globicephala melas*) at a brain weight of about 3500 g. This would be almost twice as much as the value found by Pakkenberg and Gundersen for the human cortex (23,000 million in men). However, Herculano-Houzel et al. (2014) question these values mostly for methodological reasons. In addition, they argue that cetaceans, as descendants of ungulates—characterized by relatively thin cortices with low NPD—should rather strictly follow their general brain-cortex pattern, and they predict that the number of cortical neurons even in large-brained cetaceans would be well below 10,000 million and, thus, rather conforms to the cell counts made in elephants.

Apart from these discrepancies, in mammals, the number of cortical neurons correlates better with intelligence than absolute or relative brain size or EQ. Carnivores appear to be more intelligent than "ungulates" and these are more intelligent than "insectivores," and this fits nicely the number of cortical neurons. The reason for this is that carnivore cortices have a higher NPD and consequently contain more neurons than ungulate cortices of the same size. Within primates, we again find a positive relationship between numbers of cortical neurons and levels of intelligence. The least intelligent primates, lemurs, have the lowest number of neurons, followed by monkeys and finally by the great apes and humans. Humans appear to have the largest number of cortical neurons (if we disregard the data by Mortensen et al. 2014), because of a large cortical volume combined with a high NPD. The high number of cortical neurons in monkeys (as a consequence of extremely high NPD) reported by the Herculano-Houzel group would further speak in favor of the high degree of intelligence found in the capuchin or squirrel monkeys coming close to that of the great apes.

The opposite is found in elephants and cetaceans, both of which have much fewer cortical neurons than expected based on their large brain and cortex sizes. Apparently, this is a consequence of very low NPD as well as of their relatively thin cortices. In the case of cetaceans, this may be—as Kazu et al. (2014) argue—a consequence of phylogeny: cetaceans are descendants of artiodactyls, which, as mentioned have a much lower NPD than primates, and their strong increase in body and brain size in cetaceans probably led to a further decrease in NPD. Thus, in both elephants and cetaceans, we find the puzzling fact that they have 3–7000 million cortical neurons equal to the great apes, while having modest to moderate degrees of intelligence.

In summary, while there is a relatively good correlation between intelligence and number of cortical neurons within mammalian orders, it breaks down, when we compare numbers of neurons in cortices of the same size in ungulates, carnivores, primates, elephants and cetaceans because of huge differences in NPD. At a given cortical volume, primates including humans have much more cortical neurons than the other mammalian taxa.

6.6.5 Information Processing Capacity

Information processing capacity (IPC) of a cortex is intimately related to signal transmission speed, i.e., how fast cortical neurons "transfer signals to each other." Signal transmission speed is determined by (1) interneuronal distance, (2) axonal conduction velocity, and (3) transsynaptic transmission time. While the latter appears to be rather constant among mammals, factors (1) and (2) vary greatly. Interneuronal distance, trivially, is inversely related to NPD: the higher the NPD, the shorter the interneuronal distance. Conduction velocity is a direct consequence of the thickness of the myelin sheath. The myelin sheath is thickest in primates and thinnest in elephants and cetaceans (Zhang and Sejnowski 2000; Changizi 2001). Thus, while an elephant or dolphin/whale may have the same number of cortical neurons like a chimpanzee (e.g., 4–5 billion), the former has a much lower signal transmission speed than the latter. Thus, a combination of very low NPD and low axonal conduction velocity appears to make information processing capacity of elephants and cetaceans very slow despite a high number of cortical neurons. This could, among others, explain why their cognitive abilities are modest but, also, why in cetaceans we find that the two cerebral hemispheres often work (and sleep) independently.

Conversely, small-sized monkeys like the capuchin monkey, due to their high NPD and high axonal conduction velocity, probably has a very high information processing capacity, which could be equal to that found in larger primates—let alone non-primate mammals—with a higher number of cortical neurons but lower NPD. This could explain why small monkeys with relatively small brains containing little more than one billion cortical neurons are equal, at least in some aspects of cognition, to a chimpanzee or a gorilla with 4–6 times more cortical neurons.

6.7 Conclusions

We found that in mammals neither absolute nor relative brain size nor the EQ are good predictors for higher cognitive abilities. Large-brained ungulates (762 g or more) and even larger-brained cetaceans (8200 g or more) and elephants (up to 6000 g) are of moderate or even very modest intelligence when compared to monkeys with much smaller brains. The latter appear to be almost as intelligent as the apes, while again having much smaller brains. Taking relative brain weight or the EQ into account does not yield better correlations. The number of cortical neurons appears to be a better predictor of intelligence, but here we have the paradox of elephants and cetaceans, which have at least several billion cortical neurons (as many as the great apes), but are very modest in intelligence. The best fit is reached, when we compare degrees of intelligence with a combination of the number of cortical neurons with parameters that directly determine information processing capacity of a brain/cortex, i.e., interneuronal distance (neuron density) and axonal conduction velocity. Here, primates and especially humans excel, while the large-brained elephants and cetaceans perform poorly. Taking IPC into account, we can also explain why small-brained primates can be (almost) as smart as the great apes except humans.

The human brain appears to constitute an optimal trade-off between factors determining neuronal information processing capacity, i.e., absolute and relative brain size, number of neurons, packing density, and axonal conduction velocity (Hofman 2012; Roth 2013). However, we believe that in addition the emergence of a syntactical and grammatical language about 100,000 years ago appears to have functioned as an enormous "intelligence amplifier" making thinking, problem-solving, and communication much easier (cf. Roth 2013). Among nonhuman animals there is a variety of complex and syntactical languages like bird song or dolphin language, but so far, there is no evidence that such a syntax reaches the semantic level (Fitch and Hauser 2004; Berwick et al. 2011). The invention of writing about 5000 years ago certainly has served as a second "intelligence amplifier."

Acknowledgment We thank Prof. Suzanna Herculano-Houzel, Rio de Janeiro, for the helpful suggestions.

References

Bates LA, Byrne RW (2010) Imitation: what animal imitation tells us about animal cognition. Wiley Interdiscip Rev Cogn Sci 1:685–695

Berwick RC, Okanoya K, Beckers GJL, Bolhuis JJ (2011) Songs to syntax: the linguistics of bird song. Trends Cogn Sci 15:113–121. doi:10.1016/j.tics.2011.01.002

Horowitz A (2011) Theory of mind in dogs? Examining method and concept. Learn Behav 39(4):314–317. doi:10.3758/s13420-011-0041-7

Benson-Amram S, Dantzer B, Stricker G, Swanson EM, Holekamp KE (2015) Brain size predicts problem-solving ability in mammalian carnivores. Proc Natl Acad Sci U S A 113(9):2532–2537. doi:10.1073/pnas.1505913113

Blaisdell AP, Sawa K, Leising KJ, Waldmann MR (2006) Causal reasoning in rats. Science 311:1020–1022

Bräuer J, Call J, Tomasello M (2005) All great ape species follow gaze to distant locations and around barriers. J Comp Psychol 119:145–154

Broom DM, Sena H, Moynihan KL (2009) Pigs learn what a mirror image represents and use it to obtain information. Anim Behav 78:1037–1041

Brown DH, Norris KS (1956) Observations of captive and wild cetaceans. J Mammal 37:311–326

Byrne R (1995) The thinking ape. Evolutionary origins of intelligence. Oxford University Press, Oxford

Byrne RW, Bates LA (2011) Cognition in the wild: exploring animal minds with observational evidence. Biol Lett 7(4). doi:10.1098/rsbl.2011.0352

Byrne R, Bates L, Moss CJ (2009) Elephant cognition in primate perspective. Compar Cogn Behav Rev 4:1–15

Carlier P, Jamon M (2006) Observational learning in C57BL/6j mice. Behav Brain Res 174:125–131

Changizi MA (2001) Principles underlying mammalian neocortical scaling. Biol Cybern 84:207–215. doi:10.1007/s004220000205

Cozzi B, Povinelli M, Ballarin C, Granato A (2014) The brain of the horse: weight and cephalization quotients. Brain Behav Evol 83:9–16. doi:10.1159/000356527

Delfour F, Marten K (2001) Mirror image processing in three marine mammal species: killer whales (Orcinus orca), false killer whales (Pseudorca crassidens) and California sea lions (Zalophus Californicus). J Behav Processes 53:181–190
Emery NJ, Clayton NS (2009) Tool use and physical cognition in birds and mammals. Curr Opin Neurobiol 19:27–33
Eriksen N, Pakkenberg B (2007) Total neocortical cell numbers in the mysticete brain. Anat Rec 290:83–95
Fichtel C, Kappeler PM (2010) Human universals and primate symplesiomorphies: establishing the lemur baseline. In: Kappeler PM, Silk J (eds) Mind the gap: tracing the origins of human universals. Springer, Heidelberg, pp 395–426
Fitch WT, Hauser MD (2004) Computational constraints on syntactic processing in a nonhuman primate. Science 303:377–380
Foote AL, Crystal JD (2007) Metacognition in the rat. Curr Biol 17:551–555
Gallup GG Jr (1970) Chimpanzees: self-recognition. Science 167:86–87
Guentuerkuen O (2014) Is dolphin cognition special? Brain Behav Evol 83:177–180. doi:10.1159/000357551
Hall KRL, Schaller GB (1964) Tool-using behavior of the California sea otter. J Mammal 45:287–298
Hart BL, Hart LA, Pinter-Wollman N (2008) Large brains and cognition: where do elephants fit in? Neurosci Biobehav Rev 32:86–98
Haug H (1987) Brain sizes, surfaces, and neuronal sizes of the cortex cerebri: a stereological investigation of man and his variability and a comparison with some mammals (primates, whales, marsupials, insectivores, and one elephant). Am J Anat 180:126–142. doi:10.1002/aja.1001800203
Herculano-Houzel S, Mota B, Lent R (2006) Cellular scaling rules for rodent brains. Proc Natl Acad Sci 103(32):12138–12143
Herculano-Houzel S (2009) The human brain in numbers: a linearly scaled-up primate brain. Front Hum Neurosci 3:31. doi:10.3389/neuro.09.031.2009
Herculano-Houzel S (2012) Neuronal scaling rules for primate brains: the primate advantage. Prog Brain Res 195:325–340. doi:10.1016/B978-0-444-53860-4.00015-5
Herculano-Houzel S, Avelino-de-Souza K, Neves K, Porfírio J, Messeder D, Mattos Feijó L, Maldonado J, Manger PR (2014) The elephant brain in numbers. Front Neuroanat 8:46. doi:10.3389/fnana.2014.00046
Herculano-Houzel S, Collins CE, Wong P, Kaas JH (2007) Cellular scaling rules for primate brains. Proc Natl Acad Sci U S A 104:3562–3567. doi:10.1073/pnas.0611396104
Herculano-Houzel S, Catania K, Manger PR, Kaas JH (2015) Mammalian brains are made of these: a dataset of the numbers and densities of neuronal and nonneuronal cells in the brain of Glires, primates, scandentiam eulipotyphlans, afrotherians and artiodactyls, and their relationship with body mass. Brain Behav Evol 86:145–163. doi:10.1159/000437413
Herman LM (2012) Body and self in dolphins. Cons Cogn 21:526–545
Hobhouse LT (1915) Mind in evolution. Macmillan, London
Hofman MA (2012) Design principles of the human brain: an evolutionary perspective. Prog Brain Res 195:373–390. doi:10.1016/B978-0-444-53860-4.00018-0
Irie-Sugimoto N, Kobayashi T, Sato T, Hasegawa T (2008) Evidence of means-end behavior in Asian elephants (Elephas maximus). Anim Cogn 11:359–365
Jerison HJ (1973) Evolution of the brain and intelligence. Academic, Amsterdam
Kazu RS, Maldonado J, Mota B, Manger PA, Herculano-Houzel S (2014) Cellular scaling rules for the brain of Artiodactyla include a highly folded cortex with few neurons. Front Neuroanat 8:1–19
Kendal RL, Custance DM, Kendal JR, Vale G, Stoinski TS, Rakotomalala NL et al (2010) Evidence for social learning in wild lemurs (Lemur catta). Learn Behav 38:220–234
Kern A, Siebert U, Cozzi B, Hof PR, Oelschläger HHA (2011) Stereology of the neocortex in odontocetes: qualitative, quantitative, and functional implications. Brain Behav Evol 77:79–90. doi:10.1159/000323674

Krützen M, Mann J, Heithaus MR, Connor RC, Bejder L, Sherwin WB (2005) Cultural transmission of tool use in bottlenose dolphins. Proc Natl Acad Sci U S A 102:8939–8943
Kuczaj SA II, Yeater D (2006) Dolphin imitation: who, what, when, and why? Aquat Mamm 32:413–422
Lefebvre L (2012) Primate encephalization. Prog Brain Res 195:393–412. doi:10.1016/B978-0-444-53860-4.00019-2
Lyons DE, Santos LR (2006) Ecology, domain specificity, and the origins of theory of mind: is competition the catalyst? Philos Compass 5:481–492
Manger PR (2013) Questioning the interpretations of behavioral observations of cetaceans: is there really support for a special intellectual status for this mammalian order? Neuroscience 250:664–696. doi:10.1016/j.neuroscience.2013.07.041
Marino L (2004) Dolphin cognition. Curr Biol 14:R910–R911
Marino L, Connor RC, Fordyce RE, Herman LM, Hof PR, Lefebvre L, Lusseau D, McGowan B, Nimchinsky EA, Pack AA, Rendell L, Reidenberg JS, Reiss D, Uhen MD, Van der Gucht E, Whitehead H (2007) Cetaceans have complex brains for complex cognition. PLoS Biol 5:966–972
Marino L, Butti C, Connor RC, Fordyce RE, Herman LM, Hof PR, Lefebvre L, Lusseau D, McGowan B, Nimchinsky EA, Pack AA, Reidenberg JS, Reiss D, Rendell L, Uhen MD, Van der Gucht E, Whitehead H (2008) A claim in search of evidence: reply to Manger's thermogenesis hypothesis of cetacean brain structure. Biol Rev 83:417–440
Mayer C, Call J, Albiach-Serrano A, Visalberghi E, Sabbatini G, Seed A (2014) Abstract knowledge in the broken-string problem: evidence from nonhuman primates and pre-schoolers. PLoS One 9(10):e108597
McCowan B, Marino L, Vance E, Walke L, Reiss D (2000) Bubble ring play of bottlenose dolphins (Tursiops truncates): implications for cognition. J Comp Psychol 114:98–106
McKenzie T, Cherman T, Bird LR, Naqshbandi M, Roberts WA (2004) Can squirrel monkeys (Saimiri sciureus) plan for the future? Studies of temporal myopia in food choice. Anim Learn Behav 32(4):377–390
Miklósi A, Kubinyi E, Topa J, Gacsi M, Varnyi Z, Csanyi V (2003) A simple reason for a big difference: wolves do not look back at humans, but dogs do. Curr Biol 13:763–766
Michener GR (2004) Hunting techniques and tool use by North American badgers preying on Richardson's ground squirrels. J Mammal 85:1019–1027. doi:10.1644/BNS-102
Minuzo K, Irie N, Hiraiwa-Hasegawa M, Kutsukake N (2015) Asian elephants acquire inaccessible food by blowing. Anim Cogn 19(1):215–222. doi:10.1007/s10071-015-0929-2
Mortensen HS, Pakkenberg B, Dam M, Dietz R, Sonne C, Mikkelsen B, Eriksen N (2014) Quantitative relationships in delphinid neocortex. Front Neuroanat 8:1–10
Müller CA (2010) Do anvil-using banded mongooses understand means-end relationships? A field experiment. Anim Cogn 13:325–330
Nissani M (2004) Elephant cognition: a review of recent experiments. Gajah 28:44–52
Nissani M (2006) Do Asian elephants (Elephas maximus) apply causal reasoning to tool-use tasks? J Exp Psychol Anim Behav Process 32:91–96
Okanoya K, Tokimoto N, Kumazawa N, Hihara S, Iriki A (2008) Tool-use training in a species of rodent: the emergence of an optimal motor strategy and functional understanding. PLoS One 3:e1860
Pakkenberg B, Gundersen HJG (1997) Neocortical neuron number in humans: effect of sex and age. J Comp Neurol 384:312–320
Patterson EM, Mann J (2011) The ecological conditions that favor tool use and innovation in wild bottlenose dolphins (Tursiops sp.) PLoS One 6(7):e22243
Plotnik JM, de Waal FBM, Reiss D (2006) Self-recognition in an Asian elephant. Proc Natl Acad Sci U S A 103:17053–17057
Premack D, Woodruff G (1978) Does the chimpanzee have a theory of mind? Behav Brain Sci 4:515–526
Premack D (2007) Human and animal cognition: Continuity and discontinuity. Proc Natl Acad Sci 104(35):13861–13867

Prior H, Schwarz A, Guentuerkuen O (2008) Mirror-induced behavior in the magpie (Pica pica): evidence of self-recognition. PLoS Biol 6:1642–1650
Range F, Virányi Z (2011) Development of gaze following abilities in wolves (*Canis lupus*). PLoS One 6:e16888. doi:10.1371/journal.pone.0016888
Range F, Virányi Z (2014) Wolves are better imitators of conspecifics than dogs. PLoS One 9(1):e86559
Reiss D, Marino L (2001) Mirror self recognition in the bottlenose dolphin: a case of cognitive convergence. Proc Natl Acad Sci U S A 98:5937–5942. doi:10.1073/pnas.101086398
Rensch B, Altevogt R (1955) Zähmung und Dressurleistungen indischer Arbeitselefanten. Z Tierpsychol 11:497–510
Riemer S, Müller C, Range F, Huber L (2014) Dogs can learn to attend to connectivity in string pulling tasks. J Comp Psychol 128:31–39. doi:10.1037/a0033202
Roberts WA (2012) Evidence for future cognition in animals. Learn Motiv 43:169–180
Roth G (2013) The long evolution of brains and minds. Springer, Dordrecht
Roth G, Dicke U (2005) Evolution of the brain and intelligence. Trends Cogn Sci 9:250–257
Roth G, Dicke U (2012) Evolution of the brain and intelligence in primates. Prog Brain Res 195:413–430. doi:10.1016/B978-0-444-53860-4.00020-9
Russel S (1979) Brain site and intelligence: a comparative perspective. In: Oakley DA, Plotkin HC (eds) Brain, Behavior and Evolution. Methuen, London, pp 126–153
Seed A, Tomasello M (2010) Primate cognition. Topics Cogn Sci 2:407–419
Smith BP, Applebee RG, Litchfield CA (2012) Spontaneous tool-use: an observation of a dingo (*Canis dingo*) using a table to access an out-of-reach food reward. Behav Processes 89:219–222
Smith JD (2009) The study of animal metacognition. Trends Cogn Sci 13:389–396
Taylor AH, Hunt GR, Media FS, Gray RD (2009) Do New Caledonian crows solve physical problems through causal reasoning? Proc R Soc B Biol Sci 276:247–254
Tarver JE, dos Reis M, Mirarab S et al (2016) The interrelationships of placental mammals and the limits of phylogenetic inference. Genome Biol Evol 8:330–344. doi:10.1093/gbe/evv261
Téglás E, Gergely A, Kupán K, Miklósi A, Topál J (2012) Dogs' gaze following is tuned to human communicative signals. Curr Biol 7:209–212
Tomasello M, Hare B, Lehmann H, Call J (2007) Reliance on head versus eyes in the gaze following of great apes and human infants: the cooperative eye hypothesis. J Hum Evol 52:314–320
Udell MAR, Dorey NR, Wynne CDL (2011) Can your dog read your mind? Understanding the causes of canine perspective taking. Learn Behav 39:289–302. doi:10.3758/s13420-011-0034-6
Van der Vaart E, Hemelrijk CK (2014) 'Theory of Mind' in animals: ways to make progress. Synthese 191(335):354
van Dongen PAM (1998) Brain size in vertebrates. In: Nieuwenhuys R, ten Donkelaar HJ, Nicholson C (eds) The central nervous system of vertebrates. Springer, Berlin, pp 2099–2134
Wallis LJ, Range F, Müller CA, Serisier S, Huber L, Virányi Z (2015) Training for eye contact modulates gaze following in dogs. Anim Behav 106:27–35
Zhang K, Sejnowski TJ (2000) A universal scaling law between gray matter and white matter of cerebral cortex. Proc Natl Acad Sci U S A 97:5621–5626. doi:10.1073/pnas.090504197

Chapter 7
On the Matter of Mind: Neural Complexity and Functional Dynamics of the Human Brain

Michel A. Hofman

Abstract The evolutionary expansion of the brain is among the most distinctive morphological features of anthropoid primates. During the past decades, considerable progress has been made in explaining brain evolution in terms of physical and adaptive principles. The object of this review is to present current perspectives on primate brain evolution, especially in humans, and to examine some of the design principles and operational modes that underlie the information processing capacity of the cerebral cortex. It is shown that the development of the cortex coordinates folding with connectivity in a way that produces smaller and faster brains than otherwise would have been possible. It will be argued that in primates the complexity of the neural circuitry of the cerebral cortex is the neural correlate of higher cognitive functions, including mind-like properties and consciousness.

Keywords Brain evolution • Cerebral cortex • Cognition • Consciousness • Human mind • Information processing • Biological intelligence • Neural network • Primates

Evolution is a passage from the most automatic to the most voluntary.
John Hughlings Jackson (1884)

7.1 Introduction

Organisms are faced during their lives with an immense variety of problems, ranging from purely physical ones, such as changes in climate or geomorphic disturbances, to organism-specific problems related to food supply, predation, homeostasis, and reproduction. Problem-solving, in other words, is an essential dynamic survival mechanism, evolved to cope with disturbances in the ecological equilibrium. It can

M.A. Hofman, Ph.D.
Netherlands Institute for Neuroscience, Royal Netherlands Academy of Arts and Sciences, Meibergdreef 47, Amsterdam 1105 BA, The Netherlands
e-mail: M.Hofman@nin.knaw.nl

S. Watanabe et al. (eds.), *Evolution of the Brain, Cognition, and Emotion in Vertebrates*, Brain Science, DOI 10.1007/978-4-431-56559-8_7

therefore be looked upon as an adaptive capacity enabling organisms to adjust themselves to one another and to their physical environment (see, e.g., Macphail and Bolhuis 2001; Reader et al. 2011; Shettleworth 2012a). In fact, with the evolution of sensory systems as adaptations to specialized environments, the capacity to process large amounts of sensory information increased and, with that, the power to create more complex physical realities.

If the ability of an organism to process information about its environment is a driving force behind evolution, then the more information a system, such as the brain, receives and the faster it can process this information, the more adequately it will be able to respond to environmental challenges and the better will be its chances of survival (Macphail and Bolhuis 2001; Roth and Dicke 2012; Hofman 2015). The limit to any intelligent system therefore lies in its abilities to process and integrate large amounts of sensory information and to compare these signals with as many memory states as possible, and all that in a minimum of time. It implies that the functional capacity of a neuronal structure is inherently limited by its neural architecture and signal processing time (see, e.g., Laughlin and Sejnowski 2003; Buzsáki et al. 2013). In this chapter, some of the organizational principles and operational modes will be explored that underlie the information processing capacity of the human brain, and it will be argued that the complexity of the cortical network circuitry is the neural correlate of cognition.

7.2 Principles of Brain Evolution

If we assume that the cognitive capacity of complex organisms is the product of integrated sensory information processing and mental faculties, responsible for the planning, execution, and evaluation of intelligent behavior, variations among species in cognition must in principle be observable in the neural substrate. In mammals, especially in primates, the complexity of the neural circuitry of the cerebral cortex is considered to be the neural basis for the brain's coherence and predictive power and, thus, a correlate of higher cognitive and perceptual functions (Gazzaniga et al. 2008; Wang 2010; Chittka et al. 2012).

The evolutionary expansion of the cerebral cortex, indeed, is among the most distinctive morphological features of mammalian brains. Particularly in species with large brains, and most notably in great apes and marine mammals, the brain becomes disproportionately composed of the cortical structure (Striedter 2005; Aboitiz and Montiel 2012; Smaers et al. 2012; Hofman 2014; Lewitus et al. 2014; Fig. 7.1). The volume of cortical gray matter, for example, expressed as a percentage of total brain volume increases from about 25% for insectivores to 50% for humans (Frahm et al. 1982; Hofman 1988), whereas the relative size of the entire cerebral cortex (including white matter) goes from 40% in mice to about 80% in humans (Hofman 1988; Azevedo et al. 2009; Herculano-Houzel 2009, 2012).

On the other hand, the relative size of the cerebellum remains constant across phylogenetic groups, occupying about 10–15% of the total brain mass in different

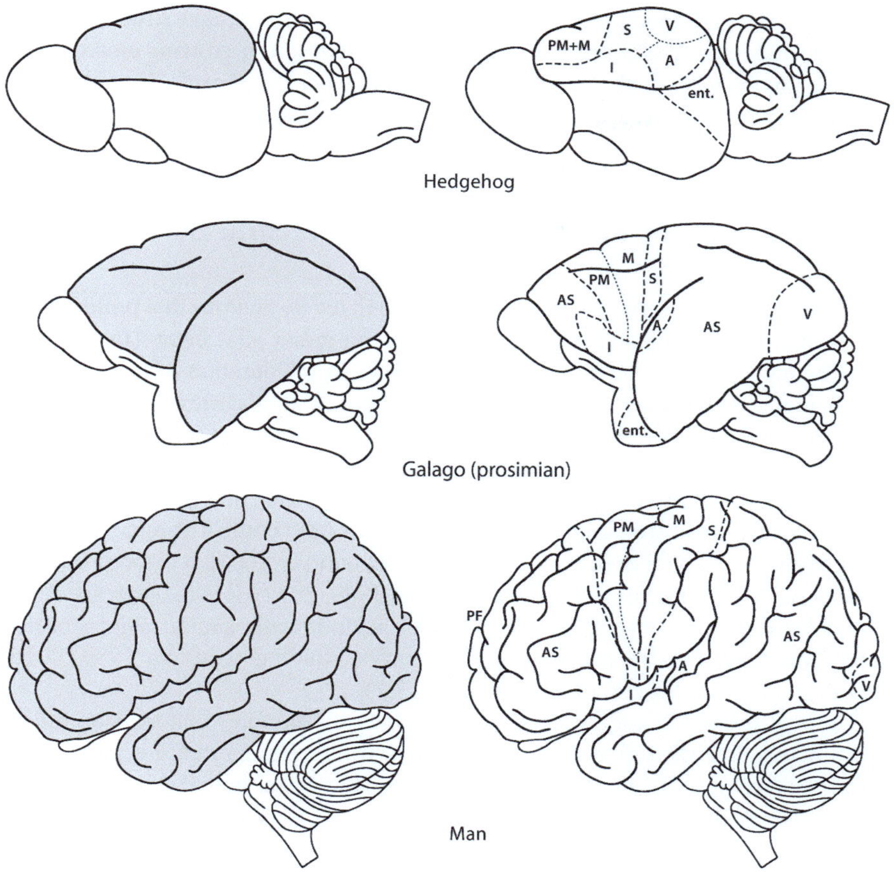

Fig. 7.1 Lateral views of the brains of some mammals to show the evolutionary development of the neocortex (*gray*). In the hedgehog almost the entire neocortex is occupied by sensory and motor areas. In the prosimian galago, the sensory cortical areas are separated by an area occupied by association cortex (AS). A second area of association cortex is found in front of the motor cortex. In human beings these anterior and posterior association areas are strongly developed. *A* primary auditory cortex, *AS* association cortex, *Ent* entorhinal cortex, *I* insula, *M* primary motor cortex, *PF* prefrontal cortex, *PM* premotor cortex, *S*, primary somatosensory cortex, *V* primary visual cortex. Modified from Nieuwenhuys (1994)

orders (Stephan et al. 1981). Comparative studies among four mammalian orders, including primates, have revealed that the absolute neuronal composition in the cerebral cortex covaries significantly with that of the cerebellum (Herculano-Houzel et al. 2008; Lent et al. 2012), showing that these two brain structures display coordinated growth during phylogenesis in mammals.

Such a coordinated evolution of the cerebral cortex and cerebellum fits well with the recent clinical and experimental evidence suggesting an important role of the cerebellum in cognitive and affective functions, in close connection with cortical associative areas (for reviews, see Schmahmann 2010; MacLeod 2012; Barton and

Venditti 2014). Although the cerebral cortex is not the only brain structure which was selected for in evolution for greater growth, as a result of growing environmental pressure for more sophisticated cognitive abilities, it has played a key role in the evolution of intelligence.

7.3 Organization and Evolution of the Cerebral Cortex

The cerebral cortex forms as a smooth sheet populated by neurons that proliferate at the ventricular surface and migrate outward along radial glial fibers (for reviews, see Cheung et al. 2007; Rakic 2009). Differences in the duration of neurogenesis, which increases more rapidly with brain size for the cerebral cortex than for subcortical areas (Charvet and Finlay 2014; Molnár et al. 2014; Suárez et al. 2014), lead to a systematic increase in the ratio of the cortical to subcortical regions. Whereas in small-brained species the cortical volume expands by virtue of a combined increase in surface area and cortical thickness, the increase of the cortical volume in species with a brain size of more than 3–4 cm^3 is almost entirely due to a disproportionate expansion of the cortical surface area (Hofman 1989, 2012). It is the increase of the cortical surface area beyond that expected for geometrically similar objects of different volumes that creates the need for cortical folding (Hofman 1989, 2016; Razavi et al. 2015; Striedter et al. 2015; Fig. 7.2).

Cortical folding is not a random process. Instead, the folds that develop reflect the functional organization of the cortex, and form patterns that are remarkably consistent across individuals and even among some species. How this happens is not well understood. Although many developmental features and evolutionary adaptations have been proposed as the primary cause of cortical folding, it is not evident that gyrification is reducible in this way (for a review, see Ronan and Fletcher 2015). There is now empirical evidence that suggests that the folding of the cerebral cortex is the product of a generalized mechanism, namely, the differential expansion of the

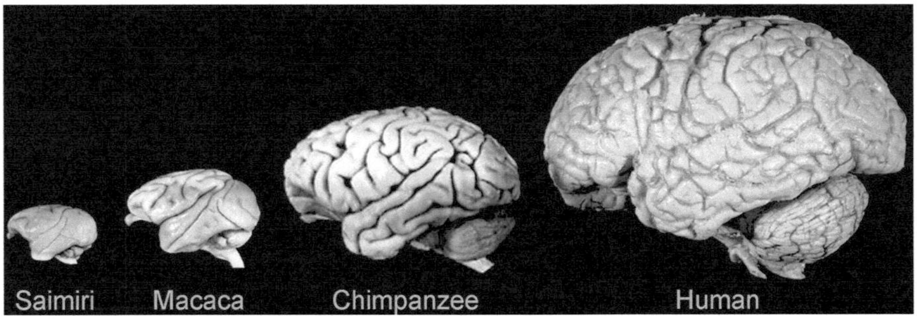

Fig. 7.2 Lateral views of the brains of some anthropoid primates showing the evolutionary expansion of the neocortex. Note the diverse configurations and gyral and sulcal patterns. *Saimiri sciureus*: E = 22 g; *Macaca mulatta*: E = 95 g; *Pan troglodytes*: E = 420 g; *Homo sapiens*: E = 1350 g. Reproduced with permission from Hofman (2007)

cortex. Mota and Herculano-Houzel (2012), for example, propose that cortical folding is driven by white matter connectivity. Specifically, they argue that the mechanical tension generated by the pattern of connectivity of fiber bundles traveling through white matter may account for the observed pattern of cortical surface convolutions. The authors propose the degree of tension, taken as directly proportional to the morphological characteristics of the fiber bundle (i.e., axonal length and average cross-sectional area and the proportion of efferent neurons), determines how much the cortical surface folds inward. This model is used to explain how surface convolutions vary with brain size and how gray matter thickness varies.

During the past decades considerable progress has been made in explaining the evolution of the cerebral cortex in terms of physical and adaptive principles (see, e.g., Macphail and Bolhuis 2001; Lefebvre 2012; Roth and Dicke 2012; Isler and Van Schaik 2014). In addition, a quantitative approach to the comparative morphology of the brain has made it possible to identify and formalize empirical regularities in the diversity of brain design, especially in the geometry of the cortex (Hofman 1989, 2012, 2016; Changizi 2001, 2007; Clark et al. 2001). Analysis of the cerebral cortex in anthropoid primates, for example, revealed that the volume of the neocortex is highly predictable from absolute brain size (Hofman 1989; Finlay and Darlington 1995; Zhang and Sejnowski 2000; Finlay et al. 2001; Hofman and Falk 2012). The volume of the cortical gray matter, containing local networks of neurons that are wired by dendrites and mostly nonmyelinated axons, is basically a linear function of brain volume, whereas the mass of long-range axons, forming the underlying white matter volume, increases disproportionately with brain size. As a result, the volume of gray matter expressed as a percentage of total brain volume is about the same for all anthropoid primates.

The relative white matter volume, on the other hand, increases with brain size, from 9% in pygmy marmosets (*Cebuella pygmaea*) to about 35% in humans, the highest value in primates (Hofman 1989). The nonlinear nature of this process is further emphasized by plotting the relative volume of white matter as a function of brain size (Fig. 7.3). The high correlation between both variables ensures that the curve, and its confidence limits, can be used for predictive purposes to estimate the volume of white matter relative to brain volume for a hypothetical primate. The model, for example, predicts a white matter volume of about 1470 cm^3 for an anthropoid primate with a brain volume of 3000 cm^3 (Hofman 2001, 2012). In other words, in such a large-brained primate, white matter would comprise about half of the entire brain volume, compared to one-third in modern man.

Volumetric measurements of gray and white matter in the neocortex of anthropoid primates have shown that the "universal scaling law" of neocortical gray to white matter applies separately for frontal and non-frontal lobes and that changes in the frontal (but not non-frontal) white matter volume are associated with changes in other parts of the brain, including the basal ganglia, a group of subcortical nuclei functionally linked to executive control (Smaers et al. 2010; Sherwood et al. 2012). These comparative analyses indicate that the evolutionary process of neocorticalization in primates is mainly due to the progressive expansion of the axonal mass that implements global communication, rather than to the increase in the number of

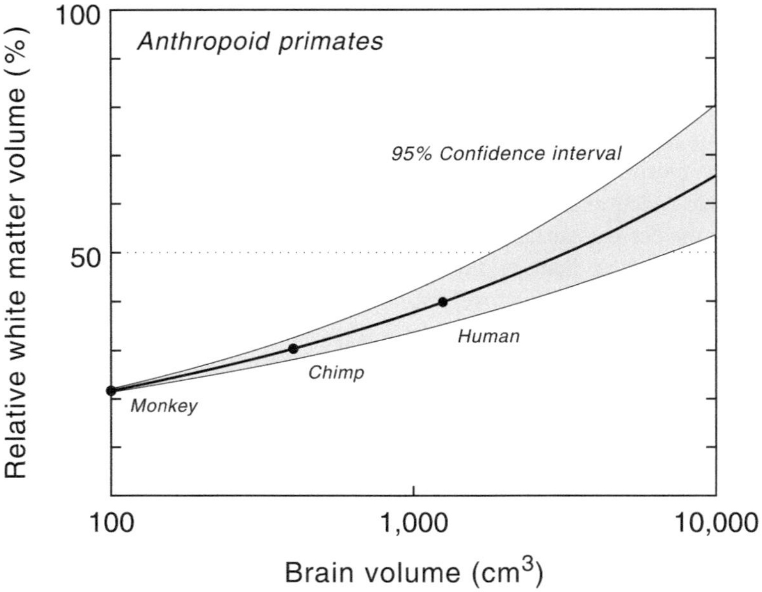

Fig. 7.3 Relative white matter volume as a function of brain volume in anthropoid primates. Semilogarithmic scale. The proportion of white matter increases with brain size, from 22% in a monkey brain of 100 cm^3 to about 65% in a hypothetical primate with a brain size of 10,000 cm^3. Modified from Hofman (2001)

cortical neurons and the importance of high neural connectivity in the evolution of brain size in anthropoid primates.

Wen and Chklovskii (2005) have shown that the competing requirements for high connectivity and short conduction delay may lead naturally to the observed architecture of the mammalian neocortex. Obviously, the brain functionally benefits from high synaptic connectivity and short conduction delays. A magnetic resonance imaging study, furthermore, focusing specifically on the prefrontal cortex, has shown that the volume of the white matter underlying prefrontal areas is disproportionately larger in humans than in other primates (Schoenemann et al. 2005). It suggests that the connectional elaboration of the prefrontal cortex, which mediates such important behavioral domains as planning, aspects of language, attention, and social and temporal information processing, has played a key role in human brain evolution.

7.4 Design Principles of Neural Organization

Evolutionary changes in the cerebral cortex have occurred mainly parallel to the cortical surface (tangentially) and have been sharply constrained in the vertical (radial) dimension, which makes it especially well suited for the elaboration of multiple projections and mapping systems. A mosaic of functionally specialized areas has indeed been found in the mammalian cortex, some of the functions being remarkably diverse (Kaas 2000, 2012; Krubitzer 2007; Schoenemann 2006). At the lower processing levels of the cortex, these maps bear a fairly simple topographical relationship to the world, but in higher areas precise topography is sacrificed for the mapping of more abstract functions. Here, selected aspects of the sensory input are combined in ways that are likely to be relevant to the animal.

Using modern anatomical tracing methods, physiological recordings and mapping studies it has been established that each sensory modality is mapped several times in different areas, with about a dozen representations of the visual world and a half a dozen each of auditory inputs and somatosensory sensations. In fact, the maps differ in the attributes of the stimulus represented, in how the field is emphasized, and in the types of computations performed. Clearly, the specifications of all these representations mean that functional maps can no longer be considered simply as hard-wired neural networks. They are much more flexible than previously thought and are continually modified by feedback and lateral interactions. These dynamic changes in maps, which seem likely to result from local interactions and modulations in the cortical circuits, provide the plasticity necessary for adaptive behavior and learning. Although primate species vary in the number of cortical areas, and in the patterns of connections within and between these areas, the structural organization of their neocortex is remarkably similar (Young 1993; Hill et al. 2010; Preuss 2011; Papo et al. 2014a, b).

The tremendous increase in the cortical surface without a comparable increase in its thickness during mammalian evolution has been explained in the context of the radial unit hypothesis of cortical development (for reviews, see Rakic 2007, 2009). According to this model, neocortical expansion is the result of changes in proliferation kinetics that increase the number of radial columnar units without changing the number of neurons within each unit significantly. Therefore the evolutionary expansion of the neocortex in primates is mainly the result of an increase in the number of radial columns.

The widespread occurrence of these neocortical columns, furthermore, qualifies them to be considered as fundamental building blocks in neural evolution (Mountcastle 1997; Rockland 2010; Buxhoeveden 2012; Opris and Casanova 2014). It has become evident that these cortical circuits integrate at higher levels of information processing, as a result of the hierarchical organization of the brain, thus enabling the system to combine dissimilar views of the world. The grouping of neurons into clusters that interact over short distances has been found to be characteristic not only of the cerebral cortex but of subcortical structures as well, such as the thalamus and basal ganglia (Pay 1980; Mountcastle 1997). An important implication of this basic plan of brain organization is that functionally related subsets of cortical areas and subcortical nuclei could evolve independently of the development of other subsystems during evolution (Sherwood et al. 2008; Lefebvre 2012; Smaers and Soligo 2013). So if we seek the neural basis of cognition, including mind-like properties and consciousness, we can hardly localize it in a specific region of the brain, but must suppose it to involve all those regions through whose activity an organism is able to construct an adequate model of its external world, perhaps it may even encompass the entire neo- and subcortical network.

7.5 Neural Network Wiring

Although the details of the interpretation of the columnar organization of the neocortex are still controversial (for critical reviews, see Da Costa and Martin 2010; Rockland 2010; Preuss 2001; DeFelipe 2015), it is evident that the potential for brain evolution results not from the unorganized aggregation of neurons but from cooperative association by the self-similar compartmentalization and hierarchical organization of neural circuits and the invention of fractal folding, which reduces the interconnective axonal distances. The human cerebral cortex, for example, contains about 20 billion neurons, which are interconnected via a massive yet highly organized network of axonal and dendritic wiring. This wiring enables both near and distant neurons to coordinate their responses to external stimulation. Understanding the organizing principles of cortical wiring, therefore, represents a central goal toward explaining human cognition and perception (see for example, Preuss 2011; Budd and Kisvárday 2013; De Reus et al. 2014; Wang and Liu 2014).

In the mammalian cerebral cortex, reciprocal connections between excitatory and inhibitory neurons are distributed across multiple layers, encompassing modular, dynamical, and recurrent functional networks during information processing. These dynamical brain networks are often organized in neuronal assemblies interacting through rhythmic phase relationships. Accordingly, these oscillatory interactions are observed across multiple brain scale levels, and they are associated with several sensory, motor, and cognitive processes. Recently Bosman and Aboitiz (2015) argued that there are functional constraints in the evolution of brain circuits and that these constraints may be the result of advantages that oscillatory activity contributes to brain network processes, such as information transmission and code reliability.

Network studies, using diffusion tensor imaging (DTI), have demonstrated that not only the neurons in the cerebral cortex are structurally and functionally highly organized but that it also holds for the wiring of the entire brain (Van den Heuvel and Sporns 2011; Wedeen et al. 2012; Van den Heuvel et al. 2016). The interconnecting white matter axonal pathways are not a mass of tangled wires, as thought for a long time, but they form a rectilinear three-dimensional grid continuous with the three principal axes of development. The topology of the brain's long-range communication network looks like a 3-D chessboard with a number of highly connected neocortical and subcortical hub regions. The development of new technologies for mapping structural and functional brain connectivity has led to the creation of comprehensive network maps of neuronal circuits and systems. The architecture of these brain networks can be examined and analyzed with a large variety of graph theory tools (for a review, see Sporns and Betzel 2016). It turns out that modularity is a key characteristic of brain networks across species and scales. Indeed, the modular organization of the primate neocortex may confer increased robustness and more flexible learning, help to conserve wiring cost, and promote functional specialization and complex brain dynamics (Gómez-Robles et al. 2014).

The competing requirements for high connectivity and short conduction delay may lead naturally to the observed architecture of the human neocortex. Obviously, the brain functionally benefits from high synaptic connectivity and short conduction delays. The design of the primate brain is such that it may perform a great number of complex functions with a minimum expenditure of energy and material both in the performance of the functions and in the construction of the system. In general there will be a number of adequate designs for an object, which, for practical purposes, will all be equivalent.

Recently I have shown that in species with convoluted brains, the fraction of mass devoted to wiring seems to increase more slowly than that needed to maintain a high degree of connectivity between the neural networks (Hofman 2014, 2016). These findings are in line with a model of neuronal connectivity (Deacon 1990; Ringo 1991) which says that as brain size increases there must be a corresponding fall in the fraction of neurons with which any neuron communicates directly. The reason for this is that if a fixed percentage of interconnections is to be maintained in the face of increased neuron number, then a large fraction of any brain size increase would be spent maintaining such degree of wiring, while the increasing axon length would reduce neural computational speed (Ringo et al. 1994). The human brain, for example, has an estimated interconnectivity of the order of 10^3, based on data about the number of neocortical columns and myelinated nerve fibers (Hofman 2012).

Herculano-Houzel et al. (2010) have shown that in primates the mass of the white matter scales linearly across species with its number of nonneuronal cells, which is expected to be proportional to the total length of myelinated axons in the white matter. Decreased connectivity in the brain is compatible with previous suggestions that neurons in the cerebral cortex are connected as a small-world network and should slow down the increase in global conduction delay in cortices with larger numbers of neurons (Sporns et al. 2004, 2007; Wang et al. 2008, De Reus et al. 2014; Wang and Liu 2014).

Once the brain has grown to a point where the bulk of its mass is in the form of connections, then further increases (as long as the same ratio in interconnectivity is maintained) will be unproductive. Increases in number of units will be balanced by decreased performance of those units due to the increased conduction time. This implies that large brains may tend to show more specialization in order to maintain processing capacity. Indeed, an increase in the number of distinct cortical areas with increasing brain size has been reported (Welker 1990; Kaas 2000, 2012; Striedter 2005). It may even explain why large-brained species may develop some degree of brain lateralization as a direct consequence of size. If there is evolutionary pressure on certain functions that require a high degree of local processing and sequential control, such as linguistic communication in human brains, these will have a strong tendency to develop in one hemisphere (Ringo et al. 1994; Aboitiz et al. 2003; Rilling 2014).

7.6 Neural Correlates of Consciousness

That consciousness depends on the function of the brain has been known from ancient times. Although detailed understanding of the neural mechanisms of consciousness has not been achieved, correlations between states of consciousness and functions of the brain are possible. The emergence of computational theories of mind and advances in the understanding of neurophysiology have contributed to a renewal of interest in consciousness. Since the 1980s, there has been a great deal of investigation of the neural correlates of consciousness, which have led to a number of plausible suggestions about how such a theory might be developed. One such suggestion is that consciousness is an internal scanning or perception. Another is that it involves an explicit higher-order thought, i.e., a thought that one is in a specific mental state.

Consciousness and affective experience may have arisen concurrently in the evolution of the nervous system, as a way to elaborate and extend the potential reach of instinctual urges, while new levels of cortical information processing and cognition promoted the ability of organisms to efficiently pursue goals essential to survival. In fact, affective experience, being an intrinsic brain function, cannot exist independent of consciousness, since in essence it is something that exists as part and parcel of conscious perception (Zeman 2001, 2005; Baars 2005; Shettleworth 2012a).

The conscious representation of the world is likely to be widely distributed over many areas of the cerebral cortex and possibly over certain subcortical structures as well (Baars 2005; Fabbro et al. 2015). Crick and Koch (1998) postulated that only some types of specific neurons will express the neural correlate(s) of consciousness and that these neurons will probably be fairly close together and will all project roughly to the same place. An alternative hypothesis is that the neural correlate of consciousness is necessarily global (Greenfield 1995; Greenfield and Collins 2005). In its most extreme form, this would mean that, at one time or another, any neuron in the cortex and associated structures could be part of the neural correlate of consciousness. Greenfield and Collins (2005) also argue that this raw, basic subjective state does not appear to be an exclusive property of the human brain. There is no obvious qualitative transformation in either the anatomy or the physiology of the central nervous system of human or nonhuman animals. Similarly, there is no clear ontogenetic line that is crossed as the brain grows in the womb, no single event or change in brain physiology, and certainly not at birth, when consciousness might be generated in an all-or-none fashion. A more plausible, and scientific, view of consciousness might be therefore that it is not a different property of the brain but that it is a consequence of a quantitative increase in the complexity of the human brain: consciousness will grow as brains grow. Hence, consciousness is most likely to be a continuously variable property of the brain, in both phylogenetic and ontogenetic terms.

The neural correlate of consciousness is defined as the minimal set of neuronal events that gives rise to a specific aspect of a conscious percept (Crick and Koch 2003; Tononi and Koch 2015). The cerebral cortex is probably the most suited part of the brain to look for this neural substrate, as it has very highly and specifically interconnected neuronal networks, many types of excitatory and inhibitory interneurons, and acts by forming transient coalitions of neurons, that is, assemblies of nerve cells, the members of which support one another. The dynamics of coalitions are not simple, as Crick and Koch (1990, 2003) have pointed out. In general, at any moment the winning coalition is somewhat sustained and embodies what an animal is conscious of. On the basis of experimental results in the macaque, Desimone and Duncan (1995) suggest that selective attention biases the competition among competing cell assemblies, but they do not explicitly relate this idea to consciousness. Edelman and Tononi (2000) presented a theory of consciousness, based on the idea of a "dynamic core," which resembles the coalition concept to a large extent. The dynamic core hypothesis, however, rejects the idea that there is a special subset of neurons that alone expresses the neural correlate of consciousness, a view which is also defended in the present essay.

Most of the theories of consciousness have the idea of competing assemblies of neurons in common. Consciousness depends on certain coalitions that rest on the properties of very elaborate neuronal networks. It is suggested that attention consists of mechanisms that bias the competition among coalitions, especially during their formation. Furthermore, the idea that the spatiotemporal dimensions of these nodes represent the neural correlates of mind is most appealing, as it suggests that consciousness, being an integral part of the species' problem-solving capacity, correlates to some extent with the degree of complexity of a nervous system. Therefore the search for the neural correlates of consciousness should be complemented by a search for its computational correlates (see, e.g., Atkinson et al. 2000; Zeman 2001, 2005).

7.7 Evolutionary Models of Mind

Considering biological intelligence as the problem-solving capacity of an organism makes it possible to speak of degrees of intelligence and of its evolution from amoeba to man (Jerison 1991; Hofman 2003; Dicke and Roth 2016). It means that there are differences in the abilities of organisms to perceive and interpret the physical world. Biological intelligence can thus be conceived as to reflect the temporal and spatial complexity of the species' niche, without referring, however, to the kinds of situations organisms encounter in everyday life (Reader et al. 2011; Roth and Dicke 2012; Isler and Van Schaik 2014). It is, in fact, a measure of capacity, independent of the way the capacity is used, and it may be treated as a trait for "anagenetic" rather than "cladistic" analysis (Gould 1976; Jerison 1985). It implies that when distantly related species are comparable in their problem-solving capacity, we should consider the species to be comparable in biological intelligence. Yet the near equality in intelligence may be based upon radically different adaptations. Since neural mechanisms and action patterns evolve in the contexts of the environments in which they are effective, and since species never occupy identical niches, many and various intelligences (in the plural) must have evolved in conjunction with evolving environments (Jerison 1985; Bouchard 2014).

In theory, each ecological niche requires its own degree of biological intelligence. That means that specific neural and sensorimotor adaptations always occur in relation to particular environments. A striking example is the mammalian brain, where the evolutionary changes in the balance of the sensory systems are the result of the adaptive radiation of species into many different ecological niches (Macphail and Bolhuis 2001; Striedter 2005; Hofman 2007; Allen 2009; Reader et al. 2011). These sensory systems, like any other biological feature, could evolve as a result of natural selection, because any subject that forms inadequate representations of outside reality will be doomed by natural selection.

In this view, cognitive systems and emotional phenomena can also be considered to be the result of interactions between genetic aptitude and natural environment, as they have a number of biologically useful functions: one is to keep track of the individual's whereabouts in the world by constructing a schematic model of reality (Churchland and Churchland 2002; Premack 2007). It is evident that the mind, as an emergent property of sufficiently complex living systems, has its evolutionary history like any other trait that increases adaptation to the environment and that its functions have increased with the evolution from simple to more complex systems (Sherwood et al. 2008; Roth 2013).

According to John Hughlings Jackson, in his famous Croonian lectures on the evolution and dissolution of the nervous system in *The Lancet* in 1884, the highest level of nervous function is the most complex, the most integrated, and the most widely interconnected expansive network of coordination of sensorimotor functions (see Gillett and Franz 2013). Hughlings Jackson therefore distanced the higher centers from purely reflex functions, as the former showed the least automatic or stereotyped of nervous processes, continually reorganizing themselves throughout life, to become "the organ of mind," a dynamic and changing set of patterns of sensitivity and response.

The notion of evolution as passage from the most simple and automatic to the most complex and voluntary implies that in highly complex organisms, such as primates, behavior rather than environmental change may be the major driving force for evolution at the organismal level (Hofman 2003, 2015). Free-moving organisms, for example, can actively explore their environment and thus generate new selection forces that can modify the structures involved. Mayr (1982, p. 612) even argues that "many if not most acquisitions of new structures in the course of evolution can be ascribed to selection forces by newly acquired behaviors." It might explain the dramatic evolutionary expansion of the human neocortex, being the region where both perception and instruction take place, where the external world is interpreted and modeled, and where concepts are formed and hypotheses tested, in short, where the physical world interacts with the mind.

One of the basic assumptions according to this hypothesis of the evolution of perceptual worlds is that each new system is modeled after a pre-existing one. Thus, human perceptual worlds involve some new developments beyond those of any other species, and these developments are logical extensions of those which evolved previously. The evolution of increasingly complex perceptual worlds may have reduced stereotyped behavior, thus enabling an animal to modify its behavior according to its experience. Jerison in fact writes (Jerison 1973, p. 23) that "if the nature of the perceptual world defines a dimension of intelligence, the evolution of intelligence is to be sought in the changes among species with respect to their perceptual worlds." However, one should be aware, when defining intelligence, of the extent to which particular sensory capacities have been developed as species-specific traits for life in a specialized niche. The implication is that integrative functions of the brain, rather than the sensory modalities themselves, may best define intelligence. In particular the enlargement of the brain in mammals beyond the basal somatic requirements may be associated with the tendency to integrate all information coming from different sensory systems by special intracortical networks.

Evolutionary psychology seeks to explain these evolved, functional characteristics of the human mind through the lense of an explanatory framework where special adaptive mechanisms are postulated to have been critical for hominin survival and reproductive success (Panksepp et al. 2002). These "adaptive modules" are theoretical constructs unique to the human lineage and should be clearly distinguished from the spatiotemporally defined neural processing units of the cerebral cortex discussed in the previous sections (for a review, see Shettleworth 2012b). The existence of a variety of genetically inherited, "adaptive modules" is dubious at best when considered simultaneously with our current understanding of mammalian brain organization. Indeed, the organization of the cerebral cortex, which is commonly assumed to be a prime anatomical substrate for unique cognitive functions, exhibits no robust signs of localized anatomical specialization above and beyond specific sensory and motor connections, and their polymodal interactions.

Although adaptation of an organism to its environment is the chief process directing biological evolution with the evolution of intelligence organisms became more and more independent of their environments, by modifying the environments according to their needs. This process culminated in the evolution of mankind, which can be understood only as a result of the interaction of two kinds of evolution, the biological and the cultural (Herrmann et al. 2007; Premack 2007; Hofman 2015). Such considerations have led various authors to argue that the human brain can acquire a large variety of epigenetically derived functions via interactions of a limited number of evolutionary conserved affective/motivational systems (situated largely in subcortical areas) with a set of plastic general-purpose learning mechanisms in the cerebral cortex (Panksepp and Panksepp 2000; Adolphs 2009). It does not mean that there are no special-purpose learning systems in the brain, such as fear learning, but the human cerebral cortex includes much more than a conglomeration of special-purpose learning mechanisms. It contains a neural architecture that can generate flexible features which may be best conceptualized as *rewritable*.

Cultural evolution, however, being the emergent result of the evolution of mind, cannot dispense with biological preconditions; it builds on biological facts and faculties. Though cultural evolution indeed presupposes biological evolution, it is not fully explicable in terms of theories and methods of the latter. In fact, cultural evolution has transgressed organic evolution and shows a certain autonomy (see, Donald 1991; Hofman 2003, 2015). The special status of cultural heredity can be derived from the fact that most cultural innovations are devised precisely in order to meet the environmental challenges or to improve our models of reality, whereas biological evolution has a mindless, random character. It is appropriate, therefore, to distinguish adaptations to the environment due to cultural selection from those that take place by the selection of genotypes. Cultural inheritance, furthermore, is an infinitely faster process than genetic inheritance, since it is based on the transmission of information through direct communication, and through books, the arts, and the media, which makes that a new scientific discovery or technical achievement, can be transmitted to the whole of mankind in less than one generation.

7.8 Human Language and Cognition

The most peculiar phenomenon in human problem-solving is the manifestation of language and, in particular, that of its newly acquired functions (Macphail 1982; Deacon 1998, 2006; Roth 2013; Hauser et al. 2014; Lieberman 2015). While animals can communicate by expressing their inner state by means of their behavior and by signaling to congeners, (e.g., in case of danger), human beings are the only animals that are able to make true and false statements and to produce valid and invalid arguments. Some of these higher functions, however, have been ascribed to some nonhuman primates as well. It is evident that the role of human language is so important and pervasive that it is not possible to estimate human general intellectual capacity independent of linguistic capacity (Macphail 1982; Passingham 2008; Schoenemann 2012; Bolhuis et al. 2014; Hauser et al. 2014; Corballis 2015).

Despite its important role in human cognition the evolutionary history of language still remains an enigma. Until recently it was widely held that language, and its left-hemispheric representation in the brain, were uniquely human, emerging quite recently in evolutionary terms after the emergence of *Homo sapiens*, some 70,000–100,000 years ago (e.g., Berwick et al. 2013; Bolhuis et al. 2014). Changing views of language, however, suggest that it was not a recent and sudden development in human evolution but that modern language is the product of a gradual coevolution of neurobiological and cultural-linguistic conditions, which took place since the genus *Pan* was separated from the hominin lineage about 4–6 million years ago (e.g., Deacon 1998; Falk 2004; Christiansen and Chater 2008, 2015; Pagel et al. 2013; Levinson 2016). New genetic evidence and the interpretation in context of fossil and artifact discoveries shed light on this controversy (see, e.g., Hillert 2015). The data indicate that premodern language might have been already spoken by *Homo erectus*. However, the protolanguage that was used by these early hominins may have been different from modern language. In fact, the complex properties of modern language require not only cortical wiring for language-specific operations, "a language-ready brain," but also a linguistic code related to complex concept formation. There are now strong indications that these cultural-linguistic features coevolved along with genetic changes over a period of several million years.

Changes in the brain that permit the advantageous supplement of language acquisition to perception and communication would have had obvious selective advantages throughout the period of hominin evolution. It has hypothesized, therefore, that the large size of the modern human brain results from demands made on the retrieval systems owing to the evolution of language (Popper and Eccles 1977; Macphail 1982; Rilling 2014; Dicke and Roth 2016). The progressive accumulation of interactions between environment (both physical and social), "conserved" subcortical systems, and the "general-purpose" cerebral cortex gave rise to a qualitatively different shade of mind—one that could communicate not merely with signs but in symbolic terms (Chomsky 2007; Bouchard 2013). On the other hand, we have seen that a language system—of the type found in humans—is not essential for consciousness. It is plausible that organisms, who do not possess a sophisticated language system, are aware of the external world. This is not to say, however, that language does not enrich consciousness or that it does not contribute to our model of reality.

If we assume that part of the basis of human speech is inherited in the DNA and that language is as much a biological as a cultural adaptation, then changes in the brain that permit the advantageous supplement of language acquisition to perception and communication would have had obvious selective advantages throughout the period of hominin evolution (Deacon 1998; Schoenemann 2012; Bouchard 2013; Bolhuis et al. 2014). We may conceive human language, therefore, as a superorganic form of adaptation, evolved not only as a cognitive adaptation contributing to the knowledge of reality of each individual but also as a means of sharing and, even more importantly, influencing states of mind among conspecifics. Indeed, because of language, human beings are not only able to construct individual representations of the external world, but they can also contribute to and learn from *collective* models of reality, that is, the cumulative experience of the whole of mankind. With its cognitive and linguistic skills, *Homo sapiens* tries to know its world and even exerts itself to the utmost to control it.

It is obvious that by virtue of language, human beings tend to have highly organized informational states of mind and, consequently, are excellent problem-solvers. But although knowledge of reality may be a necessary condition for survival, it is surely not enough: the degree of intelligence reached by a species does not determine the propensity of its reproductive success (Premack 2004; Bouchard 2013). This may be inferred from the indiscriminate elimination of millions of species through the eras, from ammonites to australopithecines. It means that though adaptability increases with the evolution of biological intelligence, environmental catastrophes can always be fatal to a species. But not only external factors can threaten the existence of organisms; *Homo sapiens*, despite its impressive intellectual capacities, might in the end become the victim of its own mind by, paradoxically, creating problems that it is then unable to solve.

7.9 Concluding Remarks

All organisms are constantly engaged in solving problems and must therefore have fitting and relevant models of their specific environments in order to enhance their chances of survival. Consequently, the problem-solving capacity of a species is assumed to reflect the temporal and spatial complexity of its ecological niche. The thesis presented here is that cognition can be considered to be a correlate of the problem-solving capacity of a species, manifesting itself in the complexity of the species' model of reality. With the evolution of sensory systems as adaptations to specialized environments, the capacity to process large amounts of sensory information increased and, with that, the power to create more complex physical realities. The processing of large amounts of information originating from the various sense organs, and the construction of complex models of reality require a neural system that selects, integrates, stores, and models. In other words, a system with mind-like properties that enables the organism to make sense of an otherwise chaotic world. But once we allow mind-like properties to come in, such as motivation, emotion, preference, and anticipation, we must allow that it is not only the hostile environment which plays an organizing or designing role in the evolution of higher cognitive functions but also the active search of an organism for a new ecological niche, a new mode of living.

Since the mind, prehuman and human, takes a most active part in evolution and especially in its own evolution, hominization and the evolution of our linguistic world may have begun as a cultural adaptation to new ecological niches. The process probably started at the time of hominin divergence a few million years ago, as part of the cognitive and manipulative adaptation to what was in essence a more complex physical reality. In other words, some of the seemingly unique higher functions of the human brain, such as language and other neuro-symbolic capacities, were not necessarily due to genetic selection and may have emerged epigenetically through learning and cultural experiences because of the dramatic expansion of the neocortex and its increased tendency to neural plasticity.

References

Aboitiz F, Montiel JF (2012) From tetrapods to primates: conserved developmental mechanisms in diverging ecological adaptations. Prog Brain Res 195:3–24

Aboitiz F, Lopez J, Montiel J (2003) Long distance communication in the human brain: timing constraints for inter-hemispheric synchrony and the origin of brain lateralization. Biol Res 36:89–99

Adolphs R (2009) The social brain: neural basis of social knowledge. Annu Rev Psychol 60:693–716

Allen JS (2009) The lives of the brain: human evolution and the organ of mind. Belknap, Cambridge

Atkinson AP, Thomas MSC, Cleeremans A (2000) Consciousness: mapping the theoretical landscape. Trends Cogn Sci 4:372–382

Azevedo FAC, Carvalho LRB, Grinberg LT, Farfel JM, Ferretti REI, Leite REP, Filho WJ, Lent R, Herculano-Houzel S (2009) Equal numbers of neuronal and nonneuronal cells make the human brain an isometrically scaled-up primate brain. J Comp Neurol 513:532–541

Baars BJ (2005) Global workspace theory of consciousness: toward a cognitive neuroscience of human experience. Prog Brain Res 150:45–53

Barton RA, Venditti C (2014) Rapid evolution of the cerebellum in humans and other great apes. Curr Biol 24:2440–2444

Berwick RC, Friederici AD, Chomsky N, Bolhuis JJ (2013) Evolution, brain, and the nature of language. Trends Cogn Sci 17:89–98

Bolhuis JJ, Tattersall I, Chomsky N, Berwick RC (2014) How could language have evolved. PLoS Biol 12:e1001934. doi:10.1371/journal.pbio.1001934
Bosman CA, Aboitiz F (2015) Functional constraints in the evolution of brain circuits. Front Neurosci 9:303. doi:10.3389/fnins
Bouchard D (2013) The nature and origin of language. Oxford University Press, Oxford
Bouchard TJ (2014) Genes, evolution and intelligence. Behav Genet 44:549–577
Budd J, Kisvárday ZF (2013) How do you wire a brain? Front Neuroanat 7:14. doi:10.3389/fnana.2013.00014
Buxhoeveden DP (2012) Minicolumn size and human cortex. Prog Brain Res 195:219–235
Buzsáki G, Logothetis N, Singer W (2013) Scaling brain size, keeping time: evolutionary preservation of brain rhythms. Neuron 80:751–764
Changizi MA (2001) Principles underlying mammalian neocortical scaling. Biol Cybern 84:207–215
Changizi MA (2007) Scaling the brain and its connections. In: Kaas JH (ed) Evolution of nervous systems, vol 3. Academic, New York, pp 167–180
Charvet CJ, Finlay BL (2014) Evo-devo and the primate isocortex: the central organizing role of intrinsic gradients of neurogenesis. Brain Behav Evol 84:81–92
Cheung AF, Pollen AA, Tavare A, DeProto J, Molnár Z (2007) Comparative aspects of cortical neurogenesis in vertebrates. J Anat 211:164–176
Chomsky N (2007) Of minds and language. Biolinguistics 1:9–27
Chittka L, Rossiter SJ, Skorupski P, Fernando C (2012) What is comparable in comparative cognition? Philos Trans R Soc Lond B 367:2677–2685
Christiansen MH, Chater N (2008) Language as shaped by the brain. Behav Brain Sci 31:489–508; discussion 509–558
Christiansen MH, Chater N (2015) The now-or-never bottleneck: a fundamental constraint on language. Behav Brain Sci 14:1–52
Churchland PS, Churchland PM (2002) Neural worlds and real worlds. Nat Rev Neurosci 3:903–907
Clark DA, Mitra PP, Wang SS-H (2001) Scalable architecture in mammalian brains. Nature 411:189–192
Corballis MC (2015) What's left in language? Beyond the classical model. Ann NY Acad Sci 1359:14–29. doi:10.1111/nyas.12761
Crick F, Koch C (1990) Towards a neurobiological theory of consciousness. Semin Neurosci 2:263–275
Crick F, Koch C (1998) Consciousness and neuroscience. Cereb Cortex 8:97–107
Crick F, Koch C (2003) A framework of consciousness. Nat Neurosci 6:119–126
Da Costa NM, Martin KAC (2010) Whose cortical column would that be? Front Neuroanat 4:16. doi:10.3389/fnana.2010.00016
De Reus MA, Saenger VM, Kahn RS, Van den Heuvel MP (2014) An edge-centric perspective on the human connectome: link communities in the brain. Philos Trans R Soc Lond B 369:20130527. doi:10.1098/rstb.2013.0527
Deacon TW (1990) Rethinking mammalian brain evolution. Am Zool 30:629–705
Deacon TW (1998) The symbolic species: the co-evolution of language and the brain. Norton and Company, New York
Deacon TW (2006) Evolution of language systems in the human brain. In: Kaas J (ed) Evolution of nervous systems. The evolution of primate nervous systems. Academic, NewYork, pp 1–26
DeFelipe J (2015) The anatomical problem posed by brain complexity and size: a potential solution. Front Neuroanat 9:104. doi:10.3389/fnana.2015.00104
Desimone R, Duncan J (1995) Neural mechanisms of selective visual attention. Annu Rev Neurosci 18:193–222
Dicke U, Roth G (2016) Neuronal factors determining high intelligence. Philos Trans R Soc Lond B 371:20150180. doi:10.1098/rstb.2015.0180
Donald M (1991) Origins of the modern mind: three stages in the evolution of culture and cognition. Harvard University Press, Cambridge

Edelman GM, Tononi GA (2000) A universe of consciousness. Basic Books, New York
Fabbro F, Aglioti SM, Bergamasco M, Clarici A, Panksepp J (2015) Evolutionary aspects of self and world consciousness in vertebrates. Front Hum Neurosci 9:157. doi:10.3389/fnhum.2015.00157
Falk D (2004) Prelinguistic evolution in early hominins: whence motherese? Behav Brain Sci 27:491–503
Finlay BL, Darlington DB (1995) Linked regularities in the development and evolution of mammalian brains. Science 268:1578–1584
Finlay BL, Darlington DB, Nicastro N (2001) Developmental structure in brain evolution. Behav Brain Sci 24:263–278
Frahm HD, Stephan H, Stephan M (1982) Comparison of brain structure volumes in Insectivora and primates. Part I Neocortex. J Hirnforsch 23:375–389
Gazzaniga MS, Ivry RB, Mangun GR (2008) Cognitive neurosciences: the biology of the mind, 3rd edn. W.W. Norton, New York
Gillett G, Franz E (2013) John Hughlings Jackson: bridging theory and clinical observation. Lancet 381:528–529
Gómez-Robles A, Hopkins WD, Sherwood CC (2014) Modular structure facilitates mosaic evolution of the brain in chimpanzees and humans. Nat Commun 5:5469. doi:10.1038/ncomms5469
Gould SJ (1976) Grades and clades revisited. In: Masterton RB, Hodos W, Jerison HJ (eds) Evolution, brain and behavior: persistent problems. Erlbaum, Hillsdale, pp 115–122
Greenfield SA, Collins TF (2005) A neuroscientific approach to consciousness. Prog Brain Res 150:11–23
Greenfield SA (1995) Journey to the centers of the mind. Freeman, New York
Hauser MD, Yang C, Berwick RC, Tattersall I, Ryan MJ, Watumull J, Chomsky N, Lewontin RC (2014) The mystery of language evolution. Front Psychol 5:401. doi:10.3389/fpsyg.2014.00401
Herculano-Houzel S (2009) The human brain in numbers: a linearly scaled-up primate brain. Front Hum Neurosci 3:31. doi:10.3389/neuro.09.031.2009
Herculano-Houzel S (2012) Neuronal scaling rules for primate brains: the primate advantage. Prog Brain Res 195:325–340
Herculano-Houzel S, Collins CE, Wong P, Kaas JH, Lent R (2008) The basic nonuniformity of the cerebral cortex. Proc Natl Acad Sci U S A 105:12593–12598
Herculano-Houzel S, Mota B, Wong P, Kaas JH (2010) Connectivity-driven white matter scaling and folding in primate cerebral cortex. Proc Natl Acad Sci U S A 107:19008–19013
Herrmann E, Call J, Hernandez-Lloreda MV, Hare B, Tomasello M (2007) Humans have evolved specialized skills of social cognition: the cultural intelligence hypothesis. Science 317:1360–1366
Hill J, Inder T, Neil J, Dierker D, Harwell J, Van Essen D (2010) Similar patterns of cortical expansion during human development and evolution. Proc Natl Acad Sci U S A 107:13135–13140
Hillert DG (2015) On the evolving biology of language. Front Psychol 6:1796. doi:10.3389/fpsyg.2015.01796
Hofman MA (1988) Size and shape of the cerebral cortex in mammals. Part II. The cortical volume. Brain Behav Evol 32:17–26
Hofman MA (1989) On the evolution and geometry of the brain in mammals. Prog Neurobiol 32:137–158
Hofman MA (2001) Brain evolution in hominids: are we at the end of the road. In: Falk D, Gibson KR (eds) Evolutionary anatomy of the primate cerebral cortex. Cambridge University Press, Cambridge, pp 113–127
Hofman MA (2003) Of brains and minds: a neurobiological treatise on the nature of intelligence. Evol Cogn 9:178–188
Hofman MA (2007) Brain evolution and intelligence in primates. In: Watanabe S, Hofman MA (eds) Integration of comparative neuroanatomy and cognition. Keio University Press, Tokyo, pp 33–53
Hofman MA (2012) Design principles of the human brain: an evolutionary perspective. Prog Brain Res 195:373–390

Hofman MA (2014) Evolution of the human brain: when bigger is better. Front Neuroanat 8:15. doi:10.3389/fnana.2014.00015
Hofman MA (2015) Evolution of the human brain: from matter to mind. In: Goldstein S, Naglieri JA, Princiotta D (eds) Handbook of intelligence: evolutionary theory, historical perspective and current concepts. Springer, New York, pp 65–82
Hofman MA (2016) The fractal geometry of the human brain: an evolutionary perspective. In: Di Ieva A (ed) The fractal geometry of the brain. Springer, New York, pp 169–186
Hofman MA, Falk D (eds) (2012) Evolution of the brain in primates. From neuron to behavior. Elsevier, Amsterdam
Hughlings Jackson J (1884) Croonian lectures on the evolution and dissolution of the nervous system. Lancet 123:555–558
Isler K, Van Schaik CP (2014) How humans evolved large brains: comparative evidence. Evol Anthropol 23:65–75
Jerison HJ (1973) Evolution of the brain and intelligence. Academic, New York
Jerison HJ (1985) Animal intelligence as encephalization. Philos Trans R Soc Lond B 308:21–35
Jerison HJ (1991). Brain size and the evolution of mind. In: The 59th James Arthur lecture on the evolution of the human brain. American Museum of Natural History, New York
Kaas JH (2000) Why is brain size so important: design problems and solutions as neocortex gets bigger or smaller. Brain Mind 1:7–23
Kaas JH (2012) The evolution of neocortex in primates. Prog Brain Res 195:91–102
Krubitzer L (2007) The magnificent compromise: cortical field evolution in mammals. Neuron 56:201–208
Laughlin SB, Sejnowski TJ (2003) Communication in neural networks. Science 301:1870–1874
Lefebvre L (2012) Primate encephalization. Prog Brain Res 195:393–412
Lent R, Azevedo FAC, Andrade-Moraes CH, Pinto AVO (2012) How many neurons do you have? Some dogmas of quantitative neuroscience under revision. Eur J Neurosci 35:1–9
Levinson SC (2016) Turn-taking in human communication—origins and implications for language processing. Trends Cogn Sci 20:6–14
Lewitus E, Kelava I, Kalinka AT, Tomancak P, Huttner WB (2014) An adaptive threshold in mammalian neocortical evolution. PLoS Biol 12:e1002000. doi:10.1371/journal.pbio.1002000
Lieberman P (2015) The evolution of language. In: Goldstein S, Naglieri JA, Princiotta D (eds) Handbook of intelligence: evolutionary theory, historical perspective and current concepts. Springer, New York, pp 47–64
MacLeod C (2012) The missing link: evolution of the primate cerebellum. Prog Brain Res 195:165–187
Macphail EM (1982) Brain and intelligence in vertebrates. Clarendon, Oxford
Macphail EM, Bolhuis JJ (2001) The evolution of intelligence: adaptive specializations *versus* general process. Biol Rev 76:341–364
Mayr E (1982) The growth of biological thought. Diversity, evolution and inheritance. Belknap, Cambridge
Molnár Z, Kaas JH, De Carlos JA, Hevner RF, Lein E, Němec P (2014) Evolution and development of the mammalian cerebral cortex. Brain Behav Evol 83:126–139
Mota B, Herculano-Houzel S (2012) How the cortex gets its folds: an inside-out, connectivity-driven model for the scaling of mammalian cortical folding. Front Neuroanat 6:3. doi:10.3389/fnana.2012.00003
Mountcastle VB (1997) The columnar organization of the brain. Brain 120:701–722
Nieuwenhuys R (1994) The neocortex: an overview of its evolutionary development, structural organization and synaptology. Anat Embryol 190:307–337
Opris I, Casanova MF (2014) Prefrontal cortical minicolumn: from executive control to disrupted cognitive processing. Brain 137:1863–1875
Pagel M, Atkinson QD, Calude AS, Meade A (2013) Ultraconserved words point to deep language ancestry across Eurasia. Proc Natl Acad Sci U S A 110:8471–8476
Panksepp J, Panksepp JB (2000) The seven sins of evolutionary psychology. Evol Cogn 6:108–131

Panksepp J, Moskal JR, Panksepp JB, Kroes RA (2002) Comparative approaches in evolutionary psychology: molecular neuroscience meets the mind. Neuro Endocrinol Lett 4(23):105–115
Papo D, Buldú JM, Boccaletti S, Bullmore ET (2014a) Complex network theory and the brain. Philos Trans R Soc Lond B 369:20130520. doi:10.1098/rstb.2013.0520
Papo D, Zani NM, Pineda-Pardo JA, Boccaletti S, Buldú JM (2014b) Functional brain networks: great expectations, hard times and the big leap forward. Philos Trans R Soc Lond B 369:20130525. doi:10.1098/rstb.2013.0525
Passingham RE (2008) What is special about the human brain? Oxford University Press, New York
Pay RG (1980) Contextual organization of unitary information processes in the cortex by the thalamus and basal ganglia and the central control of attention. Int J Neurosci 11:249–277
Popper KR, Eccles JC (1977) The self and its brain. Springer, New York
Premack D (2004) Is language the key to human intelligence? Science 303:318–320
Premack D (2007) Human and animal cognition: continuity and discontinuity. Proc Natl Acad Sci U S A 104:13861–13867
Preuss TM (2001) The discovery of cerebral diversity: an unwelcome scientific revolution. In: Falk D, Gibson KR (eds) Evolutionary anatomy of the primate cerebral cortex. Cambridge University Press, Cambridge, pp 138–164
Preuss TM (2011) The human brain: rewired and running hot. Ann N Y Acad Sci 1225(Suppl. 1):E182–E191
Rakic P (2007) The radial edifice of cortical architecture: from neuronal silhouettes to genetic engineering. Brain Res Rev 55:204–219
Rakic P (2009) Evolution of the neocortex: a perspective from developmental biology. Nat Rev Neurosci 10:724–735
Razavi MJ, Zhang T, Liu T, Wang X (2015) Cortical folding pattern and its consistency induced by biological growth. Sci Rep 5:14477. doi:10.1038/srep14477.
Reader SM, Hager Y, Laland KN (2011) The evolution of primate general and cultural intelligence. Philo Trans R Soc Lond B 366:1017–1027
Rilling JK (2014) Comparative primate neurobiology and the evolution of brain language systems. Curr Opin Neurobiol 28:10–14
Ringo JL (1991) Neuronal interconnection as a function of brain size. Brain Behav Evol 38:1–6
Ringo JL, Doty RW, Demeter S, Simard PY (1994) Time is of the essence: a conjecture that hemispheric specialization arises from interhemispheric conduction delay. Cereb Cortex 4:331–343
Rockland KS (2010) Five points on columns. Front Neuroanat 4:22. doi:10.3389/fnana.2010.00022.
Ronan L, Fletcher PC (2015) From genes to folds: a review of cortical gyrification theory. Brain Struct Funct 220:2475–2483
Roth G (2013) The long evolution of brains and minds. Springer, New York
Roth G, Dicke U (2012) Evolution of the brain and intelligence in primates. Prog Brain Res 195:413–430
Schmahmann JD (2010) The role of the cerebellum in cognition and emotion: personal reflections since 1982 on the dysmetria of thought hypothesis, and its historical evolution from theory to therapy. Neuropsychol Rev 20:236–260
Schoenemann PT (2006) Evolution of the size and functional areas of the human brain. Ann Rev Anthropol 35:379–406
Schoenemann PT, Sheehan MJ, Glotzer ID (2005) Prefrontal white matter volume is disproportionately larger in humans than in other primates. Nat Neurosci 8:242–252
Schoenemann, P.T. (2012) Evolution of brain and language. Prog Brain Res 195:443–459
Sherwood CC, Bauernfeind AL, Bianchi S, Raghanti MA, Hof PR (2012) Human brain evolution writ large and small. Prog Brain Res 195:237–254
Sherwood CC, Subiaul F, Zawidzki TW (2008) A natural history of the human mind: tracing evolutionary changes in brain and cognition. J Anat 212:426–454
Shettleworth SJ (2012a) Fundamentals of comparative cognition. Oxford University Press, New York

Shettleworth SJ (2012b) Modularity, comparative cognition and human uniqueness. Philos Trans R Soc Lond B 367:2794–2801
Smaers JB, Soligo C (2013) Brain reorganization, not relative brain size, primarily characterizes anthropoid brain evolution. Proc R Soc Lond Ser B 280:20130269. doi:10.1098/rspb.2013.0269
Smaers JB, Dechmann DK, Goswami A, Soligo C, Safi K (2012) Comparative analyses of evolutionary rates reveal different pathways to encephalization in bats, carnivorans, and primates. Proc Natl Acad Sci U S A 109:18006–18011
Smaers JB, Schleicher A, Zilles K, Vinicius L (2010) Frontal white matter volume in anthropoid primates. PLoS One 5:e9123. doi:10.1371/journal.pone.0009123
Sporns O, Betzel RF (2016) Modular brain networks. Annu Rev Psychol 67:19.1–19.28
Sporns O, Chilavo DR, Kaiser M, Hilgetag CC (2004) Organization, development and function of complex brain networks. Trends Cogn Sci 8:418–425
Sporns O, Honey CJ, Kötter R (2007) Identification and classification of hubs in brain networks. PLoS One 2:e1049. doi:10.1371/journal.pone.0001049
Stephan H, Frahm HD, Baron G (1981) New and revised data on volumes of brain structures in insectivores and primates. Folia Primatol 35:1–29
Striedter GF (2005) Principles of brain evolution. Sinauer, Sunderland
Striedter GF, Srinivasan S, Monuki ES (2015) Cortical folding: when, where, how, and why? Annu Rev Neurosci 38:291–307
Suárez R, Gobius I, Richards LJ (2014) Evolution and development of interhemispheric connections in the vertebrate forebrain. Front Hum Neurosci 8:497. doi:10.3389/fnhum.2014.00497.
Tononi G, Koch C (2015) Consciousness: here, there and everywhere? Philos Trans R Soc Lond B 370, pii: 20140167. doi:10.1098/rstb.2014.0167
Van den Heuvel MP, Sporns O (2011) Rich-club organization of the human connectome. J Neurosci 31:15775–15786
Van den Heuvel MP, Bullmore ET, Sporns O (2016) Comparative connectomes. Trends Cogn Sci 20:345–361
Wang D, Liu H (2014) Functional connectivity architecture of the human brain: not all the same. Neuroscientist 20:432–438
Wang SS-H, Shultz JR, Burish MJ, Harrison KH, Hof PR, Towns LC, Wagers MW, Wyatt KD (2008) Functional trade-offs in white matter axonal scaling. J Neurosci 28:4047–4056
Wang X-J (2010) Neurophysiological and computational principles of cortical rhythms in cognition. Physiol Rev 90:1195–1268
Wedeen VJ, Rosene DL, Wang R, Dai G, Mortazavi F, Hagmann P, Kaas JH, Tseng WY (2012) The geometric structure of the brain fiber pathway. Science 335:1628–1638
Welker W (1990) Why does cerebral cortex fissure and fold? A review of determinants of gyri and sulci. In: Jones EG, Peters A (eds) Cerebral cortex, vol 8B. Plenum, New York, pp 23–136
Wen Q, Chklovskii DB (2005) Segregation of the brain into gray and white matter: a design minimizing conduction delays. PLoS Comput Biol 1:617–630
Young MP (1993) The organization of neural systems in the primate cerebral cortex. Proc R Soc Lond B 252:13–18
Zeman A (2001) Consciousness. Brain 124:1263–1289
Zeman A (2005) What in the world is consciousness? Prog Brain Res 150:1–10
Zhang K, Sejnowski TJ (2000) A universal scaling law between gray matter and white matter of cerebral cortex. Proc Natl Acad Sci U S A 97:5621–5626

Chapter 8
Behavioural, Cognitive and Neuronal Changes in the Acquisition of Tool Use

Yumiko Yamazaki and Atsushi Iriki

Abstract Tool use is observed and shared in animals with and without phylogenetic relationships. Therefore, tool use is an excellent behavioural model to explore dynamic relationships between animals' physical and cognitive abilities and their environmental conditions. In this chapter, we will focus on tool use in primates, especially Japanese macaques and common marmosets, which have not been observed to use tools in the wild but can be trained to do so through appropriate behavioural training. This approach enables us to determine the conditions that are needed for acquisition, as well as the types of cognitive and neuronal characteristics that can be observed during and after the development of tool use. Several studies of Japanese macaques have indicated that they can be trained to use tools within a few weeks, suggesting that the behavioural and cognitive components for tool use were already present before training. The acquisition of tool use is supported by structural changes in several brain regions throughout training. Furthermore, through careful, step-by-step training, the use of tools to obtain visual cues, such as an endoscope, was successfully established for the first time in macaques. Japanese macaques exhibit different characteristics in the acquisition of tool use than common marmosets as marmosets required much longer periods of time, and they worked mainly according to their own motivation to use tools. These differences were reflected in structural changes in the brain. In the discussion, we will compare the tool use behaviours of primates and other species, particularly birds, to explore the possible physical, cognitive and environmental conditions for exhibiting tool use from a comparative perspective.

Keywords Body image • Brain structural change • Common marmoset • Japanese macaque • Motor tool • Sensory tool • Tool use • VBM

Y. Yamazaki (✉)
Advanced Research Centres, Keio University,
35 Shinanomachi, Shinjuku-ku, Tokyo 160-8582, Japan

Laboratory for Symbolic Cognitive Development, RIKEN BSI,
2-1 Hirosawa, Wako-shi, Saitama 351-0198, Japan
e-mail: yumyam@brain.riken.jp

A. Iriki
Laboratory for Symbolic Cognitive Development, RIKEN BSI,
2-1 Hirosawa, Wako-shi, Saitama 351-0198, Japan

S. Watanabe et al. (eds.), *Evolution of the Brain, Cognition, and Emotion in Vertebrates*, Brain Science, DOI 10.1007/978-4-431-56559-8_8

8.1 In What do Animals Use the Tools for?

"To use a tool is to incorporate the tool into a part of one's own body, like transferring one's thoughts from the brain to the hands. For us, the tool is the tip of our body".
Tsunekazu Nishioka
(Director of carpenters specializing in shrines and temples in Japan, 1988)

Numerous studies have documented tool use in animals, clearly illustrating that tool use is not limited to certain animal taxa (e.g. Beck 1980). Why only these animals, and not all animals, use tools is a puzzling question. To say that the environment drives them to use tools is too simple to be the answer because similar conditions do not cause all animals living in a specific environment to use tools (Baber 2003) nor can cognitive characteristics explain this issue because non-tool users in the wild can learn to use tools very quickly with training (Hihara et al. 2003a). Environmental adaptation and cognitive ability are necessary conditions to manifest tool use, but there might be other necessary conditions that are not sufficient.

Researches of tool use in non-human animals started with those of wild apes (e.g. Goodall 1964). Since then, many non-human tool users use tools for foraging (e.g. Malaivijitnond et al. 2007; Moura and Lee 2004; Inoue-Nakamura and Matsuzawa 1997), while others use tools for grooming or hygiene (e.g. Watanabe et al. 2007; Pansini and de Ruiter 2011; Deecke 2012; Hart et al. 2001) and for displays (McGrew 2013). Humans also use tools for other purposes (e.g. playing, writing, sewing, making tools and killing). Beck (1980) selected four functions of tool use by animals. First, animals use tools, such as rakes, to their reach. Second, animals use tools, such as stone tools and hammers, to amplify mechanical forces. Third, animals use tools to augment agonistic displays, for example, dropping branches. Fourth, animals use tools, such as sponges, to allow for more effective control of fluids.

Asano (1994) proposed three "classes" of tools that function in each of the following components of behaviour: antecedent stimuli, behaviour and consequences. The first is the "expansion of discriminative stimulus", which facilitates the acquisition and maintenance of the control of stimuli gained by the tools, for example, telescopes and clocks. The second class is "expansion of response topography", which expands the natural effects of motor organs, such as hammers and cars. The third class is the "expansion of reinforcing stimulus" and includes, for example, painkillers and money, which strengthen the effects of consequences.

According to this categorization, the rich variety of examples raised by Beck (1980) primarily fall into the second class of tools, i.e. "expansion of response topography". Thus, the questions arises, do non-human animals also use tools for expansion of discriminative and reinforcing stimuli? Are the functions of tool use essentially different between non-human animals and humans or are they just differences in volume and variations that share an original purpose? In this chapter, we attempt to search for some of the possible factors that enable animals to use tools by focusing on several examples of artificially acquired tool use in two primate species, namely, Japanese macaques and common marmosets. We would like to compare these species because of interesting ecological and physiological features. As we described below, both species are non-tool users in the wild, so they are suitable models for studying learning abilities of novel behaviour like tool use. Additionally, as primate species, they are much different in terms of hand dexterity, which is deeply related to the presence (macaques) and absence (marmosets) of cortico-motoneuronal system (Lemon 2008). This clear feature enables us to see how tool use is learned with different physiological backgrounds.

8.2 Tool-Use Learning in Japanese Macaques

8.2.1 Learning to Use Tools

Recent findings suggest that wild macaques can use tools, as exemplified by stone tool use in Thailand (Malaivijitnond et al. 2007; Gumert and Malaivijitnond 2013) and frothing with human hair by semi-captive macaques in the temples of Thailand (Watanabe et al. 2007) and frothing with their own hair by Japanese macaques in Arashiyama (Leca et al. 2010). In these cases, these animals use tools both to amplify mechanical forces and to extend the user's reach in a narrow space (Beck 1980).

Although reports of tool use by Japanese macaques are not abundant, these animals can be trained to use tools without difficulty. By training Japanese macaques to use rake-shaped tools, we were able to determine how they learned to use rakes and to identify the neurological processes that occur during training (e.g. Iriki et al. 1996; Hihara et al. 2006). In experimental situations, the monkeys sit in chairs and are passed a rake. The experimenters control the locations of the reinforcers (food items) on the table, beginning with a position that is slightly beyond their reach. After several hundreds of trials, the monkeys gradually learn to grasp and operate the shaft of the rake and finally become skilful in retrieving the food items from anywhere on the table. This training usually takes less than 2 weeks (e.g. Hihara et al. 2003a; Quallo et al. 2009; Yamazaki et al. 2009); thus, rake use might be a relatively easy task for Japanese macaques to acquire, although there are no reports of raking or tool use to reach food items in wild populations.

8.2.2 Changes in Neural Activity Due to the Acquisition of Tool Use

Because tool use behaviour is completely novel for Japanese macaques, the acquisition of such behaviour must involve changes in neural activity and/or the rewiring of neural connectivity. If neural activity is indeed changed, there must be neural groups that are in charge of the coding of the rake itself or the rake's movement. By recording neural activity in the somatosensory region, Iriki et al. (1996) found "bimodal" neurons that code both tactile and visual stimulation of the hand. These "bimodal" neurons, located in the intraparietal sulcus (IPS), are activated when the monkey's hand is touched by the experimenter's hand or is pointed at with something such as a laser marker. When the monkeys learned to use the rake to retrieve food from the table, the visual receptive fields of the bimodal neurons expanded to the whole rake area, including both the rake shaft and their original hands. Another type of neurons in the same intraparietal region responded to the movement of the shoulder joint and to the visual marker corresponding to the point on the table that the hand could reach. When the monkey uses the tool, this type of neuron expands its response area to include areas that are accessible with the rake. These neurons change their original receptive fields only when they are actively engaged in the raking task. The receptive fields are unchanged when the animals hold the rake passively. Thus, rake-use training reprogrammed the body image of the monkey to incorporate the rake as a body part (Maravita and Iriki 2004).

8.2.3 Changes in Brain Structure Due to Tool Use

The changes in behaviour and neural activity in tool-trained Japanese macaques are supported at the genetic, axonal and structural levels of the brain. Increased expression of brain-derived neurotrophic factor and its receptor *trk*B (Ishibashi et al. 2002) was observed in the same region in which the bimodal neurons were recorded (Iriki et al. 1996). Tracer labelling has revealed novel projections from the temporo-parietal junction and the ventrolateral-prefrontal areas to the intraparietal in tool-trained monkeys (Hihara et al. 2006). These findings strongly suggest that rake-use training induces the reorganization of brain structures.

Using voxel-based morphometry (VBM), Quallo et al. (2009) compared the volumes of brain structures before, during and after rake-use training in Japanese macaques. Two out of three subjects exhibited increases in grey matter volume of more than 10% in the secondary somatosensory area (SII), intraparietal sulcus (IPS) and superior temporal sulcus (STS), as wells as increased white matter volume in the cerebellar cortex. However, the volumes of these regions were the highest at the midpoint of the training period and decreased to their original levels at the end of the training. Nevertheless, tool-use performance scores remained steady for more than 10 days after the end of the training, suggesting that these changes in brain volume are distinct from whatever changes underlie the maintenance of tool use. Thus, these brain regions increased in volume only when the monkeys were acquiring the novel behaviour through intensive training to retrieve food items through over 20,000 trials (Quallo et al. 2009).

8.2.4 Acquisition of Sensory Tools Via Motor Tools

Japanese macaques were fully competent at using rakes after 2 weeks of training. After training, they were so skilful that they were able to retrieve food items from anywhere on the table. The monkeys also demonstrated the spontaneous use of a tool to retrieve another long tool when the food item was located beyond the reach of the original rake (Hihara et al. 2003a). These abilities were supported by neurological plasticity, including newly connected projections (Hihara et al. 2006), gene expression (Ishibashi et al. 2002) and structural changes in specific regions (Quallo et al. 2009).

Referring again to the definition of tools according to their function (Asano 1994), the tool-use behaviours acquired by Japanese macaques fall under the "expansion of the response topography": The monkeys expand the accessible area of their original motor organs, e.g. their arms and hands, by using rakes. This type of "motor tool" comprises the majority of examples of tool use by non-human animals (Beck 1980). The above studies clearly indicate that Japanese macaques are potentially capable of using this type of motor tool when properly trained.

Asano (1994) also proposed the use of tools to expand discriminative stimuli. These types of tool can be considered "sensory tools" because they serve as substitutes for sensory organs (e.g. the eyes and ears), for example, binoculars and stethoscopes. Mirrors are among the most popular and classical examples of tools that can be used to acquire visual stimuli that cannot otherwise be obtained. Some studies have reported that captive animals can use mirrors to acquire discriminative cues about their own bodies (Prior et al. 2008; de Veer et al. 2003; Plotnik et al. 2006), but such examples of sensory tool use are rare in the wild. This fact suggests that some animals are competent in the use of sensory tools but that this ability does not emerge in the wild. Gaps likely exist in the levels of necessity and motivation between motor and sensory tool use, thus raising the question: Is it possible to bridge these gaps, i.e., is it possible to create sensory tools based on motor tools?

We addressed this question by training naïve Japanese macaques to use a special rake that functions like an endoscope (Yamazaki et al. 2009). The final goal of the training was to show that monkeys can use tools to acquire visual information to guide their behaviours under conditions in which neither their hands nor the tool is seen directly, much like the way a doctor uses an endoscope to search for lesions in the stomach. To this end, a special training protocol was invented to transfer motor tool use to sensory tool use (Fig. 8.1). First, the monkeys were trained to use the standard rake to retrieve food on the table as shown in previous studies (e.g. Iriki et al. 1996). After the appropriate motor function was acquired, the training protocol then focused on separating the visual information obtained from the monkeys' eyes to that obtained from a secondary object. To obtain visual information about a food item that was hidden behind a bump on the table, we employed a tool similar to a dental mirror to capture the image of the food item behind the bump. At this point, the rake has both sensory and motor functions. Next, several types of mirrors, including standing mirrors, manual mirrors, remotely controlled mirrors and a TV monitor, were introduced to separate the sensory and motor functions. Finally, the bump was removed from the table, which was instead covered with an opaque screen to prevent the monkeys from directly seeing anything on the table. The tool used in this training step was a "camera rake": the shaft of the rake contained a small camera capable of capturing the image in front of the transparent rake tip. The monkeys were asked to retrieve the food item under the screen by looking at a monitor on which a video image from the camera rake was displayed. During the initial use of the camera rake,

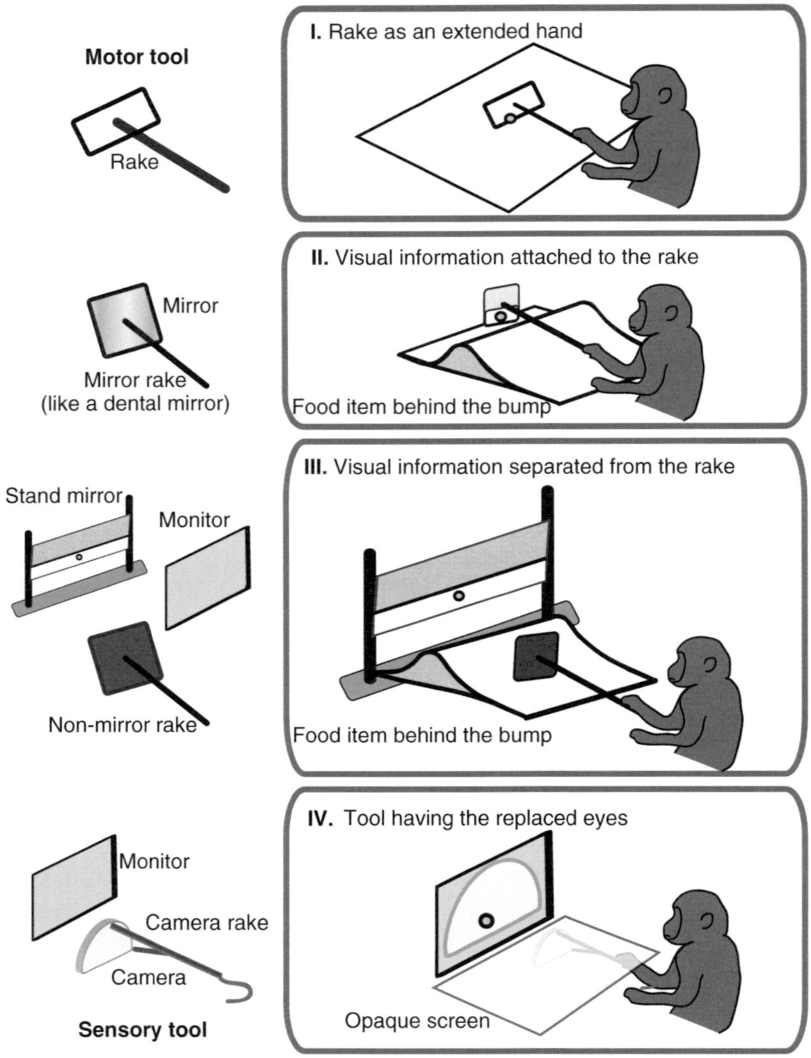

Fig. 8.1 Sensory tool training protocol for Japanese macaques. The four training phases differed in terms of the materials used. I. Rake training: the animals acquire the motor tool, and at this point, the tool is thought to become an extension of the hands. II. Mirror rake training: the subject is trained to use the rake to find and subsequently retrieve the food item behind the bump. III. Mirror and monitor use: the animals use mirrors and monitors to acquire visual information and use the rake for motor purposes. In this phase, the sensory and motor functions are separated. IV. Camera rake training: the monkey is required to search for the food item under the opaque screen by moving the camera rake and looking at the monitor that displays the image from the camera rake. The camera rake is considered a sensory tool because the animals use the rake as a substitute for their eyes. Modified from Yamazaki et al. (2009)

the monkeys made many vertical movements of the rake as though they were randomly and blindly pushing and pulling the rake. However, the monkeys eventually became capable of searching with the rake using a minimum number of circular movements to scan the entire area under the screen, as indicated by the tracking data. Thus, the monkeys successfully mastered sensory tool use to acquire a visual image that was otherwise not accessible with their eyes but that was accessible through the rake, enabling them to access a food item located beyond their reach.

Over the course of training with the sensory tool, we employed a systematic and small-step method to shape the monkeys' behaviour. The sensory tool training protocols involved two major challenges. The first was the dental mirror that was initially used to provide the monkeys with indirect visual information. The second was the camera rake, which required the monkeys to integrate information from their own movement of the rake with the visual information provided by the tip of the camera rake. Thus, these two training steps obviously required the monkeys to employ novel cognitive abilities that would not otherwise be called upon in the wild.

Humans reportedly started using mirrors approximately 8000 years ago (Enoch 2006), making mirrors the first sensory tool in human history. At a certain point, mirrors are developed into a tool to see what could not otherwise be seen. Although we do not know how early humans came to use mirrors for multiple purposes, the successful acquisition by Japanese macaques of the use of sensory tools through the use of motor tools highlights a potential route for the evolution of sensory tools in human history.

8.3 Tool-Use Learning in Common Marmosets

8.3.1 Learning to Use Tools

Given that laboratory macaques can be trained to use tools within 2 weeks and that non-laboratory macaques use several tools, a set of abilities for tool use, including both neural and cognitive abilities, is likely to be already present in these monkeys. Thus, we have several examples of tool use by old-world monkeys. Among new-world monkeys, capuchins are well known as stone tool users; however, tool use is not common among new world monkeys. Are non-tool-using new-world monkeys able to use tools if they are properly trained, similar to Japanese macaques? We addressed this question using common marmosets (*Callithrix jacchus*) as subjects.

Common marmosets have served as a "biomedical supermodel" following the successful generation of transgenic animals (Sasaki et al. 2009). The recent development of genetic manipulation techniques has enabled us to identify the genetic background of diseases and effective drugs, which is not possible in rodent models (Sasaki 2015). Thus, through the use of common marmosets, it will be possible in the future to identify genetic, neural and evolutionary factors related to tool use.

The first step in this direction was to establish a protocol for tool-use training in common marmosets, as there are no reports of these monkeys using any types of tools (Yamazaki et al. 2011). The training protocols differed between macaques and marmosets. Whereas macaques can use a precise grip to grasp the shaft of a rake without difficulty, this is not always the case with marmosets. Thus, a handle was added to the end of the shaft to make it easy for the marmosets to move the rake. Moreover, the training protocol for common marmosets consisted of small training steps that were not always necessary for the Japanese macaques (Hihara et al. 2003a; Quallo et al. 2009). Furthermore, whereas Japanese macaques must be hungry before the training session, this is not always the case with the marmosets. The training protocol was divided into four stages that included several small steps (Fig. 8.2). In the earlier stages (Stages 1 and 2), the marmosets only pulled the rake to the side. However, in the later stages (Stages 3 and 4), they were required to push and then pull the rake, and this process required a completely different motor control than that required in the earlier stages. After approximately 10,000 trials, which required nearly a year in total, the marmosets acquired tool use; i.e. they were able to retrieve the food item even when it was located behind the tool tip (Stage 4, step 3 in Fig. 8.2).

8.3.2 Changes in Brain Structure Through Tool-Use Learning

Among the differences between these two species, the clearest difference was the time required to master tool use. Whereas the Japanese macaques took approximately 2 weeks to become fully capable of retrieving the food item on the table using the rake, it took more than a year for the common marmosets to achieve a similar level of skill (Yamazaki et al. 2011). In addition, when they became capable of using the rake, the Japanese macaques tended to exhibit consistent hand use regardless of the food location. However, the common marmosets exhibited bilateral hand use even after mastering the majority of the task. The marmosets chose to use either hand depending on the position of the food item on the table. They consistently preferred adduction when retrieving the food item; i.e. when the food item was on the left side of the tool shaft, they used the right hand and vice versa. This fact reasonably corresponds to the anatomical difference in the cortico-spinal projection neurons between old- and new-world monkeys, with the exception of capuchins (Lemon 2008).

The acquisition of tool use by common marmosets is different from that of Japanese macaques; thus, are the brain regions employed for tool-use behaviour also different? When VBM was conducted to track changes in brain structures before, during and after tool-use training in common marmosets (Yamazaki et al. 2016), some of the regions of structural change were the same as those observed in the Japanese macaques, but some quite unique regions were observed in only the common marmosets. In the grey matter, volume changes in the nucleus accumbens

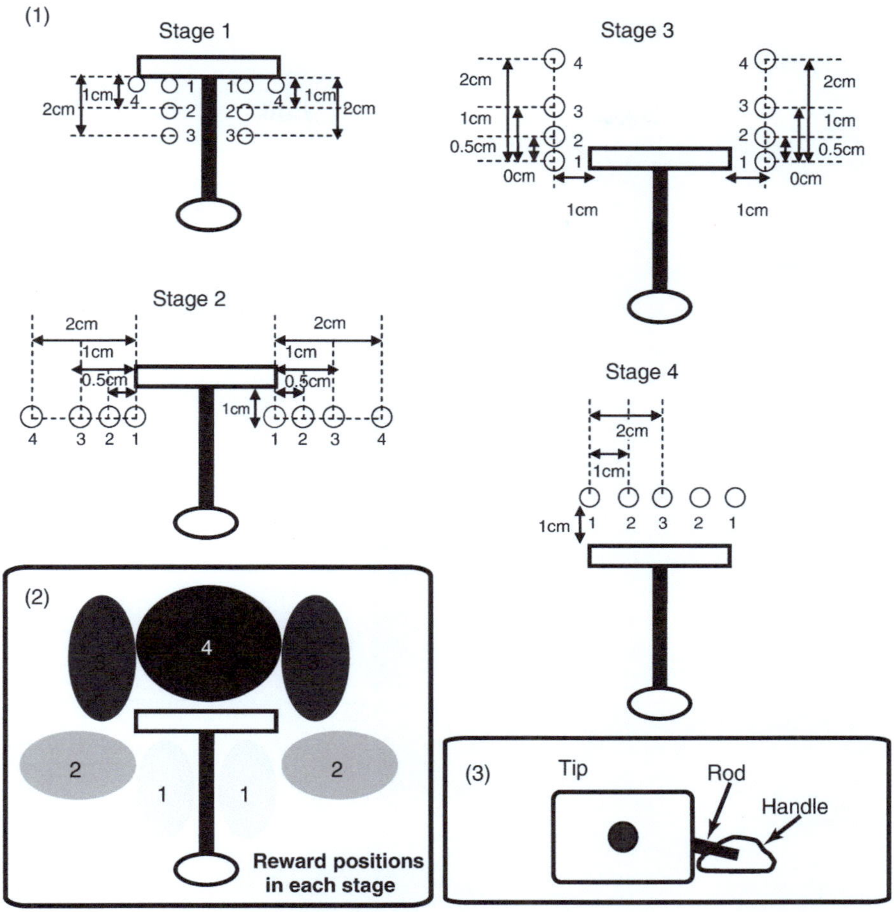

Fig. 8.2 Tool-use training of common marmosets. (1) Top view of the detailed positions of the rake (with a rectangular tip and a *black shaft*) and the food item (the small circles around the rake) in the four training phases (Stages 1–4). (2) Top view of the spatial relationships between the rake and the food item. The numbers inside each ellipse correspond to the training phases. (3) Illustration of the rake used by the common marmosets. A handle is located at the end of the shaft. Modified from Yamazaki et al. (2011)

(Acb, Fig. 8.3a) and third visual area (V3, Fig. 8.3b) were observed only during the training period. Although the increase in the Acb was prominent in the later phase of the training, the increase in V3 was consistently observed throughout training. In the white matter, the volumes of the anterior trunk of the corpus callosum (CC), the middle and inferior cerebellar peduncles (MCP and ICP) and the superior temporal sulcus (STS) increased. Interestingly, with the exception of the MCP, these white matter increases were greater in the early phase of training than in the later phase.

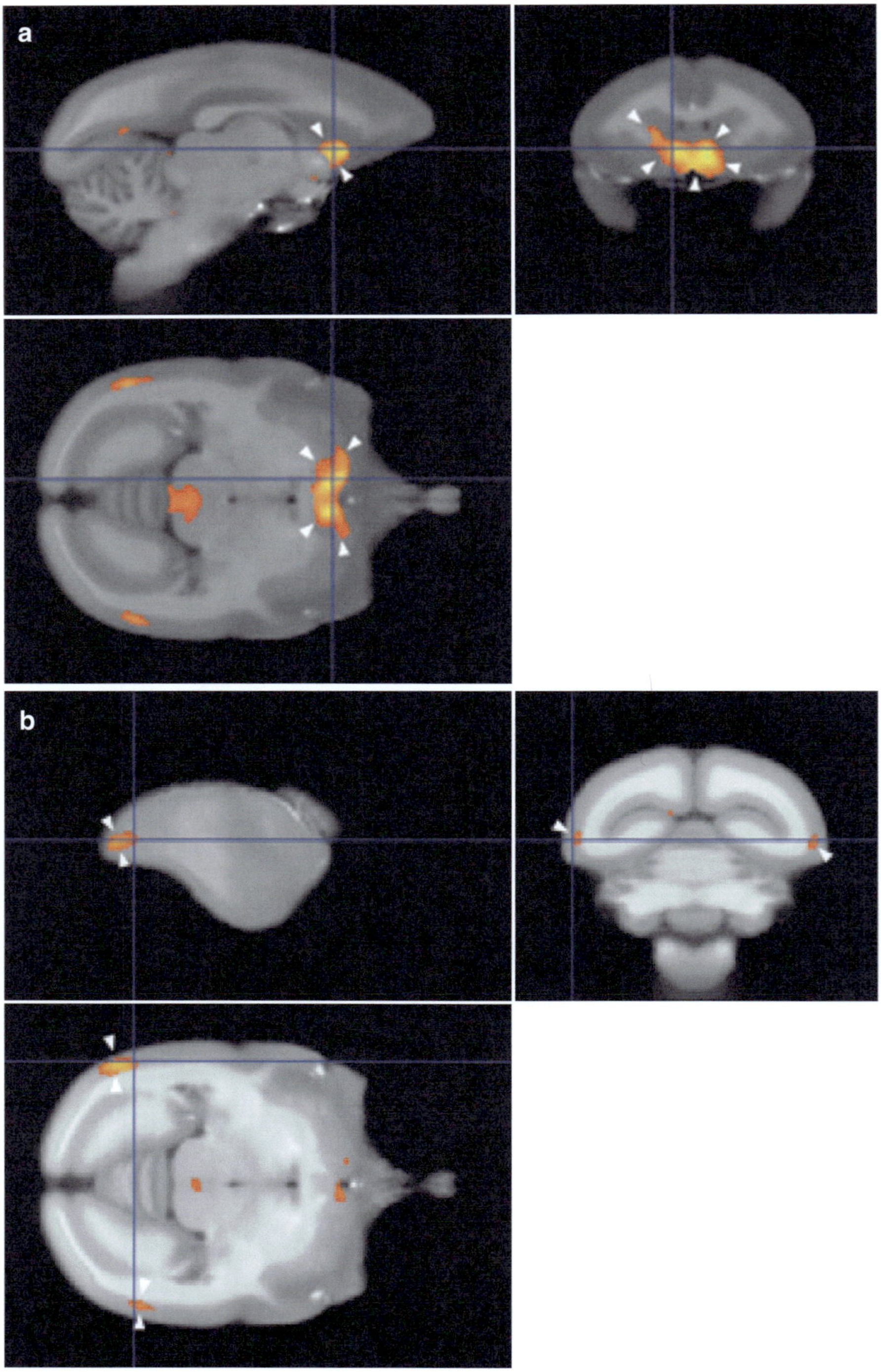

Fig. 8.3 Gray matter increases in (a) bilateral Acb and (b) bilateral lateral extrastriate cortex (V2/V3) observed only during the tool use training periods. Areas with significant increase are pointed by small triangles. Modified from Yamazaki et al. (2016)

8.3.3 How Does Tool-Use Learning Differ Between Marmosets and Macaques?

In addition to the behavioural processes of the acquisition of tool use, the differences in the altered grey and white matter regions of the brain were quite substantial when we compared Japanese macaques and common marmosets. None of the altered grey matter regions in the common marmosets (Acb and V3) corresponded to those of the macaques (STS, IPS and SII). The changes in the CC of the common marmosets were not observed in the macaques. Additionally, after the training was halted, the volumes of the changed regions decreased in the Japanese macaques, whereas these volumes were maintained in the common marmosets, although tool-use performance was maintained in both species.

What characteristics produced these differences in the two species? Overall, the difference in the length of the training period was quite prominent; approximately 2 weeks were required by the macaques, compared with 1 year for the marmosets. Not only length of the training period but also the density of training trials per day was quite different. Whereas the macaques completed several hundreds to thousands of trials per day, the marmosets completed only 30–40 trials, and sometimes they refused to work at all. This difference was the result of the food access not being limited in the case of the marmosets. In other words, the marmosets completed the task not because they were hungry but because they might have been interested in the task and the additional food resource (sweets). In contrast, it was quite difficult to make the Japanese macaques participate when they were full. Difference in hand morphology should also have contributed to the difference in learning speed. While Japanese macaques have opposable thumb and use precision grip when grasping the tool shaft, marmosets have digits with claws acting in union (Ankel-Simons 2007) and had difficulty in handling the tool shaft so that we provided a handle at the end of the shaft. The marmosets sometimes tried hard even to pick up a small piece of food by their hands when they succeeded to retrieve it.

What was the motivation for the common marmosets to participate in the task? The marmosets love eating sweetened food items, so the reinforcer (i.e. small pieces of sweetened jelly) itself obviously motivated them to complete the task. However, the attraction of a favourite food does not explain the sustained motivation for participating in the training by the common marmosets. The training protocol for the marmosets was divided into four stages, with several minor steps in each stage, and the task became increasingly difficult as they mastered the steps. Occasionally, they required many attempts before successfully retrieving the food item, particularly in the later phases of training. In these phases, the reinforcement rate per minute was lower than that in the earlier phases.

One possible factor is that the study employed only female subjects to exclude sex differences in the volumes of the brain regions. In a study involving a foraging situation (Yamamoto 2004), females exhibited faster response times than males when a familiar food was placed in an unfamiliar container, and the authors suggested that energy intake was not the main reason for the observed sex difference. Despite limited evidence that female marmosets are more motivated to forage, the results of the marmoset VBM study may be related to the original disposition of the female marmosets. If this is the case, the increased volume of the Acb in common marmosets, which was prominent in the later phases of training, is consistent with previous results reporting the activation of the Acb by tasks that require effort to obtain a reward (Salamone and Correa 2012) and when subjects perform a task skilfully (Lutz et al. 2012). Such self-driving motivation is referred to as "intrinsic motivation" in humans (Kage 1994).

The postures assumed by the subjects when engaging in tool use were also different between the two species. The macaques sat in a chair and retrieved the food items in the upright position. The marmosets were in a quadrupedal position when they used the rake to retrieve the food item from the table. Whereas the macaques maintained a constant viewpoint from the chair, the marmosets moved along the table edge and had different viewpoints depending on the positions of the food items. These visual characteristics likely affected the increase in the volume of the visual areas that were only observed in the common marmosets.

8.4 Changes in Brain Structures and Their Functions in Relation to Tool-Use Learning

The successful acquisition of tool use by Japanese macaques and common marmosets under laboratory conditions clearly indicates that tool use, in this case motor tool use, is not restricted to species that have a natural tendency to use tools in the wild. The brain structures of the two species exhibited plasticity in response to novel object use that aided the incorporation of the object into their bodies. The groups of structural changes indicate that two different processes occurred during tool-use learning.

One group of structural changes corresponded to the novel sensorimotor learning and the reconstruction of the brain network during the incorporation of tool into their own bodies. The IPS, STS, SII, CC, V3 and cerebellum belong to this group. This group is obviously important for learning new skills, and the IPS has been confirmed to be involved in sensorimotor plasticity in humans (Draganski et al. 2004). Specifically, the changes in the IPS that have been confirmed to be involved in tool-use acquisition in electrophysiological, fMRI and neuroanatomical studies (Iriki et al. 1996; Obayashi et al. 2001; Hihara et al. 2006; Ishibashi et al. 2002) are deeply related to novel skill acquisition within a relatively short time.

The other group corresponds to the motivational aspect of the sustained participation in the training. The Acb is the only structure in this group. To date, there are no reports of structural changes in the volume of the Acb due to behavioural experience. The fact that the Acb change was observed only in the marmosets and not in Japanese macaques or human subjects suggests that long-term participation in the training and the associated gradual behavioural changes are supported by Acb activation. Thus, increased activation of the Acb may be responsible for sustaining the motivation to use tools despite many failures.

The implication of the latter group of structural changes is that this type of brain change is necessary for learning to use tools that typically take years to master, regardless of whether the subjects are humans or non-human animals. As we know from examples of human tool use, e.g. pencils and chopsticks, learning is a slow process that is gradually developed and reinforced by the consequences of tool use and by the increasingly skilful use of the tool. Within this circularity, tool use is developed and refined. The evidence from common marmosets sheds light on the hidden process of how tool use is maintained, and this maintenance resulted in the further evolution of tool use.

8.5 Tool Use and Cognitive Evolution

Evidence of tool use in populations of captive primates suggests that the potential ability for tool use, particularly motor tool use, is widespread among animals. After a comprehensive review of tool use by various animals (Beck 1980), we observed many new lines of evidence in wild animals. Observations of New Caledonian crows, for example, provide surprising evidence in the sense that they not only use tools but also make tools themselves (e.g. Hunt 1996). These crows become skilful in making tools over the course of development and experience (Hunt and Gray 2004). Moreover, tool shapes are thought to evolve (Hunt 2000). Thus, the question arises as to how such evolved tool use affects the cognitive abilities of crows. Whether the cognitive ability to understand physical relations is a prerequisite for tool use or whether the emergence of tool use drove the general cognitive abilities of tool-using animals is sometimes a difficult and puzzling question. However, excellent tool use skills do not correspond to excellent performance in cognitive tasks in other physical domains (Povinelli 2000; Taylor et al. 2014; Jelbert et al. 2015).

Considering the evolution of cognition in relation to tool use, one important observation from New Caledonian crows is that they use tools for non-foraging, explorative purposes (Wimpenny et al. 2011). In these experiments, crows spontaneously brought stick tools with their beaks to explore novel objects that were potentially harmful or dangerous. The crows may use stick tools as extended versions of their beaks, as suggested by the neural activities of Japanese macaques (Iriki et al. 1996). However, the crows differentiate between the stick and their actual beak because they can avoid dangerous objects by maintaining distance between these objects and their beaks. Such multifunctional tool use has been observed in chimpanzees (McGrew 2013). Archaeological evidence suggests that prehistoric humans frequently used their manufactured tools for secondary purposes (Saito 2011). All of these species can manufacture tools. Originally, these tools were manufactured for specific purposes, but after a while, the tools came to serve a variety of purposes.

Studies of tool manufacturing suggest that tool-using species influence their environments through the use of tools and that they acquire feedback from the environment, with which they modify the tools for better use. This circular process allows both the tool and the underlying cognitive ability to evolve in a specific direction. However, during the modification of the original tool, the animals may find another purpose for the tool, and they may develop another process that can broaden the environments to be modified. At this point, tool use becomes context-free and can be associated with anything far removed from its original purpose. Because the number of animals that manufacture tools is limited (Beck 1980), tool manufacture must represent a substantial challenge for animals in general. However, when animals acquire the technology, unexpected interactions between the tool and the environment emerge and lead to additional paths for tool evolution.

Neither the Japanese macaques nor the common marmosets exhibited tool manufacture or differential tool use during the course of the experimental sessions. However, the tool-trained monkeys spontaneously differentiated vocalizations depending on the different conditions of the experiments and exhibited sequential raking to acquire an appropriate tool to retrieve the food item (Hihara et al. 2003a, b) and showed novel tool use for getting sensory information like endoscope (Yamazaki et al. 2009). These observations represent the first step of the evolution of cognitive ability via tool use. The altered brain areas (Quallo et al. 2009) and newly connected networks (Hihara et al. 2006) support such novel behaviour and its further evolution.

However, in the case of common marmosets, such novel behaviour has not yet been observed. Rather, marmosets exhibit structural changes in brain areas that are known to be related to reward (Yamazaki et al. 2016) in addition to areas that are related to sensorimotor learning. Thus, the results suggest that sustained tool-use training had an emotional aspect, which may include the experience of pleasure (Berridge and Kringelbach 2015). The fact that sustained tool use was supported by positive emotions contributes to the primary step for the maintenance of tool use.

Schaik et al. (1999) argued that several conditions, such as extractive foraging, dexterous manipulation, intelligence related to imitation and insight, would be the keys for tool use to emerge. These conditions explain well about the absence of tool use in Japanese macaques and common marmosets in different levels. However, Japanese macaques showed emergence of novel cognitive abilities after tool use, and common marmosets showed motivational change during the extensive tool learning, which would propose another hidden sets of conditions for tool use. Such new perspectives of cognitive abilities related to tool use cannot be found without artificial training on primate species which originally do not use tool in the wild.

8.6 Conclusion

We have reviewed the artificial training projects that have been employed to teach two different types of monkeys to use tools, and these species exhibited surprising abilities that otherwise cannot be observed in wild situations. However, there are huge gaps between these monkeys and tool-manufacturing animals, such as chimpanzees and New Caledonian crows, in the sense that the former species use tools when they are shaped to do so, whereas the latter voluntarily make tools to manipulate their environment. It remains unknown why Japanese macaques seldom use tools even though they acquire rake skills easily when taught.

Why animals use tools, how do they begin using them, and how they manufacture tools are all difficult questions to address, and it is difficult to empirically reconstruct the developmental paths. However, training animals to use tools can be used to approximately reconstruct the evolution of tool use. Tracking the changes in brain structure, neural activities and networks and observing the drastic behavioural changes in the subjects will provide an effective framework to reconstruct tool-use behaviour from a comparative perspective.

Acknowledgements The authors wish to acknowledge Dr. Watanabe in Keio University for the support and advice of the studies presented here.

References

Ankel-Simons F (2007) Chapter 8 - postcranial skeleton. In: Ankel-Simons F (ed) Primate anatomy, 3rd edn. Academic, Burlington, pp 283–390

Asano T (1994) Tool using behavior and language in primates. In: Hayes S, Hayes L, Sato M et al (eds) Behavior analysis of language and cognition. Context, Reno, pp 145–148

Baber C (2003) Cognition and tool use: forms of engagement in human and animal use of tools. CRC, Florida

Beck BB (1980) Animal tool behavior. Garland, New York

Berridge KC, Kringelbach ML (2015) Pleasure systems in the brain. Neuron 86(3):646–664. doi:10.1016/j.neuron.2015.02.018

de Veer MW, Gallup GG Jr, Theall LA et al (2003) An 8-year longitudinal study of mirror self-recognition in chimpanzees (*Pan troglodytes*). Neuropsychologia 41(2):229–234. doi:S0028393202001537 [pii]

Deecke VB (2012) Tool-use in the brown bear (*Ursus arctos*). Anim Cogn 15(4):725–730. doi:10.1007/s10071-012-0475-0

Draganski B, Gaser C, Busch V et al (2004) Neuroplasticity: changes in grey matter induced by training. Nature 427(6972):311–312. doi:10.1038/427311a

Enoch JM (2006) History of mirrors dating back 8000 years. Optom Vis Sci 83(10). doi:10.1097/01.opx.0000237925.65901.c0

Goodall J (1964) Tool-using and aimed throwing in a community of free-living chimpanzees. Nature 201:1264–1266

Gumert MD, Malaivijitnond S (2013) Long-tailed macaques select mass of stone tools according to food type. Philos Trans R Soc Lond Ser B Biol Sci 368(1630):20120413. doi:10.1098/rstb.2012.0413

Hart BL, Hart LA, McCoy M et al (2001) Cognitive behaviour in Asian elephants: use and modification of branches for fly switching. Anim Behav 62:839–847. doi:10.1006/anbe.2001.1815

Hihara S, Obayashi S, Tanaka M et al (2003a) Rapid learning of sequential tool use by macaque monkeys. Physiol Behav 78(3):427–434

Hihara S, Yamada H, Iriki A et al (2003b) Spontaneous vocal differentiation of coo-calls for tools and food in Japanese monkeys. Neurosci Res 45(4):383–389. doi:S0168010203000117 [pii]

Hihara S, Notoya T, Tanaka M et al (2006) Extension of corticocortical afferents into the anterior bank of the intraparietal sulcus by tool-use training in adult monkeys. Neuropsychologia 44(13):2636–2646. doi:10.1016/j.neuropsychologia.2005.11.020

Hunt GR (1996) Manufacture and use of hook-tools by New Caledonian crows. Nature 379(6562):249–251

Hunt GR (2000) Human-like, population-level specialization in the manufacture of pandanus tools by New Caledonian crows *Corvus moneduloides*. Proc Biol Sci 267(1441):403–413. doi:10.1098/rspb.2000.1015

Hunt GR, Gray RD (2004) The crafting of hook tools by wild New Caledonian crows. Proc Biol Sci 271(Suppl 3):S88–S90. doi:10.1098/rsbl.2003.0085

Inoue-Nakamura N, Matsuzawa T (1997) Development of stone tool use by wild chimpanzees (*Pan troglodytes*). J Comp Psychol 111(2):159–173

Iriki A, Tanaka M, Iwamura Y (1996) Coding of modified body schema during tool use by macaque postcentral neurones. Neuroreport 7(14):2325–2330

Ishibashi H, Hihara S, Takahashi M et al (2002) Tool-use learning selectively induces expression of brain-derived neurotrophic factor, its receptor trkB, and neurotrophin 3 in the intraparietal multisensory cortex of monkeys. Brain Res Cogn Brain Res 14(1):3–9

Jelbert SA, Taylor AH, Gray RD (2015) Reasoning by exclusion in New Caledonian crows (*Corvus moneduloides*) cannot be explained by avoidance of empty containers. J Comp Psychol 129(3):283–290. doi:10.1037/a0039313

Kage M (1994) A critical review of studies on intrinsic motivation. Jpn J Educ Psychol 42(3):345–359

Leca JB, Gunst N, Huffman MA (2010) The first case of dental flossing by a Japanese macaque (*Macaca fuscata*): implications for the determinants of behavioral innovation and the constraints on social transmission. Primates 51(1):13–22. doi:10.1007/s10329-009-0159-9

Lemon RN (2008) Descending pathways in motor control. Annu Rev Neurosci 31:195–218. doi:10.1146/annurev.neuro.31.060407.125547

Lutz K, Pedroni A, Nadig K et al (2012) The rewarding value of good motor performance in the context of monetary incentives. Neuropsychologia 50(8):1739–1747. doi:10.1016/j.neuropsychologia.2012.03.030

Malaivijitnond S, Lekprayoon C, Tandavanittj N et al (2007) Stone-tool usage by Thai long-tailed macaques (*Macaca fascicularis*). Am J Primatol 69(2):227–233. doi:10.1002/ajp.20342

Maravita A, Iriki A (2004) Tools for the body (schema). Trends Cogn Sci 8(2):79–86. doi:S1364661303003450 [pii]

McGrew WC (2013) Is primate tool use special? Chimpanzee and New Caledonian crow compared. Philos Trans R Soc Lond Ser B Biol Sci 368(1630):20120422. doi:10.1098/rstb.2012.0422

Moura AC, Lee PC (2004) Capuchin stone tool use in Caatinga dry forest. Science 306(5703):1909. doi:10.1126/science.1102558

Nishioka T (1988). Ki ni Manabe (in Japanese, 木に学べ). Shogakkan (小学館), Tokyo

Obayashi S, Suhara T, Kawabe K et al (2001) Functional brain mapping of monkey tool use. NeuroImage 14(4):853–861. doi:10.1006/nimg.2001.0878

Pansini R, de Ruiter JR (2011) Observation of tool use and modification for apparent hygiene purposes in a mandrill. Behav Process 88(1):53–55. doi:10.1016/j.beproc.2011.06.003

Plotnik JM, de Waal FB, Reiss D (2006) Self-recognition in an Asian elephant. Proc Natl Acad Sci U S A 103(45):17053–17057. doi:10.1073/pnas.0608062103

Povinelli D (2000) Folk Physics for apes: The chimpanzee's theory of how the world works. Oxford University Press, Oxford

Prior H, Schwarz A, Gunturkun O (2008) Mirror-induced behavior in the magpie (Pica pica): evidence of self-recognition. PLoS Biol 6(8):e202. doi:10.1371/journal.pbio.0060202

Quallo MM, Price CJ, Ueno K et al (2009) Gray and white matter changes associated with tool-use learning in macaque monkeys. Proc Natl Acad Sci U S A 106(43):18379–18384. doi:10.1073/pnas.0909751106

Saito M (2011) The Archaeology of diversion. Bull Nagoya Univ Arts 32:143–154

Salamone JD, Correa M (2012) The mysterious motivational functions of mesolimbic dopamine. Neuron 76(3):470–485. doi:10.1016/j.neuron.2012.10.021

Sasaki E (2015) Prospects for genetically modified non-human primate models, including the common marmoset. Neurosci Res 93:110–115. doi:10.1016/j.neures.2015.01.011

Sasaki E, Suemizu H, Shimada A et al (2009) Generation of transgenic non-human primates with germline transmission. Nature 459(7246):523–527. doi:10.1038/nature08090

Taylor AH, Cheke LG, Waismeyer A et al (2014) Of babies and birds: complex tool behaviours are not sufficient for the evolution of the ability to create a novel causal intervention. Proc Biol Sci 281(1787). doi:10.1098/rspb.2014.0837

van Schaik CP, Deaner RO, Merrill MY (1999) The conditions for tool use in primates: implications for the evolution of material culture. J Hum Evol 36(6):719–741. doi:10.1006/jhev.1999.0304
Watanabe K, Urasopon N, Malaivijitnond S (2007) Long-tailed macaques use human hair as dental floss. Am J Primatol 69(8):940–944. doi:10.1002/ajp.20403
Wimpenny JH, Weir AA, Kacelnik A (2011) New Caledonian crows use tools for non-foraging activities. Anim Cogn 14(3):459–464. doi:10.1007/s10071-010-0366-1
Yamamoto ME, Domeniconi C, Box H (2004) Sex differences in common Marmosets (Callithrix jacchus) in response to an unfamiliar food task. Primates 45(4):249–254. doi:10.1007/s10329-004-0088-6
Yamazaki Y, Namba H, Iriki A (2009) Acquisition of an externalized eye by Japanese monkeys. Exp Brain Res 194(1):131–142. doi:10.1007/s00221-008-1677-1
Yamazaki Y, Echigo C, Saiki M et al (2011) Tool-use learning by common marmosets (*Callithrix jacchus*). Exp Brain Res 213(1):63–71. doi:10.1007/s00221-011-2778-9
Yamazaki Y, Hikishima K, Saiki M et al (2016) Neural changes in the primate brain correlated with the evolution of complex motor skills. Scientific Reports 6:31084

Chapter 9
Great Ape Social Attention

Fumihiro Kano and Josep Call

Abstract Recent advances in infrared eye-tracking technology have allowed researchers to examine social attention in great apes in great detail. In this chapter we summarize our recent findings in this area. Great apes, like humans, exhibit spontaneous interest in naturalistic pictures and movies and selectively attend to socially significant elements such as faces, eyes, mouth, and the targets of others' actions. Additionally, they follow the gaze direction of others and make anticipatory looks to the targets of others' actions; the expression of these behaviors is adjusted flexibly according to the social contexts, and the viewers' memories and understandings of others' goals and intentions. Our studies have also revealed systematic species differences in attention to eyes and gaze following, particularly between bonobos and chimpanzees; several lines of evidence suggest that neural and physiological mechanisms underlying gaze perception, which are related to the individual differences within the human species, are also related to the species differences between bonobos and chimpanzees. Overall, our studies suggest that cognitive, emotional and physiological underpinnings of social attention are well conserved among great apes and humans.

Keywords Action anticipation • Anticipatory look • Eye contact • Eye movement Eye tracking • Gaze following • Great ape • Memory

---

F. Kano (✉)
Kumamoto Sanctuary, Wildlife Research Center, Kyoto University, Kumamoto, Japan

Department of Developmental and Comparative Psychology, Max-Planck Institute for Evolutionary Anthropology, Leipzig, Germany
e-mail: fkanou@gmail.com

J. Call
Department of Developmental and Comparative Psychology, Max-Planck Institute for Evolutionary Anthropology, Leipzig, Germany

School of Psychology and Neuroscience, University of St. Andrews, St. Andrews, UK
e-mail: call@eva.mpg.de

S. Watanabe et al. (eds.), *Evolution of the Brain, Cognition, and Emotion in Vertebrates*, Brain Science, DOI 10.1007/978-4-431-56559-8_9

Imagine the following situation. While you are walking down the office corridor, a stranger in necktie catches your attention. You are a little surprised because you don't usually see people wearing necktie in this office. You then notice that the stranger is working at a printer, inspecting the inside of the machine. Now one of your colleagues approaches him and starts talking to him. Although you witnessed only a few events in a short time period, you can probably make a good guess about the identity and the job of the stranger. This person is likely to be a specialist from some other company, and he is here to fix your colleague's printer.

Events like this are common in our everyday lives and despite the challenge that they pose, we can cope with them almost effortlessly thanks to our cognitive system. One key component of this system is our ability to selectively orient to essential social information in our surroundings, such as the appearance of other people and their focus of attention (Birmingham and Kingstone 2009; Klein et al. 2009). Using this information, we are able to identify people, understand their intentions, and infer their social relations with others. Our ability to spontaneously orient to and decode social information is one of our most fundamental cognitive abilities.

Nonhuman primates also seem to utilize similar skills. For example, juvenile apes and monkeys learn from their close kin members about consumable foods (Ueno and Matsuzawa 2005; van De Waal et al. 2013). Chimpanzees learn from others how to obtain out-of-reach or hard-to-process foods [e.g., termite fishing, nut cracking (Biro et al. 2003; Lonsdorf 2006)]. In a food competition context, a subordinate individual carefully avoids taking the foods that are being watched by a dominant (Hare et al. 2000). In a dominance competition, an alpha male keeps watching over his opponent so as to prevent him from making an alliance with the other dominants (De Waal 1982). Previous studies using observation methods to examine their attentional foci revealed an essentially similar nature of social attention in human and nonhuman primates. As in humans, nonhuman primates tend to look at the same foods and tools that the other individuals are handling [i.e., stimulus enhancement, joint attention; (Emery 2000; Tomasello 1995; Whiten 1992)]. Nonhuman primates also tend to look at the same objects and locations at which others are looking [i.e., gaze following; (Rosati and Hare 2009; Shepherd 2010; Tomasello et al. 1998)]. They were also observed to use eye contact to regulate social interactions among individuals (De Waal 1990a; Gomez 1996).

Traditional methods have examined social attention indirectly by recording an individual's body and head orientation as a way to assess their attentional foci, but recent advances in infrared, non-invasive eye-tracking technology have allowed researchers a much more direct and precise analysis of social attention. More specifically, eye-tracking technology offers greater spatial and temporal resolution compared to body and head orientation measures. On the one hand, a greater spatial resolution means that eye tracking can distinguish the individual's focus of attention with great precision, for instance, whether the viewer is looking at the eyes or the mouth in a face. On the other hand, a greater temporal resolution means that eye tracking can measure for how long individuals maintain focus on certain targets and how they shift attention between them over time, for instance, when the viewer is anticipating the target of an individual's action.

This method has been commonly used with human adults and infants, but a series of studies by us and our colleagues showed that the same technique can be applied to our closest relatives, the great apes (henceforth apes) who exhibit spontaneous interest in naturalistic pictures and movies. They systematically scan pictures and movies, and their gaze movements inform us about what aspects they perceive as relevant about the depicted objects, actions, and stories. In this chapter we summarize recent studies using eye tracking to examine social attention in apes. The similarities and differences between species will constitute a key aspect for inferring the cognitive and emotional mechanisms of social attention and how they may have evolved over time. The chapter is organized as follows. We first give a brief introduction on the application of this method. In the next section we explore the social elements that apes prefer to attend and its determinants such as the context dependency and the species difference. The third section is devoted to explore how they coordinate thier attention with others, and how they adjust these behaviors according to the social contexts, their own memories and understandings of others' goals. We will close the chapter with a summary of the main findings and future research directions.

9.1 A Method for Tracking the Gaze of Apes

Eye tracking is a method that directly measures the participants' gaze location with a high temporal and spatial resolution (Duchowski 2007). Figure 9.1 presents our basic experimental setup. Stimuli are presented on a monitor, and a table-mounted remote eye tracker records the ape's gaze as indicated by the position of her eyes and pupils (corneal and retinal reflections). The eye-tracker and the monitor are placed outside a test booth with a transparent acrylic panel separating them from the ape

Fig. 9.1 Eye-tracking with great apes. A chimpanzee sitting in front of a computer screen and eye-tracker (*left*). A juvenile gorilla sipping grape juice while watching movies on a screen (*right*)

who sits in the booth watching the stimuli presented on the screen. The ape needs to stay roughly in the same position throughout the recording, although their heads do not need to be restrained (unlike a traditional eye-tracking method) because the infrared cameras of the eye tracker can continuously detect both the eye and pupil positions. In order to keep apes as still as possible, we give them access to a dripping nozzle attached to the panel that continuously delivers a small quantity of juice.

Although apes are not rewarded contingent on looking at the stimuli, most of them show spontaneous interests in them. We even observed an occasion when one of the bonobos was so engaged with the presented movie that she stopped drinking the juice and just stared at it. Moreover, we often observe signs of engagement with the stimuli, e.g., bobbing their heads or banging the panel while watching. However, these responses have never been followed by a refusal to participate in subsequent sessions. Therefore, apes seem interested in the pictures and movies, but they do not seem to confuse them with their corresponding referents.

Despite their interest in pictures and videos, the stimuli have to be chosen carefully because they do not view all sorts with the same interest. Here are a few examples that usually fail to keep the apes' attention: animations made of simple circles and triangles, puppet plays made to entertain human children, and Hollywood-like movies with complex cinematography techniques (including quick transitions of shots and camera works). In contrast, apes maintain their interest when the contents include familiar backgrounds (e.g., home cages), familiar people (e.g., keepers), conspecific primates, simple actions (e.g., eating foods, handling objects and tools), and simple camera angles. Intense social interactions, such as aggression, are also typically successful in keeping apes engaged (Kano and Tomonaga 2010a).

Besides the content, the general pace of the scenes needs to be moderately fast to keep apes' attention. The reason for this is that compared to humans, apes fixate on one location more briefly (i.e., shorter duration of fixations) and shift their gaze to the next location earlier (i.e., more frequent saccades) (Kano et al. 2011; Kano and Tomonaga 2011). When presented with a still scene, apes typically complete the scanning faster than humans do (and then immediately quit watching). Thus, due to the different basic ape eye-movement characteristics, the transitions of scenes, actions, and events should be kept at a moderately fast pace.

9.2 Apes' Attentional Focus and Its Determinants

9.2.1 Focal Elements of Interest

Figure 9.2 presents the gaze locations of 14 bonobos looking at the face and full body pictures of conspecific (Kano et al. 2015). Similarities between humans and apes can be seen clearly here. Apes attend to the face, especially the eyes. In addition, upon stimulus presentation their first fixation is typically directed to the face and eyes. This may be due to the fact that, like humans, apes' social perception is also specialized

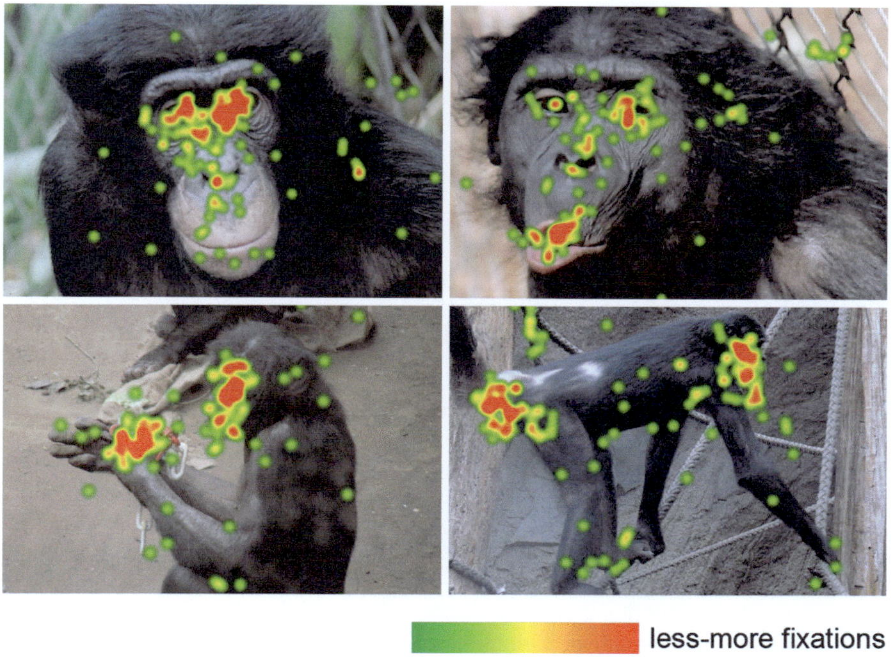

Fig. 9.2 How bonobos look at the pictures. The heat maps represent the aggregation of fixations made by 14 bonobos viewing the conspecific pictures. Those elements with greater "heat" signatures (depicted by *red areas*) attracted more fixations of a larger number of participants

for faces (Parr et al. 2009; Tomonaga and Imura 2009), and their initial fixation in an area around eyes is used for face recognition (Guo 2007; Hsiao and Cottrell 2008).

The similarities between humans and apes are not only about the attentional foci but also about the flexibility in the shift of attentional foci based on the scene contents. When the pictures include simply the face and the body, or the eyes and the mouth, without particular postures or expressions, they mostly look at the face and the eyes. When the pictures include facial expressions (Fig. 9.2, top-right), they look longer at the mouth than the eyes. When the individual in the pictures holds an object (Fig. 9.2, bottom-left), they look longer at the objects than the face. Not surprisingly, when ano-genital areas are visible (Fig. 9.2, bottom-right), these areas also attract more attention than the face.

Figure 9.3 depicts another example showing the apes' ability to shift attention based on an agent's action (Kano and Call, unpublished data). In the presented video, an actor ape faced an out-of-reach grape; he then got a stick and used it to rake in the grape. When the chimpanzees were presented with this video, they initially focused on the actor ape, but once the action started they shifted their focus of attention to the grape and sustained it there until the whole action completed. This means that apes do not merely focus on basic scene elements, but they can dynamically shift their attention to different elements of the scene as the action unfolds.

Fig. 9.3 How chimpanzees look at the model's action. In the video, the model individual uses a stick to rake an out-of-reach grape. Heat maps represent the aggregation of fixations made by 13 chimpanzees

9.3 Context Dependency of Gaze

Like in humans, the focus of attention is sensitive to the ongoing contexts in great apes. Context-dependent gaze requires a viewer to flexibly adjust her focus of attention depending on the context, which in turn offers a glimpse of the viewer's understanding about a particular event. For instance, at the zoo, we have often observed that apes usually do not pay much attention to unfamiliar (regular) visitors in the public areas. However, when those same unfamiliar visitors enter the service areas behind the scenes, apes suddenly become interested in them; some apes may even tease them by spitting a mouthful of water. Likewise, when apes spot a familiar person in the visitor areas, they might greet them enthusiastically or tease them by throwing a branch at them.

Kano and Tomonaga (2013) examined a chimpanzee's eye movement in a context-rich real-life environment depicted in Fig. 9.4. For several days, the chimpanzee participant saw a different experimenter coming into the test room to play a simple gesture game on each day. The experimenter was either a familiar experimenter that the chimpanzee saw in the particular test room regularly (familiar, regular experimenter), a familiar experimenter that the chimpanzee did not see in that test room regularly (familiar, irregular experimenter), or a completely unfamiliar experimenter that the chimpanzee had never seen anywhere (unfamiliar, irregular experimenter). The results showed that the chimpanzee looked at the unfamiliar-irregular experimenter the longest and the familiar-regular the shortest (the familiar-irregular experimenter somewhere between). This pattern could indicate a degree of surprise in the chimpanzee. In addition, when the experimenter started a simple gesture game, the chimpanzee shifted her attention from the experimenter's face to the task-relevant elements, i.e., the experimenter's hands and the rewards. Thus, the chimpanzee's gaze was task-dependent. This study examplified a remarkably strong context-dependency of gaze when the chimpanzee was tested in a context-rich real-life environment, where (potential) interaction with the social partner matters.

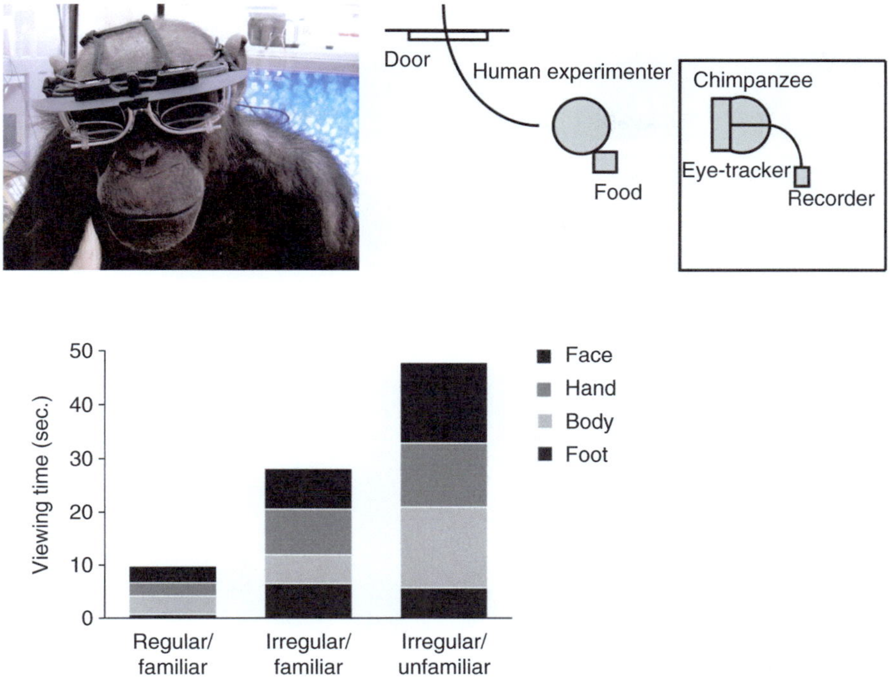

Fig. 9.4 A chimpanzee wearing a head-mounted eye-tracking device during a live interaction with an experimenter. The device recorded the chimpanzee's focus of attention, while she observed and spontaneously interacted with the experimenter who greeted the chimpanzee during 1 min

9.3.1 Species Uniqueness of Gaze

Despite their basic similarities, apes and humans also show significant differences in their gaze patterns. Yarbus's (1967) classic study showed that when presented with facial pictures, the trajectory of eye movement (i.e. scan paths) in humans tends to describe an inverted triangle, with its angles on each of the eyes and the mouth. In contrast, chimpanzees, orangutans, and gorillas tend to look shorter to the eye region and longer to the mouth region than humans do (Kano et al. 2012; Kano and Tomonaga 2010b). Thus their scan paths tend to describe lines connecting the eye and mouth, with the connection between the eyes being less pronounced (but see bonobos below). This is a robust pattern since such a difference in scan path pattern can be observed relatively independently from the stimulus variation. For example, when presented with the pictures of facial expressions instead of neutral faces (Fig. 9.2 top-right), apes looked longer at the mouth, while humans still devoted some time to looking at the eyes (Kano and Tomonaga 2010b).

However, interestingly, bonobos are somewhat exceptional among nonhuman great ape species. In an experiment, we presented the two species with a series of pictures containing the images of both species (Kano et al. 2015).

Species	Name	Sex	Age	Eye (ms)	Mouth (ms)	Face (ms)	Genital (ms)	Target (ms)	Missclassified	Discriminant Score
Bonobo	Vijay	M	10	1368	302	1232	178	455		-3.20
	Ulindi	F	21	972	489	1034	400	292		-2.40
	Luiza	F	9	1158	689	1269	450	478		-2.33
	Lenoire	F	32	681	678	799	256	175		-2.17
	Ikela	F	23	1461	356	1006	562	379		-1.82
	Louise	F	42	1093	398	807	226	408		-1.61
	Kuno	M	18	1142	508	827	391	436		-1.14
	Jasongo	M	24	965	694	1078	532	615		-0.80
	Fimi	F	6	831	547	825	503	451		-0.68
	Junior	M	19	821	457	748	758	295		-0.50
	Joey	M	32	1253	857	815	479	553		-0.27
	Loto	M	5	519	1106	922	550	563	*	-0.03
	Yasa	F	17	924	522	896	596	669	*	0.14
	Lolita	F	25	163	343	456	570	316	*	0.37
Chimpanzee	Ulla	F	37	1455	471	1098	739	474	*	-1.28
	Riet	F	39	1247	675	863	811	370	*	-0.51
	Zamba	M	18	249	152	663	601	392		-0.11
	Kofi	M	9	361	923	492	318	403		0.25
	Alex	M	13	816	1027	1004	340	901		0.43
	Kara	F	9	406	752	565	251	606		0.57
	Iroha	F	5	679	413	602	673	474		0.59
	Lome	M	13	461	878	650	560	506		0.62
	Jahaga	F	21	978	481	722	830	523		0.66
	Sandra	F	21	751	1232	896	867	535		0.66
	Getrudia	F	21	417	394	337	375	421		0.71
	Lobo	M	9	543	897	735	555	611		0.72
	Misaki	F	15	581	272	645	729	557		0.87
	Robert	M	39	546	987	634	818	490		1.24
	Mizuki	F	17	764	370	699	470	859		1.31
	Hatsuka	F	5	466	546	544	496	664		1.32
	Bangolo	M	5	420	941	585	284	771		1.35
	Fraukje	F	38	819	968	616	811	657		1.87
	Fifi	F	21	379	458	324	530	664		2.17
	Natsuki	F	8	883	725	620	1097	809		2.99

Fig. 9.5 Bonobo-chimpanzee differences in social orienting. Heat maps represent the aggregation of fixations made by participant bonobos (*left*) and chimpanzees (*right*). The table shows each participant's viewing time to each scene element, averaged over a total of 60 trials presenting both bonobo and chimpanzee pictures. Each picture was presented for 3 s. The discriminant scores show the classification rates based on those viewing times

Figure 9.5 illustrates the main viewing differences that we observed between them. In general, bonobos viewed longer the face and the eyes compared to chimpanzees, whereas chimpanzees viewed longer the mouth, the ano-genital region, and the objects manipulated by the models than bonobos. A discriminant analysis clearly distinguished the two species (Fig. 9.5): 29 out of 34 participant apes were correctly classified into their own species.

One plausible explanation for the species differences between bonobos and chimpanzees may be related to their species-typical temperaments. In humans, looking into another individual's eyes——eye contacts——predicts the viewer's temperament and the interpersonal relationships (Argyle and Dean 1965; Kleinke 1986); people with a higher need for affiliation showed an increased level of eye contact. In the case of bonobos and chimpanzees, they are known for their differences in the patterns of aggression and affiliation. Bonobos exhibit a larger repertoire of affiliative behaviors toward conspecifics than chimpanzees do, such as non-conceptive sexual behaviors, frequent play among adults, and non-aggressive encounters with strangers (De Waal 1990b; Furuichi 2011; Palagi 2006; Tan and Hare 2013). It has been hypothesized that these species-typical behaviors of bonobos have evolved in part as a response to the relaxation of intra-group competition and the selection against male aggression (Furuichi 2011; Hare et al. 2012). As a result, bonobos may have developed more relaxed attitudes toward others, and such temperamental characteristics may allow them to look into the eyes of others without experiencing as much a social stress as chimpanzees do.

Several lines of evidence suggest that the differences in gaze perception between bonobos and chimpanzees are related to their neural and physiological differences. First, although the general brain structure is largely conserved across the species, the local gray matter in the amygdala and the anterior insula differ between bonobos and chimpanzees (Rilling et al. 2012). As a region related to social interaction and stress, the amygdala was found to respond to the eyes selectively, when the viewers were fixating on the eyes of stimulus faces (Dalton et al. 2005; Mosher et al. 2014). Second, the oxytocin-receptor gene is also reported to be different between bonobos and chimpanzees (Staes et al. 2014). Oxytocin is found to be related to attention to the eyes. Oxytocin administration increased eye orientation in humans and macaques (Ebitz et al. 2013; Guastella et al. 2008). Third, it has been hypothesized that prenatal androgens may be higher in chimpanzees than bonobos, as suggested by their differences in a 2D-4D (digit) ratio (McIntyre et al. 2009). The level of prenatal androgens has been associated to the level of eye contact: in human children, a higher level of prenatal androgens is associated with a decreased level of eye contact (Lutchmaya et al. 2002). These studies suggest that neural and physiological mechanisms underlying gaze perception, which are related to the individual differences within the human species, are also related to the species differences between bonobos and chimpanzees.

9.4 Co-orienting Attention with Others

Now let's turn our attention from simple orienting behavior to a more dynamic aspect of social attention. Social attention is not only about how long one looks at each social element but also about how well one coordinates attention with others. Imagine one simple reaching action performed by an agent (Fig. 9.6). When an agent is performing an action, the action sequence can be divided into several small subactions (Land et al. 1999). The agent first looks at the target object, moves his/her hand toward the object, and then grabs the object. If the viewer follows the action of

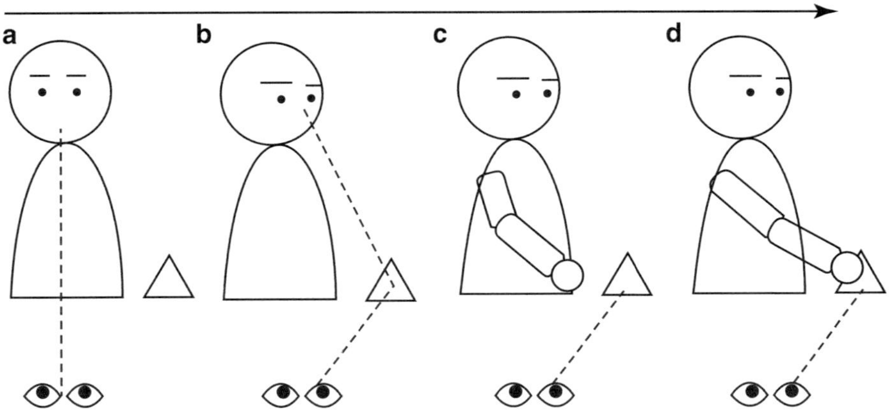

Fig. 9.6 A schematic drawing showing "attention coordination"; (**a**) face viewing, (**b**) gaze following, (**c**) action anticipation, and (**d**) action viewing

an agent, the viewer could anticipate the action goals and look to the target objects before the final grabbing action happens. Co-orientation often constitutes a first crucial step for engaging in joint activities with the agent and gauging the agent's perception, goals and intentions. The viewer's gaze responses in Fig. 9.6b (agent looking at the object) and c (agent reaching toward the object) are typically referred respectively as gaze following and action anticipation. The following three sections discuss our studies investigating apes' abilities on gaze following and action anticipation including the social context and the viewers' memories and understandings of others' goals and intentions that modulate the expression of these behaviors.

9.5 Anticipating the Actions of Others

An anticipatory look is defined as looking at the target objects of an ongoing action before the action sequence is completed. Only recently did researchers start to examine this behavior, thanks to the advance in eye-tracking technology that allows researchers to capture the subtle eye movement happening in just a fraction of a second. Like humans (Cannon and Woodward 2012; Falck-Ytter et al. 2006; Flanagan and Johansson 2003), apes also make reliable anticipatory looks in response to an agent's action (Kano and Call 2014b; Myowa-Yamakoshi et al. 2012).

The anticipatory looks made by humans are cognitively sophisticated; they attend to a reaching action based on the agent's goals and intentions (Cannon and Woodward 2012; Senju et al. 2011; Southgate et al. 2007; Woodward 1998). We conducted a study to examine whether apes' anticipatory looks could also be based on an agent's current action goal. In this study, we tested bonobos, chimpanzees, and orangutans using a paradigm designed by Cannon and Woodward (2012) (see Fig. 9.7). In the original paradigm, human infant participants first saw a human

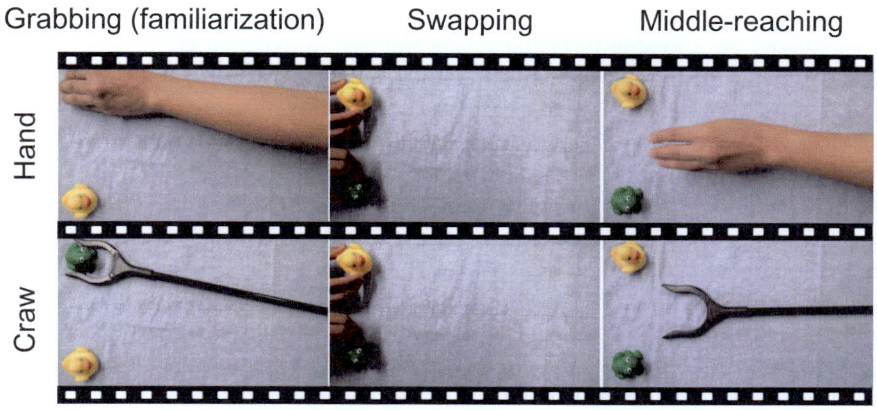

Fig. 9.7 Videos used in a goal anticipation study. Either a human hand or a mechanical claw moved toward one of the two different objects three times (familiarization), and then after the location of the objects were switched, the hand or claw moved incompletely between the two objects

agent's hand repeatedly reaching toward and grabbing the same one of two objects. Then the locations of the two objects were switched, and this time, the hand made an incomplete reach action toward the objects stopping halfway between the two objects. In the control condition, human infant participants saw instead an inanimate mechanical claw performing the same actions. Infants presented with the incomplete hand reach made anticipatory looks to the familiar object (the object that had been grabbed before) instead of the familiar location (where the object used to be). In contrast, such a result was not found in the control condition with the claw. Our apes performed just like human infants: they directed more looks at the familiar object in the hand condition but not in the claw condition. Thus, we confirmed that, like in humans, anticipatory looks in great apes are based on an agent's action goals, rather than mere location or direction of the agent's hand movements.

We found no significant species differences in this study. Bonobos and chimpanzees, and orangutans displayed a similar frequency and pattern of anticipatory looks. Thus, it is possible that species differences are only limited to gaze perception and do not extend to action perception or goal understanding. Relatedly, autistic children, known to differ from typical developing children in attention to eyes and gaze following, did not display a different pattern of anticipatory looks to an agent's reaching action from typically developing children (Falck-Ytter 2010).

9.6 The Role of Memory in Anticipatory Looking

Goal-based anticipatory looks in apes possess a memory component; after apes learned the agent's goal in the familiarization trials (see Fig. 9.7), they used that memory to anticipate the reaching goal in the test trial. This memory is most likely to be a short-term memory or possibly a procedural memory that has been learned through repeated presentations of the same actions. Can apes also make anticipatory looks based on a long-term memory of a single-trial experience? Traditionally, it has been considered that nonhuman animals lack the ability to remember single events for a long time period, except in very limited contexts [e.g., taste aversion, (Garcia et al. 1955)]. However, recent studies on episodic-like memory in nonhuman animals (Babb and Crystal 2006; Clayton et al. 2003; Kart-Teke et al. 2006; Martin-Ordas et al. 2010) revealed that it is not necessarily the case. Thus it is possible that apes' anticipatory looks to an agent's action could also be based on long-term memory after a single-trial exposure of that action.

Kano and Hirata (2015) tested this possibility in bonobos and chimpanzees by presenting them movies twice with one-day delay and measuring their anticipatory gaze patterns (Fig. 9.8). The movie story started with an ape-like character attacking a human agent. The agent then reached ambiguously toward two different objects (weapons for revenge) for a few seconds, grabbed one of them, and then hit the ape-like character with it. Apes watched this movie only once on the first day, and on the second day (24 h later), they watched the same movie again except one change; the location of objects was switched. This change, the same technique applied to the study mentioned above, enabled us to examine whether or not they remembered the content rather than the location of the objects. As apes are typically engaged in and encode better an intense social interaction such as aggression than they do to neutral events (Kano et al. 2008), the original agent's reaching events were enhanced with aggression events in this study. In addition, to create a novel event in a familiar situation that could best engage ape participants, the study used the background and a human actor that were familiar to the participant apes, in combination with objects (weapons) and an antagonist (the ape-like character) unfamiliar to the apes. If the ape participants remembered which objects the agent chose, they should anticipate the target object on the second but not on the first day of watching. Our results con-

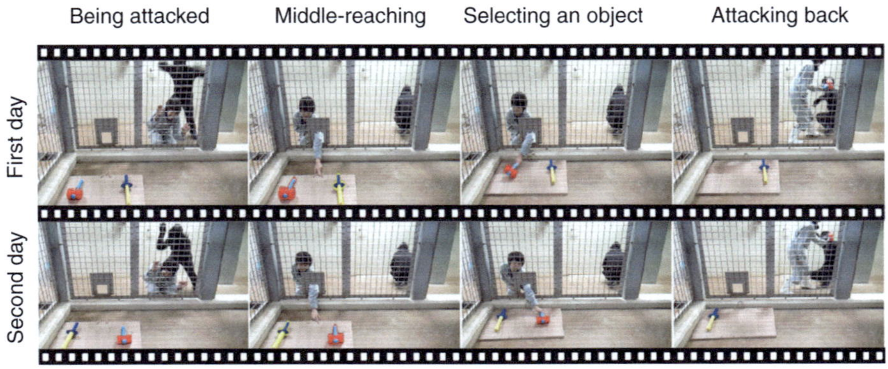

Fig. 9.8 Videos used to investigate whether apes would anticipate an agent's action based on long-term memory after a single-trial exposure of the action. The video started with a human agent being attacked by an ape-like character. The human agent then reached between the two objects, grabbed one of them, and attacked the ape-like character using the selected object. The same video was presented on the second day (24 h later), except that the object locations were switched

firmed this prediction. Importantly, their anticipatory looks to the target object were observed only just before the agent grabbed the object but not throughout the second presentation, suggesting that apes remembered the timing of event occurrence, or the relation between the agent's action and object (rather than the object per se).

9.7 Following Gaze of Others

Gaze following is defined as looking in the same direction as other individuals after seeing their gaze direction. Gaze following is probably the best studied aspect of social attention in both human and nonhuman primates. Previous studies have shown that humans, apes, monkeys, and lemurs (Ruiz et al. 2009; Sandel et al. 2011; Shepherd and Platt 2008) as well as other non-primate animals [e.g., dogs, ravens, red-footed tortoises; (Bugnyar et al. 2004; Téglás et al. 2012; Wilkinson et al. 2010)] follow the gaze of others in a real-life situation. When the situation was presented in pictures and movies, humans, apes, monkeys, and dogs still reliably follow the gaze of the models (Deaner and Platt 2003; Hattori et al. 2010; Kano and Call 2014a; Téglás et al. 2012). In humans and monkeys, gaze following occurs reflexively within a fraction of a second and relatively independently of any task demand (Deaner and Platt 2003; Shepherd 2010). Gaze following could also occur non-reflexively under a sophisticated cognitive modulation. For example, when a model's line of sight was blocked by a barrier, humans, apes, monkeys, and ravens looked around the barrier rather than simply followed the model's gaze (Amici et al. 2009; Bräuer et al. 2005; Bugnyar et al. 2004; Moll and Tomasello 2004).

Although all primates tested follow gaze, species differences emerge depending on the experimental conditions. In a study, bonobos more frequently followed the gaze of human models than chimpanzees did (Herrmann et al. 2010); stump-tail macaques followed the gaze of conspecific models more than rhesus macaques did (Tomasello et al. 1998); and human children followed the gaze of human models more than apes did (Herrmann et al. 2007). In an eye-tracking experiment, chimpanzees followed the gaze of only conspecific model but not that of human model, while human participants followed the gaze of both models (Hattori et al. 2010). Most previous studies used a pairwise comparison of species and/or a human experimenter as a model, leaving it unclear whether the reported species differences reflected particular responses to the conspecifics or general sensitivity to the gaze signals.

We thus conducted an experiment using a crossed experimental design with participant species and depicted species, in which bonobos, chimpanzees, orangutans, human adults, and infants (1-year-olds) were presented with both conspecific and allospecific models (Fig. 9.9). Movies depicted a model, either a conspecific or an allospecific, turning his/her head to one of two identical objects repeatedly. Our results showed a species difference in both conspecific preference and the general gaze sensitivity. Consistent with the previous study (Hattori et al. 2010), chimpanzees followed the gaze of a conspecific chimpanzee but not that of allospecifics

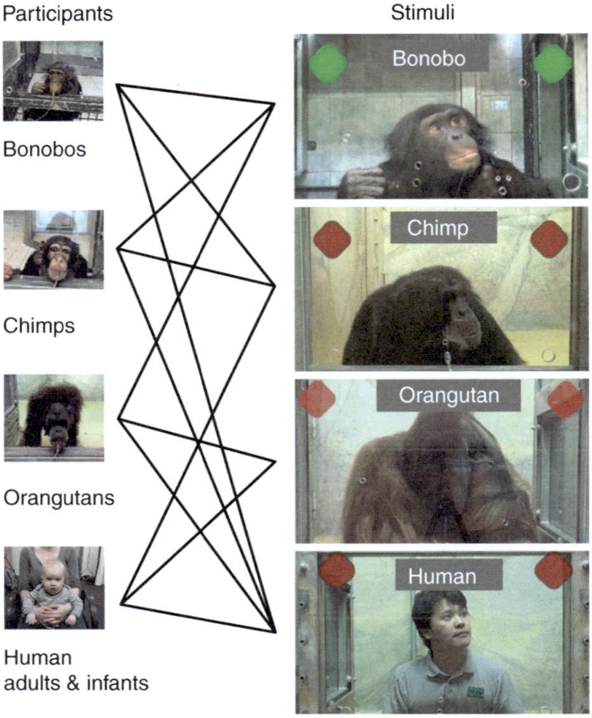

Fig. 9.9 A study on gaze following using a cross experimental design. For each participant species (*left*), both conspecific and allospecific models (*right*) were presented. Participants observed the model repeatedly looking at one of two identical objects

(human or bonobo). Bonobos also followed the gaze of both conspecific and allospecific models, and overall more frequently, with shorter latencies, than chimpanzees did. Orangutans showed somewhat intermediate responses: they followed the gaze of conspecifics and bonobos but not the gaze of humans, with overall intermediate frequencies among the ape species. Human adults followed the gaze of both conspecific and allospecific models, and overall more frequently than the ape species. Human infants, unlike human adults, followed the gaze of a human model (as frequently as bonobos did) but not the gaze of allospecific ape models.

Thus, our results showed that these species differed in both general sensitivity to the gaze signals and particular responses to the conspecifics. It is particlarly interesting to find that, among the tested participants, only chimpanzees and human infants followed the gaze of only conspecifics. One interpretation of this result is that chimpanzees and human infants are more motivated to follow the gaze of the most relevant others, the conspecifics. They may preferentially acquire information from only the most relevant others to learn about the environment efficiently. In this regard, our results may be consistent with the previous report of selective social referencing and behavioral copying by chimpanzees and human children (Dindo et al. 2009; Haun et al. 2012; van De Waal et al. 2013). The bonobo-chimpanzee difference in gaze following is again remarkable. We found that both bonobos and chimpanzees followed the gaze of conspecifics, but bonobos, but not chimpanzees, followed the gaze of allospecifics. Bonobos followed the gaze of both conspecific and allospecific models overall more frequently and with shorter latencies than chimpanzees did.

The observed bonobo-chimpanzee differences may be related to thier differences in temperaments and neural and physiological mechianisms, and the associated differences in gaze percption. In support of this idea, previous studies suggested that, in macaques and humans, the relative strength of reflexive and voluntary components in gaze following may be modulated by androgen-related mechanisms (Shepherd 2010). For example, subordinate monkeys followed the gaze more reflexively than dominant monkeys (Shepherd et al. 2006); human females followed the gaze of a familiar person more reflexively than males (Deaner et al. 2007). In the case of bonobos and chimpanzees, previous studies have shown that they are different in the levels of prenatal and adulthood androgen (Sannen et al. 2003; McIntyre et al. 2009).Thus, in bonobos, gaze following may be modulated by a reflexive (nonselective) rather than voluntary (selective) process to a larger extent than in chimpanzees; that is, bonobos may have followed the gaze of both conspecific and allospecific models more reflexively and therefore less selectively than chimpanzees did in our study.

One remaining puzzle is that, although we showed that chimpanzees do not follow the gaze of human models in our experiment, chimpanzees do follow the gaze of human experimenter in real-life behavioral experiments (Bräuer et al. 2005; MacLean and Hare 2012). It is conceivable that the context-rich real-life environment may better motivate the chimpanzees to follow human gaze. For example, when a chimpanzee faces a human experimenter in such situations, she may be more likely to attend to the human because he could provide some food. Besides, humans tend to add ostensive cues to the gaze cues during interaction (Csibra and Gergely 2009), making an eye contact and calling the chimpanzee's name before looking at the target object. Such ostensive cues increase the likelihood that human infants will follow the adult's gaze (Senju and Csibra 2008). In this regard, the ostensive cues might better motivate the chimpanzee to follow the experimenter's gaze as well. Our recent study partly replicated the effect of human ostensive cues on gaze following in chimpanzees (Kano, Moore, Krupenye, Tomonaga, Call, in prep).

9.8 Conclusions and Future Directions

In this chapter, we have discussed the characteristics of apes' social attention. We have described both similarities and differences between the species and discussed the possible mechanism underlying them. Consistent with the previous studies using observational (non-eye-tracking) methods, we found a number of similarities in the expression of social attention between apes and humans. All species attend to the same social elements including the face, eyes, and mouth of others as well as the targets of others' actions. They follow the gaze direction of others, and they make anticipatory looks to the targets of others' actions. In addition, they flexibly adjust these behaviors according to the social contexts, and the viewers' memories and understandings about others' goals and intentions. We also found significant differences between the species. Bonobos (and humans) look longer the eyes of others than chimpanzees. Bonobos may follow the gaze direction of others more reflexively and less selectively than chimpanzees do. These species differences seem to be derived from each species' unique temperaments. Several lines of evidence suggest that neural and physiological mechanisms underlying gaze perception, which are related to the individual differences within the human species, are also related to the species differences between bonobos and chimpanzees. Therefore, overall, the current data suggest that the cognitive, emotional and physiological underpinnings of social attention are well conserved across humans and all great ape species.

There are at least two important future research directions. The first is a more detailed examination of species and individual variation in social attention. We found that the pattern of social orienting is highly species-specific and related to each species' unique temperaments. We could thus predict further diversity of social orienting in primate phylogeny, including non-ape species, such as Old-World monkeys (e.g., rhesus and stumptail macaques) and New-World monkeys (e.g., capuchin and spider monkeys). Those closely related species may differ in their social systems and temperaments, similarly as bonobos and chimpanzees do. In addition, it is also important to examine how individuals' unique experiences affect the pattern of social attention, as the differences in early social experiences are known to affect the social and cognitive abilities of great apes (e.g. "encultured" chimpanzees reared by humans in human environment, Carpenter and Tomasello, 1995). Such an examination of species and individual variation should further help us to understand the mechanism and evolution of social attention in primates.

The other research direction is a more detailed examination of cognitive functions, such as a theory of mind and memory, in apes. We believe that an anticipatory looking paradigm could be particularly useful to reveal the cognitive underpinnings of social responses. Whether or not apes make anticipatory looks based on an agent's intention is a goal worth pursuing. To show that one truly understands not only the current action goals of an agent but also the intentions of an agent, ultimately we need to find an evidence of false-belief understanding in the participant (Gergely and Csibra 2003). If apes are able to make correct anticipatory looks based on agent's intention even when the agent carries a false belief about the location of the desired object, then we can be confident that apes do make anticipatory looks based on agent's intentions. Such results have already been found with infants as young as 2 years old by measuring their anticipatory looking (Senju et al. 2011; Southgate et al. 2007). The use of an appropriate eye-tracking version for apes may offer a possibility that apes possess at least certain implicit form of false-belief representation (see Krupenye, Kano, et al., 2017 for the update).

Anticipatory looks are, by definition, based on the participant's knowledge, memory, or understanding about the regularities of events and agents' actions. In this sense they differ fundamentally from simple orienting responses which are likely under the control of both low-level perceptual and high-level cognitive functions. We believe that these findings from our studies will open up further possibilities to explore the unrevealed cognitive potentials as well as the diversity of indivdiual minds in nonhuman animals.

Acknowledgments We thank Drs. Masaki Tomonaga and Satoshi Hirata for their invaluable supports in conducting the studies presented here. Financial support came from Ministry of Education, Culture, Sports, Science and Technology (K-CONNEX to FK), Japan Society for Promotion of Science (KAKENHI 26885040, 16 K21108 to FK) and European Research Council (Synergy grant 609819 SOMICS to JC).

References

Amici F, Aureli F, Visalberghi E, Call J (2009) Spider monkeys (*Ateles geoffroyi*) and capuchin monkeys (*Cebus apella*) follow gaze around barriers: evidence for perspective taking? J Comp Psychol 123(4):368–374

Argyle M, Dean J (1965) Eye-contact, distance and affiliation. Sociometry 28(3):289–304

Babb SJ, Crystal JD (2006) Episodic-like memory in the rat. Curr Biol 16(13):1317–1321

Berg DJ, Boehnke SE, Marino RA, Munoz DP, Itti L (2009) Free viewing of dynamic stimuli by humans and monkeys. J Vis 9(5):1–15

Birmingham E, Kingstone A (2009) Human social attention. Ann N Y Acad Sci 1156(1):118–140

Biro D, Inoue-Nakamura N, Tonooka R, Yamakoshi G, Sousa C, Matsuzawa T (2003) Cultural innovation and transmission of tool use in wild chimpanzees: evidence from field experiments. Anim Cogn 6(4):213–223

Bräuer J, Call J, Tomasello M (2005) All great ape species follow gaze to distant locations and around barriers. J Comp Psychol 119(2):145–154

Bugnyar T, Stöwe M, Heinrich B (2004) Ravens, *Corvus corax*, follow gaze direction of humans around obstacles. Proc R Soc Lond Ser B Biol Sci 271(1546):1331–1336

Cannon EN, Woodward AL (2012) Infants generate goal-based action predictions. Dev Sci 15(2):292–298

Carpenter M, Tomasello M (1995) Joint attention and imitative learning in children, chimpanzees, and enculturated chimpanzees. Social Development 4(3):217–237

Clayton NS, Bussey TJ, Dickinson A (2003) Can animals recall the past and plan for the future? Nat Rev Neurosci 4(8):685–691

Csibra G, Gergely G (2009) Natural pedagogy. Trends Cogn Sci 13(4):148–153

Dalton, K. M., Nacewicz, B. M., Johnstone, T., Schaefer, H. S., Gernsbacher, M. A., Goldsmith, H. H.,. .. Davidson, R. J. (2005). Gaze fixation and the neural circuitry of face processing in autism. Nat Neurosci, 8, 519–526.

De Waal FBM (1982) Chimpanzee politics: power and sex among apes. Harper & Row, New York

De Waal FBM (1990a) Peacemaking among primates. Harvard University Press, Cambridge

De Waal FBM (1990b) Sociosexual behavior used for tension regulation in all age and sex combinations among bonobos *Pedophilia*. Springer, New York, pp 378–393

van De Waal E, Borgeaud C, Whiten A (2013) Potent social learning and conformity shape a wild Primate's foraging decisions. Science 340(6131):483–485

Deaner RO, Platt ML (2003) Reflexive social attention in monkeys and humans. Curr Biol 13(18):1609–1613

Deaner RO, Shepherd SV, Platt ML (2007) Familiarity accentuates gaze cuing in women but not men. Biol Lett 3(1):64–67

Dindo M, Whiten A, De Waal FB (2009) In-group conformity sustains different foraging traditions in capuchin monkeys (*Cebus apella*). PLoS One 4(11):e7858

Duchowski AT (2007) Eye tracking methodology: theory and practice. Springer, New York

Ebitz RB, Watson KK, Platt ML (2013) Oxytocin blunts social vigilance in the rhesus macaque. Proc Natl Acad Sci 110(28):11630–11635

Emery NJ (2000) The eyes have it: the neuroethology, function and evolution of social gaze. Neurosci Biobehav Rev 24(6):581–604

Falck-Ytter T (2010) Young children with autism spectrum disorder use predictive eye movements in action observation. Biol Lett 6(3):375–378

Falck-Ytter T, Gredebäck G, Von Hofsten C (2006) Infants predict other people's action goals. Nat Neurosci 9(7):878–879
Flanagan JR, Johansson RS (2003) Action plans used in action observation. Nature 424(6950):769–771
Furuichi T (2011) Female contributions to the peaceful nature of bonobo society. Evol Anthropol 20(4):131–142
Garcia J, Kimeldorf DJ, Koelling RA (1955) Conditioned aversion to saccharin resulting from exposure to gamma radiation. Science 122:157–158
Gomez JC (1996) Ostensive behavior in great apes: the role of eye contact. In: Russon AE, Bard KA, Parker ST (eds) Reaching into thought: the minds of the great apes. Cambridge University Press, New York, pp 131–151
Guastella AJ, Mitchell PB, Dadds MR (2008) Oxytocin increases gaze to the eye region of human faces. Biol Psychiatry 63(1):3–5
Guo K (2007) Initial fixation placement in face images is driven by top-down guidance. Exp Brain Res 181(4):673–677
Hare B, Call J, Agnetta B, Tomasello M (2000) Chimpanzees know what conspecifics do and do not see. Anim Behav 59(4):771–785
Hare B, Wobber V, Wrangham R (2012) The self-domestication hypothesis: evolution of bonobo psychology is due to selection against aggression. Anim Behav 83(3):573
Hattori Y, Kano F, Tomonaga M (2010) Differential sensitivity to conspecific and allospecific cues in chimpanzees and humans: a comparative eye-tracking study. Biol Lett 6(5):610–613
Haun D, Rekers Y, Tomasello M (2012) Majority-biased transmission in chimpanzees and human children, but not orangutans. Curr Biol 22(8):727–731
Herrmann E, Call J, Hernandez-Lloreda MV, Hare B, Tomasello M (2007) Humans have evolved specialized skills of social cognition: the cultural intelligence hypothesis. Science 317(5843):1360–1366
Herrmann E, Hare B, Call J, Tomasello M (2010) Differences in the cognitive skills of bonobos and chimpanzees. PLoS One 5(8):e12438
Hsiao JH, Cottrell G (2008) Two fixations suffice in face recognition. Psychol Sci 19(10):998–1006
Kano F, Call J (2014a) Cross-species variation of gaze following and conspecific preference among great apes, human infants and adults. Anim Behav 91:137–150
Kano F, Call J (2014b) Great apes generate goal-based action predictions: an eye-tracking study. Psychol Sci 25(9):1691–1698
Kano F, Hirata S (2015) Great apes make anticipatory looks based on long-term memory of single events. Curr Biol 25(19):2513–2517
Kano F, Tomonaga M (2010a) Attention to emotional scenes including whole-body expressions in chimpanzees (*Pan troglodytes*). J Comp Psychol 124(3):287–294
Kano F, Tomonaga M (2010b) Face scanning in chimpanzees and humans: continuity and discontinuity. Anim Behav 79(1):227–235
Kano F, Tomonaga M (2011) Species difference in the timing of gaze movement between chimpanzees and humans. Anim Cogn 14(6):879–892
Kano F, Tomonaga M (2013) Head-mounted eye tracking of a chimpanzee under naturalistic conditions. PLoS One 8(3):e59785
Kano F, Tanaka M, Tomonaga M (2008) Enhanced recognition of emotional stimuli in the chimpanzee (*Pan troglodytes*). Anim Cogn 11(3):517–524
Kano F, Hirata S, Call J, Tomonaga M (2011) The visual strategy specific to humans among hominids: a study using the gap-overlap paradigm. Vis Res 51(23):2348–2355
Kano F, Call J, Tomonaga M (2012) Face and eye scanning in gorillas, orangutans, and humans: unique eye-viewing patterns in humans among hominids. J Comp Psychol 126(4):388–398
Kano F, Hirata S, Call J (2015) Social attention in pan: bonobos exhibit more eye contacts than chimpanzees. PLoS One 10(6):e0129684
Kart-Teke E, Silva MADS, Huston JP, Dere E (2006) Wistar rats show episodic-like memory for unique experiences. Neurobiol Learn Mem 85(2):173–182

Klein JT, Shepherd SV, Platt ML (2009) Social attention and the brain. Curr Biol 19(20):958–962
Kleinke CL (1986) Gaze and eye contact: a research review. Psychol Bull 100(1):78–100
Krupenye C, Kano F, Hirata S, Call J, Tomasello M (2016) Great apes anticipate that other individuals will act according to false beliefs. Science 354(6308):110–114
Land MF, Mennie N, Rusted J (1999) The roles of vision and eye movements in the control of activities of daily living. Perception 28(11):1311–1328
Lonsdorf EV (2006) What is the role of mothers in the acquisition of termite-fishing behaviors in wild chimpanzees (*Pan troglodytes schweinfurthii*)? Anim Cogn 9(1):36–46
Lutchmaya S, Baron-Cohen S, Raggatt P (2002) Foetal testosterone and eye contact in 12-month-old human infants. Infant Behav Dev 25(3):327–335
MacLean EL, Hare B (2012) Bonobos and chimpanzees infer the target of another's attention. Anim Behav 83(2):345–353
Martin-Ordas G, Haun D, Colmenares F, Call J (2010) Keeping track of time: evidence for episodic-like memory in great apes. Anim Cogn 13(2):331–340
McIntyre, M. H., Herrmann, E., Wobber, V., Halbwax, M., Mohamba, C., de Sousa, N.,. .. Hare, B. (2009). Bonobos have a more human-like second-to-fourth finger length ratio (2D: 4D) than chimpanzees: a hypothesized indication of lower prenatal androgens. J Hum Evol, 56(4), 361–365.
Moll H, Tomasello M (2004) 12- and 18-month-old infants follow gaze to spaces behind barriers. Dev Sci 7:(1), 1–(1), 9
Mosher CP, Zimmerman PE, Gothard KM (2014) Neurons in the monkey amygdala detect eye contact during naturalistic social interactions. Curr Biol 24(20):2459–2464
Myowa-Yamakoshi M, Scola C, Hirata S (2012) Humans and chimpanzees attend differently to goal-directed actions. Nat Commun 3:693. doi:10.1038/ncomms1695
Palagi E (2006) Social play in bonobos (*Pan paniscus*) and chimpanzees (*Pan troglodytes*): Implicationsfor natural social systems and interindividual relationships. Am J Phys Anthropol 129(3):418–426
Parr LA, Hecht E, Barks SK, Preuss TM, Votaw JR (2009) Face processing in the chimpanzee brain. Curr Biol 19(1):50–53
Rilling JK, Scholz J, Preuss TM, Glasser MF, Errangi BK, Behrens TE (2012) Differences between chimpanzees and bonobos in neural systems supporting social cognition. Soc Cogn Affect Neurosci 7(4):369–379
Rosati AG, Hare B (2009) Looking past the model species: diversity in gaze-following skills across primates. Curr Opin Neurobiol 19(1):45–51
Ruiz A, Gómez JC, Roeder JJ, Byrne RW (2009) Gaze following and gaze priming in lemurs. Anim Cogn 12(3):427–434
Sandel AA, MacLean EL, Hare B (2011) Evidence from four lemur species that ringtailed lemur social cognition converges with that of haplorhine primates. Anim Behav 81(5):925–931
Sannen A, Heistermann M, Van Elsacker L, Mohle U, Eens M (2003) Urinary testosterone metabolite levels in bonobos: a comparison with chimpanzees in relation to social system. Behaviour 140(5):683
Senju A, Csibra G (2008) Gaze following in human infants depends on communicative signals. Curr Biol 18(9):668–671
Senju A, Southgate V, Snape C, Leonard M, Csibra G (2011) Do 18-month-olds really attribute mental states to others? A critical test. Psychol Sci 22(7):878–880
Shepherd SV (2010) Following gaze: gaze-following behavior as a window into social cognition. Front Integr Neurosci 4:e5
Shepherd SV, Platt ML (2008) Spontaneous social orienting and gaze following in ringtailed lemurs (*Lemur catta*). Anim Cogn 11(1):13–20
Shepherd SV, Deaner RO, Platt ML (2006) Social status gates social attention in monkeys. Curr Biol 16(4):119–120
Southgate V, Senju A, Csibra G (2007) Action anticipation through attribution of false belief by 2-year-olds. Psychol Sci 18(7):587–592

Staes N, Stevens JM, Helsen P, Hillyer M, Korody M, Eens M (2014) Oxytocin and vasopressin receptor gene variation as a proximate base for inter-and intraspecific behavioral differences in bonobos and chimpanzees. PLoS One 9(11):e113364
Tan J, Hare B (2013) Bonobos share with strangers. PLoS One 8(1):e51922
Tatler BW (2007) The central fixation bias in scene viewing: selecting an optimal viewing position independently of motor biases and image feature distributions. J Vis 7(14):1–17
Téglás E, Gergely A, Kupán K, Miklósi Á, Topál J (2012) Dogs' gaze following is tuned to human communicative signals. Curr Biol 22(3):209–212
Tomasello M (1995) Joint attention as social cognition. In: Moore C, Dunham P (eds) Joint attention: its origins and role in development. Lawrence Erlbaum, Mahwah, pp 103–130
Tomasello M, Call J, Hare B (1998) Five primate species follow the visual gaze of conspecifics. Anim Behav 55(4):1063–1069
Tomonaga M, Imura T (2009) Faces capture the visuospatial attention of chimpanzees (*Pan troglodytes*): evidence from a cueing experiment. Front Zool 6(1):14
Ueno A, Matsuzawa T (2005) Response to novel food in infant chimpanzees: do infants refer to mothers before ingesting food on their own? Behav Process 68(1):85–90
Whiten A (1992) On the nature of evolution of imitation in the animal kingdom: reappraisal of a century of research. Adv Study Behav 21:239–283
Wilkinson A, Mandl I, Bugnyar T, Huber L (2010) Gaze following in the red-footed tortoise (*Geochelone carbonaria*). Anim Cogn 13(5):765–769
Woodward AL (1998) Infants selectively encode the goal object of an actor's reach. Cognition 69(1):1–34
Yarbus AL (1967) Eye movements and vision. Plenum, New York

Chapter 10
Evolution of Vocal Communication: From Animal Calls to Human Speech

Kazuo Okanoya

Abstract The evolution of speech poses a dilemma when viewed from the signaling theory in evolutionary biology. This is because the content of speech is not honest, yet the expression of speech is mostly honest. The creativity of language rests on the dishonesty of language, but why did such a dishonest signal evolve at all? In this chapter, I will try to lay out a set of hypotheses. Animal acoustical communication perhaps started as noises contingent with breathing and jaw or respiratory gestures associated with predation or feeding. These noises and movements were gradually ritualized, forming fixed patterns of motor actions indicating intentions or emotions. Especially, vocalizations associated with respiratory action became indicators of emotional states such as fear or contentment, or that of intentions such as attack or copulation. Animal calls were thus established as an honest indicator of the internal state. In certain species, stochastic combinations of such calls were used by young animals to induce parental behavior, perhaps because they reminded parents of immature articulation. This effect was then utilized by male animals to attract females. Extremes of such vocalizations are songs, used for mate attraction and/or territorial defense by many species of birds and some species of whales and primates. Songs are an honest signal of vigor, since singing is costly. Gradually, songs came to be utilized also in nonsexual social contexts seen in, for example, gibbons. Sequences of song syllables and behavioral contexts were associated through a mutual segmentation process over generations and proto-words emerged. Arbitrary combinations of proto-words referred to non-existing entities and thus the dishonesty of speech started. However, speech was always associated with honest signals such as emotions in voice or facial expressions. Thus, the receiver could mostly judge the honesty of the speech content. Furthermore, the dishonesty of linguistic expression produced creativity and cumulative culture. In this way, emotion and language evolved in humans because of the mutual dependency between dishonesty and creativity.

K. Okanoya
The University of Tokyo, Komaba, Meguro, Tokyo, Japan
e-mail: cokanoya@mail.ecc.u-tokyo.ac.jp

S. Watanabe et al. (eds.), *Evolution of the Brain, Cognition, and Emotion in Vertebrates*, Brain Science, DOI 10.1007/978-4-431-56559-8_10

Keywords Call • Speech • Evolution • Song • Ritualization • Honest signal • Sexual selection • Domestication • Mutual segmentation (emotion, vocal learning, chunking)

10.1 Introduction

In this chapter, I propose a set of new hypotheses that explain the evolution of acoustic communication including human speech (Fig. 10.1). In doing so, I rely on the theory of honest signals in evolutionary biology (Bradbury and Vehrencamp 2012; Searcy and Nowicki 2005). I argue that animal calls evolved as a ritualization of physiological noises associated with breathing and emotional conditions. I propose that repetition of calls evoked stronger responses from the receiver resulting in sequential vocalizations and songs. Such needs most likely occurred in infant animals to elicit care from the mother or parents. Some species of infant mammals including humans emit specific "isolation" calls or "cries" when separated from the mother (Newman 2007). Begging calls are also emitted in altricial birds to beg for food (Wright and Leonard 2007).

Songs presumably were first used as a calming signal by the male to reduce aggressive behavior of the female in the copulation sequence, because songs mimicked the isolation calls or begging calls of young animals (Holy and Guo

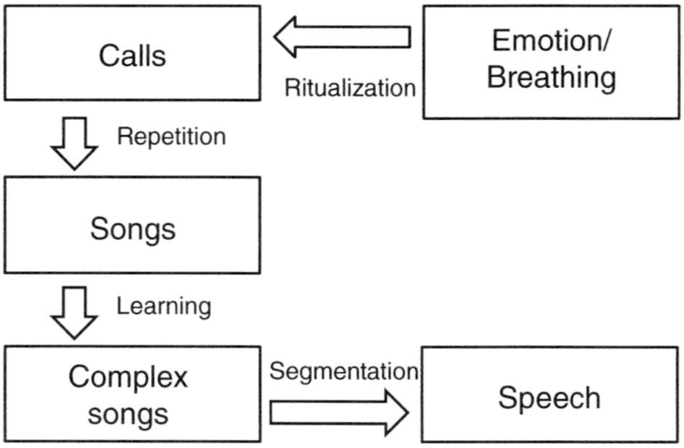

Fig. 10.1 Schematic account of the set of hypotheses I developed in this chapter. Acoustic communication started as an expression of emotion associated with breathing. Such signals were ritualized and the action patterns were fixed as calls. Repetition of calls was used by infant animals to intensify signal value for mothers or parents. Similar signals were evolutionarily mimicked by adult males to suppress escape response of females in copulation contexts. These signals are songs. Most animals sang innate songs and receivers started to extract honest information about individual vigor from these songs. Songs then became sexually selected traits. In some species, complexity was preferred as a signal of vigor and songs became a learned trait allowing further complexity. Such complex learned songs were shared in the societies of proto-humans. Mutual segmentation of behavioral contexts and song phrases led to the emergence of speech

2005). Songs then became sexually selected signals that conveyed the fitness of the singer. To be effective as a sexual signal, songs became a learnable trait in some species, because receivers preferred complexity as a signal of fitness. As complexity increased, it became more effective for songs to become an acquired trait (Nowicki and Searcy 2004).

In the hominids that eventually produced *Homo sapiens sapiens*, songs were used in various social settings. Mutual segmentation of social contexts and song phrases resulted in proto-words and became the precursor for spoken language (Merker and Okanoya 2007). When humans acquired speech, it was always used in face-to-face communication so that signal honesty was secured by emotional signals that were associated with speech production (Pentland and Heibeck 2010). But as language acquired combinatory creativity, the content of the speech itself became decoupled from signal honesty. This dual characteristic of signal honesty and dishonesty in language became a strength of language because information can be accumulated. I conclude the chapter by considering the future of human linguistic communication.

10.2 Emergence of Communicative Signals

Communication in the context of biology is defined as "the transmission of a signal from one animal to another such that the sender benefits, on average, from the response of the recipient" (Slater 1983). Since the definition does not include an intention of the signaler or a benefit to the receiver, it is useful to avoid anthropomorphic interpretations of animal behavior. The anthropomorphic views include the false notion that communication is mutually beneficial. Communication signals could have evolved as long as they were beneficial to the sender. Similarly, birds singing courtship songs do not need to have the intention to seduce females (Bradbury and Vehrencamp 2012). In this chapter, I restrict myself to vertebrate communication using acoustic signals, but allow myself to extend discussion into the emergence of human language and linguistic communication.

10.2.1 Ritualization of Respiratory Movements

Communicative acoustic signals have always started as secondary traits in vertebrate animal behavior. Acoustic signals often originate from respiratory actions because respiratory organs in vertebrates function as an air passage. Because respiration is an action that is absolutely necessary for animals to survive, the use of respiratory energy has a low physiological cost. Because the respiratory tract, the pipe connecting the bilateral lungs and the mouth opening, extends to the inside of the body, physiological conditions affect its acoustical characteristics. Coughing is associated with infection and inflammation of the respiratory tract. Strong exhalation produces noise associated with the length of the respiratory tract. Because opening the mouth is preparatory behavior to biting or attacking in predatory animals, the exhalative noise associated with the mouth opening could signal the attack. In this way, respiratory noise correlated with subsequent behavior by the signaler.

When such signals change the behavior of the receiver so that the change benefits the sender, the signals gain communicative value. For example, pup isolation calls of rodents are short and repeating ultrasonic calls that are easy to localize because there are many onset-offset cues with phase information available for the small rodent heads due to the high ultrasonic pitch. Upon detecting the isolation call, the mother quickly approaches to retrieve the pup, the sender of the call (Ehret 2005). These calls must have originated from the respiratory noises arising from the short and shrunken trachea of infant animals whose body temperature quickly fell because of isolation from the mother. Calls then must have undergone natural selection for localizability. During the process, the noise originating due to hypothermia must have become the isolation call.

10.2.2 Honest Signals

For the receivers of the signal, it is crucial that the signal reflects the true behavioral state of the sender. If not, the signal loses its value and it gradually ceases to function. The behavioral states include emotional, intentional, nutritional, and genetic (Brudzynski 2014; Searcy and Nowicki 2005). When the signal conveys enough information as to the behavioral state of the sender, the signal is defined to be "honest" in evolutionary biology (Searcy and Nowicki 2005). The honest signal bears "costs" to produce including physiological, temporal, and social. For example, birdsong requires costs in terms of neural resources, metabolism, risk of being located by a predator, and cost in terms of time (e.g., reduced time to forage or other alternative behaviors). Thus, singing can be an honest signal to indicate the singer's resourcefulness and fitness.

When considering signal honesty, human speech poses an interesting problem: the benefit of speech comes from the fact that what we say does not need to reflect what we have, resulting in signal dishonesty. We need to consider prosodic aspects of speech and content of speech separately to consider evolution of speech.

10.2.3 The Motivation-Structure Rule

Eugen Morton (1977) noticed that there are common features of vocal signals in land vertebrates. When aggressive motivation is low, calls are narrow banded and high pitched, but when it is high, calls become wide banded and low pitched. When fear is low, calls are low pitched and modulated, but when it is high, calls become

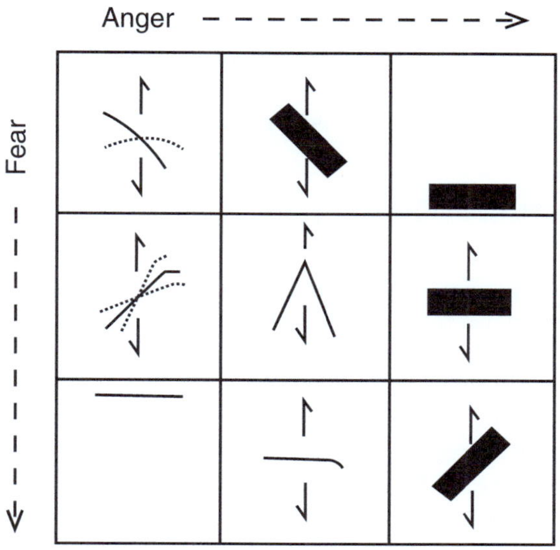

Fig. 10.2 Schematic diagrams of the motivation-structure rule suggested by Morton. Modified from Morton (1977). Each diagram is a schematic sound spectrogram. Duration is represented horizontally and pitch is represented vertically. Slanted lines represent frequency modulation and boxes represents wide-banded noise. These diagrams are arranged horizontally based on the strength of anger and vertically based on that of fear. For more details, see the text

high pitched and flat. Figure 10.2 is a schematic representation of the motivation-structure (MS) rule proposed by Morton. It is possible to provide a proximate account of this rule. First, consider the horizontal axis. Since vocalization is produced by the airway including the trachea, it reflects body size. Larger animals produce lower-pitched and wide-banded voices than smaller animals. Larger animals are more dominant than smaller ones. Thus, lower-pitched and wide-banded voices might indicate dominance and aggression. Next, consider the vertical axis. Modulation requires muscle relaxation to produce. With an increase in tension, the vocal tract will shrink and pitch will be higher.

Although the MS rule might account for some of the common features of animal vocalizations, it has a limitation. Humans, when in tension, produce more modulated voices than when they are relaxed (Aucouturier et al. 2016). Rats, at an initial stage of learning an operant discrimination task in which the rats were required to perform different behavior to different stimuli, produce more modulated calls than when the learning is advanced (Yuki and Okanoya 2014). In the case of humans, because of the complexity of the human larynx, tension may result in unstable vibration. In the case of rats, since they produce sounds in the ultrasonic range, their voice acoustics may be different from those animals that produce within audible ranges.

10.3 Emergence of Songs

Most land animals emit "calls" specific to behavioral contexts. Calls are monosyllabic, simple vocalizations. In addition to calls, some animals emit trains of various calls. Such vocalizations are often used in mating contexts. Because of the acoustic resemblance to human singing, these vocalizations are sometimes referred to as songs. It has been an enigma how songs emerged in animals. Here I first provide hypotheses regarding the emergence of animal songs. I then provide an alternative hypothesis to account for the emergence of human songs.

10.3.1 Food-Begging and Isolation Calls

In rodents, when pups are out of the nest, they emit isolation calls that induce retrieval responses from the mother. In birds, when chicks are hungry, they emit food-begging calls to make parents bring food to them. When human babies need physiological or social care, they emit baby cries. These care-inducing signals are always in the form of repeated calls. This is true in rodents, birds, and humans (Wright and Leonard 2007). Repeated signals may increase the chance of being detected, but also may increase a risk of being habituated. Infant pigmy marmosets produce repeated vocalizations when seeking care from adult animals, but they do so by combining different calls (Elowson et al. 1998). These call-repeating behaviors in young animals might be a preadaptation of songs in adults. Because these behaviors could mimic infantile characteristics, a tendency to produce randomly repeated calls may induce strong reaction from female listeners.

Supporting evidence for this infantile mimicry hypothesis comes from a neuroanatomical study in the songbird brain (Liu et al. 2009). Chicks of chipping sparrows produce variable sequences of food-begging calls. When an expression of an immediate early gene, *c-fos*, was examined in the brain of these chicks, the area corresponding to the adult RA (the song motor nucleus) showed strong activation. Partial lesions of the same area resulted in reduction of variability in food-begging calls. Results indicate that food-begging and adult songs may utilize the same neural resources. This finding supports the hypothesis that food-begging calls may be a preadaptation to songs in birds.

Another line of evidence includes neurophysiological studies with mammalian isolation calls including human cries. In rats and squirrel monkeys, lesioning the anterior cingulate cortex resulted in changes in acoustic structures of isolation calls. In human babies, crying induced activity in the same brain area (Newman 2007). In adult mice, lesioning the anterior cingulate cortex resulted in changes in temporal and acoustical structures in courtship songs (Ariaga, unpublished observation). On the other hand, a mutant mouse that lacked the formation of neocortical and hippocampal areas sung normal songs suggesting only a part of the cortex may be necessary for courtship songs (Hammerschmidt et al. 2015).

Taken together, the idea that isolation calls and food-begging calls might be precursors for adult mating songs are consistent with current data about neural mechanisms for vocal productions. Further studies are necessary to relate isolation and food-begging calls and adult mating songs in birds and mammals.

10.3.2 Singing Together to Repel Predators

While the infantile mimicry hypothesis may account for some aspects of the relationship between adult songs and infant calls, there are many questions to be answered. For example, why it is that among primates only human babies emit cries so often and so strongly. Several authors suggested that human babies became able to cry because predation pressure was suppressed and crying to obtain care became more adaptive than it was risky (Falk 2009; Nonaka et al. 2006). Tool making could be one of the factors that enabled us to prevent predation. However, tool making is a rather recent invention in hominid evolution to be useful to account for the origin of language.

There are radically new hypotheses to account for the emergence of singing proposed by a comparative musicologist, Joseph Jordania. One of his hypotheses states that *Homo sapiens* started communal singing to repel predators by synchronized, loud vocalizations (Jordania 2009). Additionally, he thinks group singing had an effect of increasing production of enkephalin and oxytocin, preparing group members for fighting. The hypothesis was based on the fact that group singing is more widespread than solo singing, and he therefore considered that the origin of human singing was group singing. The hypothesis explains why we are the only primate to sing and to have an extraordinary capacity to synchronize with external rhythm sources (Patel et al. 2009), because synchronizing vocalizations enables louder sound production and higher anti-predation effects. This hypothesis allows enough time (millions of years) for the development of songs from which speech might have arisen.

10.3.3 Songs as Honest Signals

Sequential vocalizations, especially in birds, function as honest signals of the singer's vigor. Several experimental facts support this assertion. I shall list the costs associated with song-like vocalizations. Because of these costs, singing remains an honest signal. First, it is time consuming. While singing, the singer cannot engage in other behavior such as foraging, eating, and nursing. Second, singing requires neural resources. In songbirds, males dedicate a large brain space to store neural circuits required to learn and perform singing. Because of this cost, some species like canaries adaptively change the volume of song control nuclei so that it is at a maximum when they are in breeding season, but the nuclei shrink when the season ends (Nottebohm et al. 1986). Third, the act of singing involves metabolic costs. The average metabolic rate for singing is 1.05–1.07 times higher than for not singing, and it is 2.2–2.6 times the basal metabolic rate (Ward et al. 2003). Fourth, singing enables predators to spot the singer more easily. Chaffinches singing at a risky perch, when actually encountering a risk, will sing at a less risky perch (Krams 2001). By singing at a risky perch, the male can demonstrate his vigor. Fifth, singing has a cognitive cost and it might lead to predation risk. When scared by flashlights, male Bengalese finches sometimes stop singing (Seki et al. 2008). When singing stereotyped sequences, birds stopped less often than when singing variable sequences. Thus, crystallized, stereotyped songs are less amenable to attention shifting, and this might increase predation risk.

10.4 Signal Complexity

Once a song appeared in an animal's vocal repertoire, the song might increase in complexity to fulfill its function. There are several ways in which songs increase in complexity. Here I use a case study of Bengalese finches, on which my colleagues and I had been working more than 25 years, to show these factors. The two factors are sexual selection and domestication. Bengalese finches are a domesticated form of wild white-rumped munias (Washio 1996). The process of domestication started in 1764 when a feudal lord (Daimyo) of Kyushu, Japan, imported white-rumped munias from China. The birds were initially imported for their strong parental instinct, by which they can incubate and rear the eggs of other birds. The white-rumped munia is a brownish bird with a white patch on its back. About 130 years ago, a mutation appeared in its feather pigment and birds with brown and black patches on a white base appeared. This mutation was highly valued by aviculturists and the finch became a popular bird known as the Bengalese finch. Bengalese finches have been bred for their tameness, white color pattern, and reproductive behavior. Moreover, their songs became much more complex than their wild ancestors, although there has been no record of artificial selection for their song. I attribute the evolution of song complexity in Bengalese finches to the following two factors.

10.4.1 Sexual Selection

Evolution of song complexity in Bengalese finches may be related to female choice of song complexity in the domesticated environment. This consideration led us to examine female responses to song complexity. Two experiments were designed and conducted using female Bengalese finches and munias. First, we tested female preference for song complexity by counting nesting materials carried by females when stimulated with songs of varying complexity. In this study, not only female Bengalese finches but also female white-rumped munias preferred syntactically complex songs over simple ones (Okanoya 2004). In another choice experiment in female Bengalese finches, four out of eight birds approached toward the complex song, one to the simple song and three for both songs (Morisaka et al. 2008). Song preference in female Bengalese finches varied individually, but the overall tendency detected in this experiment could lead to the evolution of complex songs in male Bengalese finches as shown by a simple population dynamics model. We then examined the physical correlates of song complexity in male Bengalese finches. We found the principal component of body size correlated well with the song bout length (Soma et al. 2008) suggesting that healthier males could produce longer songs that have more room for complexity. Because these finches were under domestication for over 250 years, female choice must have been indirect. That is, females were not actively selecting males for song complexity, but when paired with complex singers, females devoted more of their reproductive effort (Soma et al. 2009).

10.4.2 Domestication

Humans have domesticated many species for various purposes. Some of these animals show curious similarities with each other: loss of pigmentation on a part of the body surface, round face and weak biting force, and decreased aggressiveness associated with decreased cortisol level. These changes are collectively called "domestication syndrome," and a recent theoretical analysis proposed these may be due to delayed migration of neural crest cells during embryogenesis (Wilkins et al. 2014). This hypothesis states that the adrenal medulla (that secrets hormones related with stress coping), pigmentation cells, and jawbones are all descended from neural crest cells produced during embryogenesis. By selecting tame individuals, humans may have been selecting individuals with slower migration of these neural crest cells.

Because Bengalese finches originated from white-rumped munias imported from China some 250 years ago, there should have been selection for tameness that secured survival during the course of long travel. Munias were initially used as foster parents for foreign birds, and this required further tameness and forbearance, and aggressive individuals would have been removed from the stock. Parenting in a small cage requires stress tolerance, and levels of stress hormones needed to decline. These requirements probably led to selection for the delayed migration of neural crest cells, which resulted in an overall white appearance and tameness in domesticated Bengalese finches.

To examine whether the neural crest hypothesis of domestication syndrome (Wilkins et al. 2014) applies to Bengalese finches, we examined several socio-emotional factors in Bengalese finches and, when possible, compared them with white-rumped munias. Biting force was examined in both strains by holding the bird and challenging it with a stick equipped with a piezo-electric sensor (Suzuki et al., unpublished). Munias bit twice as often, and the biting force was twice as strong as for the Bengalese finches. This showed that Bengalese finches are less aggressive and their bill muscles are less developed. Fearfulness was examined by a tonic-immobility test (Suzuki et al. 2013). The bird was held on its back for 15 s and then released. The time to move and the time to fly away were measured. White-rumped munias took three times as long as Bengalese finches to fly away. Results indicated that Bengalese are less fearful than munias. Lastly, fecal corticosterone level was measured in both strains of birds. Bengalese showed corticosterone levels half that of munias, indicating lower stress levels in domesticated Bengalese finches (Suzuki et al. 2012).

These results are consistent with the neural crest hypothesis and further explain some other data. For example, we found that in a free-flight cage, Bengalese chicks learned songs not only from their fathers but also from other males (Takahasi et al. 2010). A similar experiment was conducted with munias, but they only learned from their fathers (Kagawa et al., unpublished data). This may, in part, be due to decreased fearfulness and increased tameness in Bengalese finches. Furthermore, the smaller neural structure in munias than in Bengalese might be explained by the decreased corticosterone level in Bengalese finches. Mineralo- and gluco-corticosterone receptors coexist in the HVC (the brain region necessary to learn song) of Bengalese finches. It is known in rodents that with a higher degree of corticosterone level, these receptors function to suppress neural growth (Abdanipour et al. 2015). Similar mechanisms might be functioning in the brains of Bengalese and munias, resulting in more developed neural tissues in Bengalese finches.

It is well accepted that hominoids "domesticated" themselves over millions of years (Wells et al. 2009), and the process of domestication included the selection of calm, sociable individuals to enable group living and strengthen collaborative activities. By protecting themselves from predators, hominoids could have spared more energy for sexual rituals (Miller 2000). Vocal plasticity might be a feature that is either sexually selected or that evolved through relaxation (Wells et al. 2009). Vocal plasticity prepared a behavioral vehicle on which thought could be organized. Like many domesticated animals, some of these traits might be related to the cells descended from the neural crest (Wilkins et al. 2014).

10.5 Emergence of Speech

Here my challenge is to place the emergence of human speech in a continuous evolutionary line with the emergence of songs and evolution of song complexity in nonhuman animals. Earlier, I raised two hypotheses in relation to this: the infantile mimicry hypothesis and the predator repelling hypothesis. Both have pros and cons, but both could account for the emergence of song-like vocalizations in humans. However, it still remains puzzling that humans are the only primate to have learned songs. To demonstrate the continuum of development with other primates, I will first examine songlike behavior in nonhuman primate, and then, I will propose a hypothesis related to the emergence of speech out of songs.

10.5.1 Gibbon Songs and Gelada Lip-Smacking

Gibbons are one of the five ape groups including humans, chimpanzees, gorillas, and orangutans. Gibbons, because they are not great apes, are the most distant from humans among apes. Gibbons do have song-like vocalizations

(Geissmann 2002) but they are not learned, as indicated by cross-fostering studies (Merker and Cox 1999). Nevertheless, gibbon songs are quite diverse (Clarke et al. 2006) and not only used in a mating context but also in many other social contexts (Inoue et al. 2012). In Muller's gibbons, male calls consist of two simple types: a frequency modulated "wa" call and a constant "o" call. Combinations of these calls and behavioral contexts were correlated, meaning that gibbons might exchange contextual information via the combination of calls.

The gelada is a species of primate with a rich vocal repertoire. They also make a facial expression with lip-smacking of 3–8 Hz used as an affiliative signal. On some occasions, their lip-smacking is presented with vocal sounds, making this behavior highly similar to human speech production (Bergman 2013). Other primates including macaques also show lip-smacking, and this behavior might be one of the precursors to human speech (Ghazanfar et al. 2012).

Both of these behaviors, if combined with the bird-like ability of vocal learning, would provide a basis for the emergence of human speech. In the next section, I will advance a hypothesis based on the assumption that protohumans were singing and vocal-learning primates.

10.5.2 Mutual Segmentation Hypothesis

How might song-like behavior in some primate species be connected with speech in humans? We proposed a conceptual model for this process (Merker and Okanoya 2007). In our model, each behavioral context is denoted by a particular song in a protohuman society. Consider the hypothesis that prior to language, protohumans developed singing behavior associated with several social contexts. If songs became a learned property as they are in some species of birds and whales, a phrase of syllables may be shared by more than one song. Then, likewise, parts of behavioral contexts in which a song is sung may also be shared by more than one song. For example, a song sung when hunting (song H) and a song sung when dining (song D) might have shared the same phrase h&d. Furthermore, song H and song D shared the context of doing something together. After a while, by singing the shared phrase h&d, the singer could have specified the context of "let's do that together." By repeating this process, holistic songs might have been decomposed into specific phrases and these phrases might have become proto-words. I call this the mutual segmentation hypothesis of song phrases and song contexts (Merker and Okanoya 2007; Okanoya and Merker 2007) (Fig. 10.3). Once the process of mutual segmentation started and segmented short utterances became associated with segmented restricted contexts, rudimentary forms of speech communication could have started. After this, nonbiological, cultural processes could then regulate the emergence of syntactical structures.

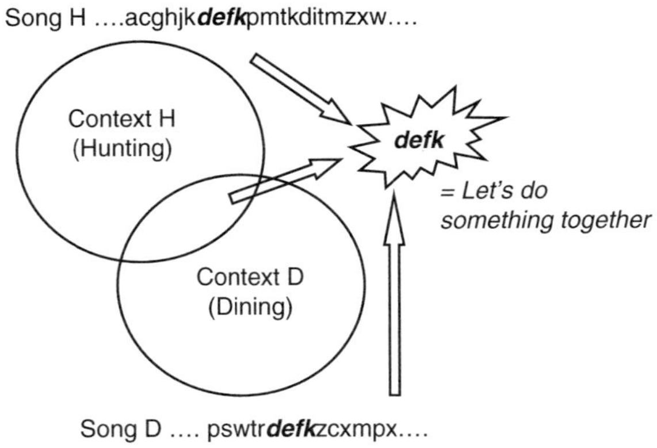

Fig. 10.3 Mutual segmentation of song phrases and behavioral contexts. When two songs had a common phrase and contexts under which songs were sung had a common context, song (part) phrase and (part) context are mutually segmented and associated. The segmented very short song phrase came to denote the segmented specific context

10.5.3 Neural Substrates of the Mutual Segmentation Hypothesis

For the above process to function, it is important to demonstrate that the faculty of segmentation is widespread and supported by a general neural mechanism among vertebrates. We showed this by observations of natural song learning in Bengalese finches (Takahasi et al. 2010) and by electroencephalogram recording in human neonates (Kudo et al. 2011).

Eleven male and ten female adult Bengalese finches were kept in a large aviary with 11 pot-nests attached for breeding. Thirty-two male chicks were born and reared in that environment and the resulting songs were compared with those of adult males. Male chicks learned, on average, from two male tutors by segmenting parts of tutor songs and putting the segmented parts together to yield individually distinctive songs. Transition probabilities from one song note to another were among the cues for segmentation: birds segmented where the probability was low. Another cue was inter-note interval: birds segmented parts where inter-note intervals were longer. Looking from the other side, birds were chunking song notes based on higher transition probabilities and shorter inter-note intervals. In adult Bengalese finches, lesioning the premotor area or the basal ganglia (Okanoya 2004) resulted in changes in song sequence, suggesting that these areas are involved in statistical segmentations.

Electroencephalograms were recorded from human neonates within 72 h of birth. We played tone sequences that were statistically organized so that four triplets each lasted 550 ms were played in random order without inter-triplet spaces. Event-related potentials (ERP) were obtained from the medial-frontally placed electrode (FCz). After exposed to the triplet-tone sequences repeatedly for 5 min, only the first sound of the triplet evoked a significantly strong positive response, suggesting that the neonates were segmenting the triplets based on transition probabilities. The same ERPs were recorded from adult participants and we were able to show the N400 associated with the first sound of the triplets again suggesting that participants were segmenting the stimulus stream based on transition probabilities (Abla et al. 2008). Further studies of adult participants using functional magnetic resonance imaging suggested that the basal ganglia were activated during the initial phase of statistical learning, but in the later phase, the premotor areas were activated (Katahira et al. unpublished).

Results in birds, human adults, and newborns all suggest that the premotor area and the basal ganglia are involved in segmentation and chunking of sound sequences. Since these areas are common in the vertebrate brain, the faculty for sequence segmentation/chunking must be also general in vertebrates. Thus, segmentation and chunking, which are basic faculties to produce sequential complexity, are shared by a wide range of animals. Based on this system, sexual selection of sequential signals in birds and perhaps in humans produced signal complexity on which thought could be implemented.

For mutual segmentation to occur, we need also to prepare contextual segmentation. We routinely segment behavioral contexts based on multiple environmental and internal cues and this would be a rudimentary form of semantics. A somewhat simpler example is spatial navigation. The idea that the hippocampus may function as a spatial map was first presented by O'Keefe and Dostrovsky (1971) with some experimental evidence. When a rat navigates a novel environment freely, a particular neuron in the hippocampus fires when the rat is exploiting a particular part of the environment. This finding was later expanded to include modeling research.

In general, it was shown that with Hebbian learning and lateral inhibition, a network of suitably interconnected neurons begins to fire when a specific class of inputs is present. Such a network can establish attractor dynamics and several attractors can specify environmental locations. Not only that, since the hippocampus receives sensory, emotional, and reinforcement information from different brain areas, this structure is suitable for segmenting a multidimensional behavioral context (Gluck and Myers 1993). In fact, an experiment by O'Keefe's group showed that these cells not only respond to specific locations but also begin to behave as attractors for specific environmental shapes (Wills et al. 2005). This is in good agreement with our current hypothesis that the hippocampus could function as a bottom-up categorizer for behavioral contexts in general. In common with many other learning networks, such a categorizing network will show the capacity for generalization (Ghirlanda and Enquist 2003), supplying the other essential prerequisite for the type of historical structural transformation.

While the hippocampus functions as a bottom-up categorizer, top-down effects are probably governed by the prefrontal cortex. While the learning rate of the neocortex is slow, the hippocampus learns rapidly based on concurrent, local, and time-limited information (O'Reilly and Rudy 2000). Thus, the hippocampus can segment contextual parameters quickly, and the result of segmentation is tested by statistical or rule-based prediction by the prefrontal cortex. The amygdala, another limbic structure, might also function in segmentation of emotional contexts (Fenker et al.

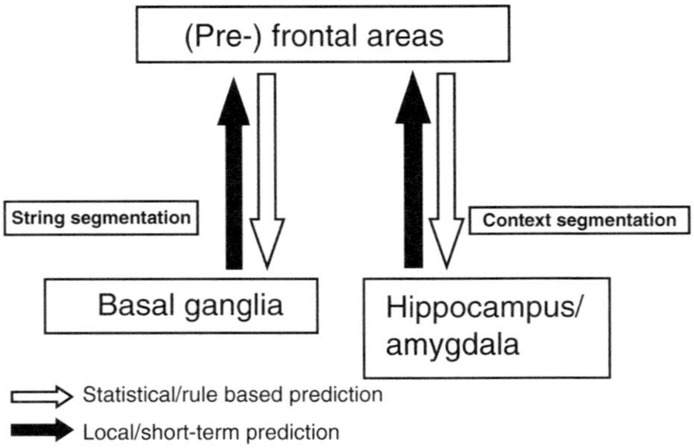

Fig. 10.4 Neural substrate for mutual segmentation. For song string segmentation, the loop structure between the prefrontal cortex and the basal ganglia is functioning. For context segmentation, the loop structure between the prefrontal cortex and the hippocampus and amygdala are functioning. From the prefrontal cortex, statistical or rule-based prediction of what is coming next is provided, while from the basal ganglia and the limbic area local and short-term prediction is provided

2005).This is similar to the case of string segmentation in which basal ganglia give rise to short-term statistics of ongoing stimuli; the prefrontal cortex is more likely to function at a slower rate with a longer time constant (Fig. 10.4).

10.6 Specificity of Linguistic Communication

As explained, I hypothesize proto-speech emerged by the process of mutual segmentation of song string and behavioral contexts (Merker and Okanoya 2007; Okanoya and Merker 2007). After speech gained the combinatory property by which new expressions became possible, the speech signal could now point to non-existing or imaginary entities. This was the beginning of imagination. By freely combining concepts that were not associated, humans came to develop their imagination and creative thinking. However, at the same time, this was the start of manipulative communication, because with language, anything could be expressed without grounding that with the trait of the speaker. This also made language a dishonest signal in the sense of signal honesty in evolutionary biology (Bradbury and Vehrencamp 2012). Nevertheless, humans continued using language once it was evolutionarily acquired. Why is this so?

10.6.1 Speech: Honest and Dishonest Components

One of the reasons why language, a dishonest signal, survived could be because language as expressed speech has multiple components. Speech is vocal behavior used in face-to-face contexts. This means that speech, in its original mode, is used

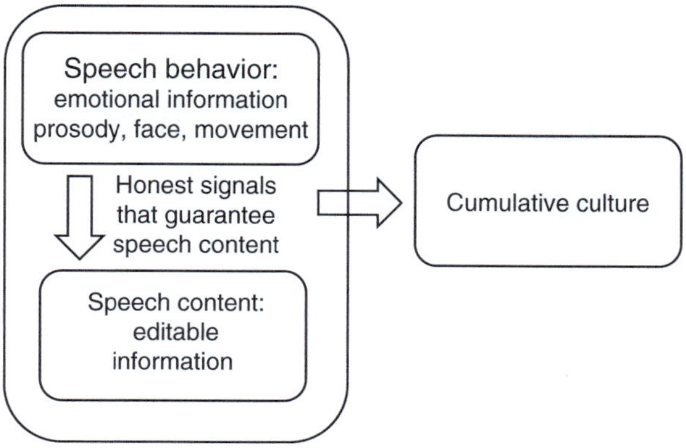

Fig. 10.5 Speech content is supported by the honesty of speech behavior. See text for details

in real time, in proximity, and with visual information. Speech behavior includes emotional information such as prosody, facial expression, and bodily movement. This emotional information mostly consists of honest signals, because they cannot be manipulated intentionally (Zuckerman et al. 1979). At the same time, of course, the content of speech is editable information. In face-to-face communication, if the speech content contained false information intentionally, prosodic or facial emotion would convey that the content was not true. Honesty of speech content was thus guaranteed by honesty of speech behavior (Fig. 10.5). In this way, human speech was used and evolved as a useful tool to accumulate knowledge.

10.6.2 The Future of Human Communication

The above scenario might account for the evolution of speech up to the invention of telecommunication. As electrical devices for telecommunication advanced, the face-to-face mode of speech communication started to lose its position as the primary mode of communication (Fig. 10.6). In modern society, a great deal of work is done through telecommunication devices in which most of the information is text based. We examined how emotional content could be transmitted in telecommunication devices and found that the sense of emotion transmission is very low in text-based communication (Arimoto and Okanoya 2015).

Although text-based communication is efficient in terms of time, cost, and accuracy of both parties, it lacks the signal honesty necessary for fruitful communication. Additionally, since devices develop much more quickly than one generation of humans, different generations are imprinted with different means of information transfer (Kelly 2016). This creates difficulties in developing mutual understanding between generations, and long-lasting traditions of respecting elders for their experiences are no longer considered valid (i.e., not efficient). Most current social problems are rooted in these simple facts. Now is the time to consider how we should design future means of communication.

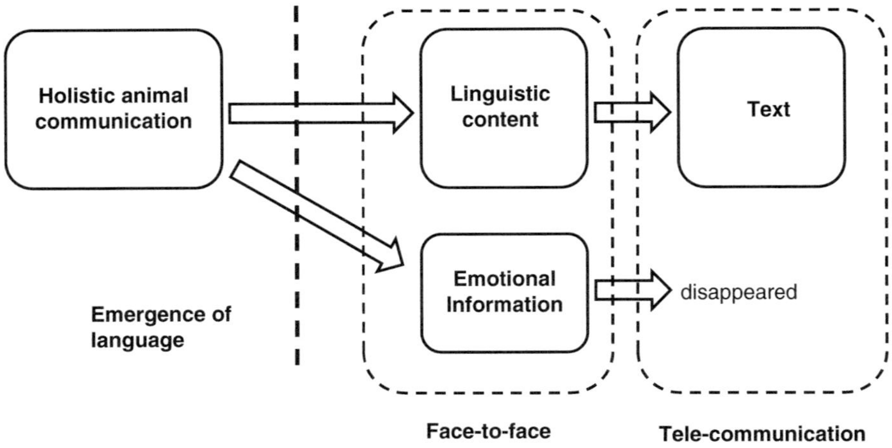

Fig. 10.6 Since the emergence of language, communication content had been divided into linguistic content and emotional information. Since the invention of telecommunication, text content alone is often conveyed, but the accompanying emotional information is often left behind

10.7 Conclusion

In this chapter, I reviewed the literature on the evolution of acoustic communication in animals. I developed a set of hypotheses to account for the emergence of human speech and language in line with the evolution of animal communication. I found that a discontinuity occurred when humans started to use devices for telecommunication, since these remove emotional information supporting the honesty of linguistic content. I considered that this might change the way humans use language.

References

Abdanipour A, Sagha M, Noori-Zadeh A, Pakzad I, Tiraihi T (2015) In vitro study of the long-term cortisol treatment effects on the growth rate and proliferation of the neural stem/precursor cells. Neurol Res 37(2):117–124

Abla D, Katahira K, Okanoya K (2008) On-line assessment of statistical learning by event-related potentials. J Cogn Neurosci 20(6):952–964

Arimoto Y, Okanoya K (2015) Mutual emotional understanding in a face-to-face communication environment: how speakers understand and react to listeners' emotion in a game task dialog. Acoust Sci Technol 36(4):370–373

Aucouturier J-J, Johansson P, Hall L, Segnini R, Mercadié L, Watanabe K (2016) Covert digital manipulation of vocal emotion alter speakers' emotional states in a congruent direction. Proc Natl Acad Sci U S A 113(4):948–953

Bergman TJ (2013) Speech-like vocalized lip-smacking in geladas. Curr Biol 23(7):R268–R269

Bradbury JW, Vehrencamp SL (2012) Principles of animal communication. Sinauer Associates, Inc., Sunderland, MA

Brudzynski SM (2014) Social origin of vocal communication in rodents. In: Biocommunication of animals. Springer, Dordrecht, pp 63–79

Clarke E, Reichard UH, Zuberb hler K (2006) The syntax and meaning of wild gibbon songs. PLoS One 1(1):e73

Ehret G (2005) Infant rodent ultrasounds–a gate to the understanding of sound communication. Behav Genet 35(1):19–29

Elowson AM, Snowdon CT, Lazaro-Perea C (1998) Infant'babbling'in a nonhuman primate: complex vocal sequences with repeated call types. Behaviour 135(5):643–664

Falk D (2009) Finding our tongues: mothers, infants and the origins of language. Basic Books, New York

Fenker DB, Schott BH, Richardson-Klavehn A, Heinze HJ, Düzel E (2005) Recapitulating emotional context: activity of amygdala, hippocampus and fusiform cortex during recollection and familiarity. Eur J Neurosci 21(7):1993–1999

Geissmann T (2002) Duet-splitting and the evolution of gibbon songs. Biol Rev 77(1):57–76. doi:10.1017/s1464793101005826

Ghazanfar AA, Takahashi DY, Mathur N, Fitch WT (2012) Cineradiography of monkey lip-smacking reveals putative precursors of speech dynamics. Curr Biol 22(13):1176–1182

Ghirlanda S, Enquist M (2003) A century of generalization. Anim Behav 66(1):15–36

Gluck MA, Myers CE (1993) Hippocampal mediation of stimulus representation: a computational theory. Hippocampus 3(4):491–516

Hammerschmidt K, Whelan G, Eichele G, Fischer J (2015) Mice lacking the cerebral cortex develop normal song: insights into the foundations of vocal learning. Sci Rep 5:8808

Holy TE, Guo Z (2005) Ultrasonic songs of male mice. PLoS Biol 3(12):e386

Inoue Y, Sinun W, Yoshida S, Okanoya K (2012) Male gibbons change the note order according to the behavioral situations. In: The evolution of language: proceedings of the 9th international conference (EVOLANG9), Kyoto, Japan, 13–16 March 2012. World Scientific, Singapore, p 458

Jordania J (2009) Times to fight and times to relax: singing and humming at the beginnings of human evolutionary history. Kadmos 1:252–276

Kelly K (2016) The inevitable: understanding the 12 technological forces that will shape our future. Viking Press, New York

Krams I (2001) Perch selection by singing chaffinches: a better view of surroundings and the risk of predation. Behav Ecol 12(3):295–300

Kudo N, Nonaka Y, Mizuno N, Mizuno K, Okanoya K (2011) On-line statistical segmentation of a non-speech auditory stream in neonates as demonstrated by event-related brain potentials. Dev Sci 14(5):1100–1106. doi:10.1111/j.1467-7687.2011.01056.x

Liu W-c, Wada K, Nottebohm F (2009) Variable food begging calls are harbingers of vocal learning. PLoS One 4(6):e5929

Merker B, Cox C (1999) Development of the female great call in *Hylobates gabriellae*: a case study. Folia Primatol 70(2):97–106

Merker B, Okanoya K (2007) The natural history of human language: bridging the gaps without magic. In: Emergence of communication and language. Springer, London, pp 403–420

Morisaka T, Katahira K, Okanoya K (2008) Variability in preference for conspecific songs with syntactical complexity in female Bengalese finches: towards an understanding of song evolution. Ornithol Sci 7(1):75–84

Morton ES (1977) On the occurrence and significance of motivation-structural rules in some bird and mammal sounds. Am Natur 111, 855–869

Newman JD (2007) Neural circuits underlying crying and cry responding in mammals. Behav Brain Res 182(2):155–165

Nonaka Y, Kudo N, Okanoya K (2006) Acoustical development of the human baby cry: anatomical and social factors. J Acoust Soc Am 120(5):3190–3190

Nottebohm F, Nottebohm ME, Crane L (1986) Developmental and seasonal changes in canary song and their relation to changes in the anatomy of song-control nuclei. Behav Neural Biol 46(3):445–471

Nowicki S, Searcy WA (2004) Song function and the evolution of female preferences: why birds sing, why brains matter. Ann N Y Acad Sci 1016(1):704–723. doi:10.1196/annals.1298.012

Okanoya K (2004) Song syntax in Bengalese finches: proximate and ultimate analyses. Adv Study Behav 34:297–346
Okanoya K, Merker B (2007) Neural substrates for string-context mutual segmentation: a path to human language. In: Emergence of communication and language. Springer, London, pp 421–434
O'Keefe J, Dostrovsky J (1971) The hippocampus as a spatial map: preliminary evidence from unit activity in the freely-moving rat. Brain Res 34(1):171–175
O'Reilly RC, Rudy JW (2000) Computational principles of learning in the neocortex and hippocampus. Hippocampus 10(4):389–397
Patel AD, Iversen JR, Bregman MR, Schulz I (2009) Experimental evidence for synchronization to a musical beat in a nonhuman animal. Curr Biol 19(10):827–830. doi:10.1016/j.cub.2009.03.038
Pentland A, Heibeck T (2010) Honest signals: how they shape our world. MIT Press, London
Searcy WA, Nowicki S (2005) The evolution of animal communication: reliability and deception in signaling systems. Princeton University Press, Princeton
Seki Y, Suzuki K, Takahasi M, Okanoya K (2008) Song motor control organizes acoustic patterns on two levels in Bengalese finches (*Lonchura striata* var. domestica). J Comp Physiol A 194(6):533–543. doi:10.1007/s00359-008-0328-0
Slater PJB (1983) The study of communication. In: Halliday TR, Slater PJB (eds) Communication, vol 2. Blackwell Scientific, New York, pp 9–42
Soma M, Hiraiwa-Hasegawa M, Okanoya K (2008) Early ontogenetic effects on song quality in the Bengalese finch (*Lonchura striata* var. domestica): laying order, sibling competition, and song syntax. Behav Ecol Sociobiol 63(3):363–370. doi:10.1007/s00265-008-0670-9
Soma M, Hiraiwa-Hasegawa M, Okanoya K (2009) Song-learning strategies in the Bengalese finch: do chicks choose tutors based on song complexity? Anim Behav 78(5):1107–1113. doi:10.1016/j.anbehav.2009.08.002
Suzuki K, Yamada H, Kobayashi T, Okanoya K (2012) Decreased fecal corticosterone levels due to domestication: a comparison between the white-backed Munia (*Lonchura striata*) and its domesticated strain, the Bengalese finch (*Lonchura striata* var. domestica) with a suggestion for complex song evolution. J Exp Zool A Ecol Genet Physiol 317(9):1–9
Suzuki K, Ikebuchi M, Okanoya K (2013) The impact of domestication on fearfulness: a comparison of tonic immobility reactions in wild and domesticated finches. Behav Process 100:58–63
Takahasi M, Yamada H, Okanoya K (2010) Statistical and prosodic cues for song segmentation learning by Bengalese finches (*Lonchura striata* var. domestica). Ethology 116(6):481–489. doi:10.1111/j.1439-0310.2010.01772.x
Ward S, Speakman JR, Slater PJ (2003) The energy cost of song in the canary, *Serinus canaria*. Anim Behav 66(5):893–902
Washio K (1996) Enigma of Bengalese finches (in Japanese). Kindai-Bungei-Sha, Tokyo
Wells T, Dunn A, Sergeant MJ, Davies MN (2009) Multiple signals in human mate selection: a review and framework for integrating facial and vocal signals. J Evol Psychol 7(2):111–139
Wilkins AS, Wrangham RW, Fitch WT (2014) The "domestication syndrome" in mammals: a unified explanation based on neural crest cell behavior and genetics. Genetics 197(3):795–808
Wills TJ, Lever C, Cacucci F, Burgess N, O'Keefe J (2005) Attractor dynamics in the hippocampal representation of the local environment. Science 308(5723):873–876
Wright J, Leonard ML (2007) The evolution of begging: competition, cooperation and communication. Springer Science & Business Media, Berlin
Yuki S, Okanoya K (2014) Behavioral correlates of 50-kHz ultrasonic vocalizations in rats: progressive operant discrimination learning reduces frequency modulation and increases overall amplitude. Anim Behav Cogn 1(4):452–463
Zuckerman M, DeFrank RS, Hall JA, Larrance DT, Rosenthal R (1979) Facial and vocal cues of deception and honesty. J Exp Soc Psychol 15(4):378–396
Miller G (2000) The mating mind: how sexual choice shaped the evolution of human nature. Doubleday Books, New York

Chapter 11
Integration Hypothesis: A Parallel Model of Language Development in Evolution

Shigeru Miyagawa

Abstract There are generally two views of how language emerged in evolution: emergent and gradual. The emergent view holds that language appeared relatively rapidly within the last 100,000 years, possibly due to some minor mutation. The gradualist view postulates stages of "protolanguage" that began as a simple system that progressively developed into ever-complex systems until language as we know it emerged. The original protolanguage may have been singing, as Darwin conjectured, or lexical in nature as proposed by a number of linguists. Human language is enormously rich and complex, which makes it difficult to imagine that all the components of it emerged somehow out of the blue in recent evolutionary time, yet there is no evidence for such a system earlier in evolution. The Integration Hypothesis holds that language is an integration of two independently occurring systems in nature that underlie communication. One system, exemplified by the alarm calls of primates, is the lexical system, which is composed of isolated units of utterance that typically have a specific referent, such as "leopard," "snake," and "eagle," we see in the calls of vervet monkeys. The expression system, associated with birdsong, creates patterns without the use of lexical items. Each system has developed over a long span of time, millions and possibly hundreds of millions of years. At some point in recent evolutionary time, the two systems, L and E, integrated uniquely in humans to give rise to language, which gives the appearance of rapid emergence. I will speculate on how the integration may have been triggered.

Keywords Language • Darwin • Birdsong • Alarm calls • Gibbons • Integration hypothesis • Finite-state grammar • Duality of semantics • Movement • Gradualist view of language evolution • Emergent view of language evolution

---

S. Miyagawa
Department of Linguistics and Philosophy, MIT, 32D-836, Cambridge, MA 02139, USA
e-mail: miyagawa@mit.edu

S. Watanabe et al. (eds.), *Evolution of the Brain, Cognition, and Emotion in Vertebrates*, Brain Science, DOI 10.1007/978-4-431-56559-8_11

11.1 Introduction

Language is a defining trait of who we are as human beings. We use it constantly and in many ways, it is fundamental to our unique successes as a species. How did human language emerge in evolution? Our best estimation is that it appeared within the past 100,000 years, before our ancestors started to migrate out of Africa into the Eurasian continent. Because language doesn't fossilize, we have no direct evidence for what precisely happened to give rise to language. One might wonder if it is even worth asking the question about language in evolution if we cannot come up with direct proof.

Indeed, back in 1866, the Linguistic Society of Paris banned all debate about evolution of language because of lack of empirical evidence. It isn't suitable for serious study, they said. But we are in a very different world today, a world in which we know a great deal more about our brain and the brain of other animals, primate communication, birdsong, and, most importantly, human language, some or all of which will form a promising foundation for exploring the question of how language emerged in evolution.

The great amount of knowledge we have amassed over the past 150 years could help to solve many of the puzzles that form the mystery behind the emergence of language. But that is not the only reason why we should pursue this question. The emergence of language is one of the great mysteries of evolution and, some say, one of the hardest problems in science (Christiansen and Kirby 2003). This great mystery about language and the mind that produces it is all the more perplexing when we note there is an enormous gap between our closest relatives, primates like the chimpanzees and the bonobos, and humans in terms of our cognitive abilities. Our closest relatives don't produce art, they don't cook with fire, they don't fight for ideals, and, of course, they don't have language like we do. It is a trait unique to humans and not shared with other branches of the same monophyletic group (Tallerman and Gibson 2012). Other members of the group do have modes of communication, but human language possesses richness and complexity far beyond anything else that we see in the animal kingdom. The mind that makes language possible is so much more powerful than what we see elsewhere in the animal world that Darwin himself commented, "the difference between the mind of the lowest man and that of the highest animal is immense" (Darwin 1871: 100). In a similar vein, Alfred Russel Wallace, a leading evolutionary thinker who was a contemporary of Darwin, noted that evolution should have endowed the humans with a brain a little better than that of an ape, yet what we ended up with is far more powerful than what a gradual evolutionary adaption would predict.

The problem of how language emerged in evolution is complex, just as complex as language itself. Not surprisingly, just as there is debate about every aspect of human language including its core function—is it primarily for internally representing thought, is it for communicating ideas and emotions externally, or is it for some other purpose?—there is debate about what evolutionary steps contributed to the development of human language. I will review some of the major proposals from

Darwin (1871) to more recent proposals by linguists, particularly by Bickerton (1990, 2000, 2014) and Jackendoff (1999, 2002), and an extension of it by Progovac (2010, 2015). I will argue that there are promising components to these proposals but also shortcomings. I will then present the Integration Hypothesis of language evolution (Miyagawa et al. 2013, 2014), which incorporates Darwin's idea that language began as song much like birdsong, and also the proposal by linguists that there was a phylogenetic precursor of language, commonly referred to as "protolanguage" (e.g., Hewes 1973). An important question that has been posed about language is how quickly did it emerge? Was it gradual, traversing through possibly many stages of protolanguage before arriving at language as we know it today? This is the so-called gradualist view, and it has many prominent proponents (e.g., Bickerton 1990, 1995, 1998, 2000, 2014; Pinker and Bloom 1990; Newmeyer 1991, 1998; Pinker 1994; Jackendoff 1999, 2002; Tallerman 2007; Hurford 2012; Progovac 2006, 2008, 2009, 2010, 2012, 2015; Progovac and Locke 2009). Or did the essence of today's language form rapidly? This is the so-called emergent view (Berwick 1998; Hauser et al. 2002; Chomsky 2005, 2008, 2010, 2012, 2013, 2017; Berwick and Chomsky 2011; Di Sciullo 2011, 2013, 2014; Bolhuis et al. 2014; Miyagawa et al. 2013, 2014; Nóbrega and Miyagawa 2015). The rapid-development hypothesis comes in two versions, the better known being that something happened, possibly a mutation, that led to a rapid emergence of language where nothing like it existed before (e.g., Berwick and Chomsky 2011). The other view is that language arose rapidly from integrating preadapted systems that existed independently in nature (Miyagawa et al. 2013, 2014). The latter is the Integration Hypothesis, which I will take up to show its advantages and also a shortcoming, a shortcoming shared with the gradualist view proposed by Bickerton, Jackendoff, and Progovac. I will speculate on a possible solution to the problem, which will allow us to utilize Darwin's "song" idea and combine it with the proposals by linguists that are based on a sophisticated knowledge of human language.

11.2 Did Protolanguage Exist?

Homo sapiens emerged in Africa some 200,000 years ago, and their brain continued to evolve. It was perhaps some 60,000–100,000 years ago that a number of things appeared that demonstrate enhanced cognitive capability, including refined tools, carved and painted art, and sophisticated weapons. It isn't clear whether language emerged concurrent with these other achievements of high-order cognitive capability, but certainly language requires immense cognitive capability that is consistent with these other achievements of the *Homo sapiens*. Another piece of evidence for when language emerged is the discovery of the so-called ochres in the Blombos Caves in South Africa. Ochre is an iron-rich mineral, and more than 8000 pieces of ochre-like material have been found in the Blombos Caves, and they have been dated back to 75,000–100,000 years ago. Some, like this one, have engravings and incisions (Fig. 11.1).

Fig. 11.1 Ochres from the Blombos Caves (Evolutionary Studies Institute University of the Witwatersrand)

Though some have questioned it, Tattersall (2009) and others have argued that these patterns represent early abstract or symbolic depiction, possibly similar to language. It is certainly within the approximate time span when it is believed that human language arose.

Although there is disagreement on just about every aspect of human language, one point that would be difficult to deny is that it is an enormously complex system composed of a countless number of elements—sounds, words and subwords, syntax, and meaning. Given this complexity, it would be natural to posit that human language must have developed gradually from an earlier, simpler protolanguage system. I will briefly review two proposals for protolanguage, the song protolanguage of Darwin (1859, 1871) and the lexical protolanguage of Bickerton, extended by Jackendoff. For other proposals, see Fitch (2010), who gives an excellent summary and critique of the major protolanguage proposals. For the development of cognition in *Homo sapiens*, see, for example, Mithen (2014).

11.2.1 Song Protolanguage

Darwin (1859, 1871) was the first to note that there may be a connection between birdsong and language; he described it as the "nearest analogy to language." In the same way that humans have an instinct to speak, songbirds have an instinct to sing, and just as language is learned, so are birdsongs learned, in both cases the learning process opening up the possibility of variety, as in dialects in both human language and birdsong. In both language and birdsong, there is a stage of acquisition that precedes mastery; in humans this stage is infant babbling, and in juvenile birds, it is the singing of "subsongs." On the last point of acquisition, recent studies show a surprising parallel between language and birdsong beyond simply sharing a premastery stage. A birdsong, such as that of zebra finch, is composed of notes that are combined to form syllables and syllables combined to form motifs, which are then combined to form a complete song. Liu et al. (2004) identified two learning styles, the "serial repetition" strategy and the "motif" strategy. For a juvenile bird that adopts the serial repetition strategy, an approximation to one syllable of the model is repeated many times, and because the unit of repetition is small, each syllable is clearly articulated. In the motif strategy, the juvenile bird adopts a global imitation of its father's song (only males sing), and because the unit of repetition is larger, the articulation is noisy and imprecise. The following spectrograms show the striking

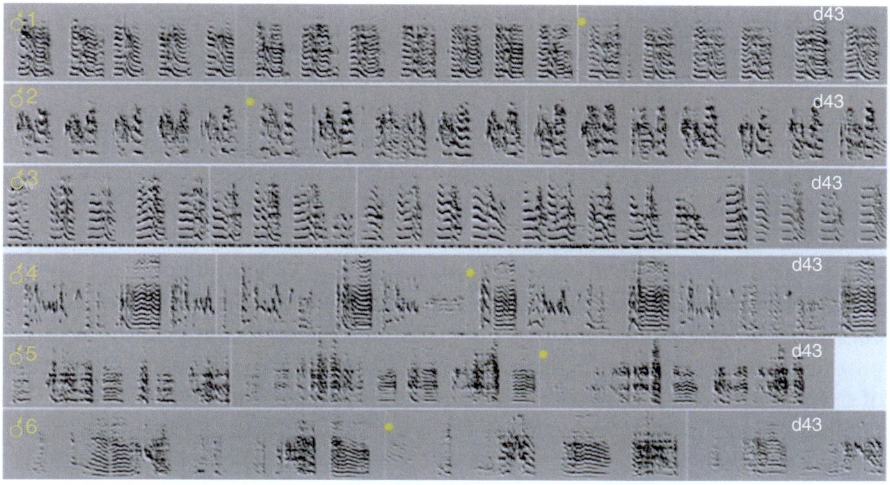

Fig. 11.2 Serial repetition and motif strategies of song learning (Liu et al. 2004)

difference in these two strategies among two groups of 41-day-old zebra finch juvenile males (Fig. 11.2).

In the upper half, we see three spectrograms of juveniles that adopt the serial repetition strategy, with each syllable clearly sung. In the bottom half are spectrograms of juveniles with motif strategy; given the larger unit being sung, the overall song lacks the precision of the other strategy. Regardless of the strategy that a juvenile adopts, by the time it matures, it is able to sing the song perfectly. As it turns out, the two strategies for acquisition are also found in human language acquisition. O'Grady (2005) (see also Lipkind et al. 2013) points out that a human infant may adopt one of two styles of learning: the "analytic" style, which focuses on breaking speech into its smallest component parts to produce short, clearly articulated, one-word utterances in the early stages and the *gestal* style, in which the infant memorizes and produces relatively large chunks of speech that are often poorly articulated that correspond to entire sequences of words in the adult language. Just as with songbirds, the human infant, regardless of the style of acquisition it adopts, ultimately acquires the language perfectly.

11.2.2 Lexical Protolanguage

One idea that a number of linguists have suggested is that before human language became the full-fledged system that we are familiar with, it went through a simpler linguistic stage, what Fitch (2010) calls "lexical" protolanguage.[1] According to Bickerton (1990, 2000, 2014), this earlier stage was composed typically of a simple

[1] "Lexical protolanguage" is called by some as "compositional protolanguage" (e.g., Tallerman and Gibson 2012).

utterance that had no structure. One way to think about this, as Jackendoff (1999) suggests, is that protolanguage was today's language minus syntax, a point also suggested by Bickerton (1990). So, it is just made up of words. In other words, for linguists such as Bickerton and Jackendoff, protolanguage was the capacity to use unstructured symbolic units. How did such a system arise, if, indeed, it did exist? Bickerton (2014) suggests that one possibility for the emergence of such a system of communication is to convey the location of a dead prey in scavenging. This was a time when our ancestors were hunter-gatherers. Why would our scavenging ancestors require such a system? It is surmised that this system allowed the hunter-gatherers to communicate the location of dead prey at a distance. To be successful in claiming the dead prey from competitors, these prehumans had to recruit members of their community in large number and communicate to these members the location of the prey and some idea of its size and the desirability of being able to claim it. This requires displacement, which is the ability to communicate about things that are not in the immediate vicinity of the speaker-hearer. Human language is adept at displacement, so are the systems employed by honeybees and ants for similar resource-locating purposes. The idea is that, in humans, protolanguage emerged to make displacement possible. This is an intriguing idea. Did language actually go through such a protolanguage stage? It is hard to say, of course, because it happened so long ago. We will return to both the song and lexical protolanguages later in the chapter.

11.3 *Both* Gradual and Rapid

The gradualist viewpoint based on protolanguage makes sense if one were only focused on language and how it may have evolved over time. Given the complexity of human language, it would be reasonable to assume that this complex system began as a simpler mechanism that over a span of evolutionary time came to take on the kind of complexity and richness we are familiar with. But there is a crucial point here: evolution is about living organisms, but language is not a living organism. From this perspective, languages may change, but they don't evolve (Chomsky 2017).[2] But clearly something happened in evolution to give rise to language. Was it an event, or series of events, that took place gradually over a long span of evolutionary time, or was it fairly rapid in nature? My suggestion is that it was *both*. I begin with the gradualist part of my idea, leaving the rapid part until the next section.

The time span I have in mind for the gradual development needed for the eventual emergence of language is much longer than the evolutionary time implied in the various models of protolanguage. In fact, it is millions of years and, in one case, hundreds of millions of years. It is our brain that provides the cognitive capacity to produce language, and the brain we have inherited has been growing ever larger

[2] Chomsky makes the same point earlier, in a 2012 interview: http://www.phenomenologyandmind.eu/wp-content/uploads/2012/12/16_Intervista-CHOMSKY.pdf

long before *Homo sapiens* appeared in Africa 200,000 years ago. *Homo habilis*, which appeared on the scene some two million years ago, started to show an increase in brain size (Deacon 1997). Why is this? Some attribute the growth in brain size to eating habits. The two systems, or organs, that consume the most energy in our body are the digestive system and the brain. As our ancestors became adept at preparing food, including grinding it, and most importantly heating it with fire to soften it, less energy was needed for digestion, and the brain received the benefit of the excess energy and grew larger (Wrangham et al. 2000; Fonseca-Azevedo and Herculano-Houzel 2012).

The *Homo sapiens* brain is not just larger relative to the body size, but well developed in regions that contribute to higher-order cognitive functions such as the prefrontal cortex which is responsible for high-level cognitive activities.[3] Language requires a number of special components in the brain and in our physiology for speech. Within the brain, two of several important centers for language are Broca's area and Wernicke's area. Both are located almost always on the left hemisphere. Broca's area, located in the frontal lobe, supports speech production, while Wernicke's area, in the superior temporal gyrus, supports speech perception. Both are needed for language. So the question is when did these regions of the brain develop? Did they develop around the time that language arose in evolution, 100,000 years or so ago? There is evidence from study of skulls of *Homo habilus* and *Homo erectus* that these centers may have been around in the brains of our ancestors over a million years ago, long before language actually emerged (Tobias 1998; Wynn 1998). In addition, there is evidence that homologues of these structures are present in some great ape brains today (Gannon et al. 1998; Cantalupo and Hopkins 2001) suggesting they may have developed over nine million years ago.

This idea that some of the key centers of the brain that support language have existed for several million years, maybe more, is part of a picture that is emerging recently about who we are. Although it was only recently in evolutionary time that we began to see evidence of high cognitive achievements, such as painted art, sophisticated weapons, refined tools, and language, the brain circuits that make them possible apparently have been present much longer. In recent articles, it is pointed out that the human cortex, where higher-order cognitive computation is carried out, has similar cell types, patterns of wiring, and gene expression as other mammals (Calabrese and Woolley 2015; Harris 2015). Furthermore, the computations that these brain circuits make possible also occur in birds. It is as if there are only a limited number of brain microcircuit designs that nature came up with, and these circuits have been repurposed to make the higher-order cognitive achievements possible in humans.

Among the recent studies, one in particular has direct implication for Darwin's song protolanguage in which he surmised that prehumans sang, and this ultimately led to language. Pfenning et al. (2014) look at vocal learning as reflected in regions of the brain of songbirds (zebra finch) and humans, and also birds that don't sing

---

[3] Recent studies point more on expansion of the temporoparietal junction (e.g, Bruner and Iriki (2016)). Extending mind, visuospatial integration, and the evolution of the parietal lobes in the human genus. Quaternary International, 405, 98–110.

(dove, quail), and a primate that doesn't have language (macaque). Taking a computational approach, they screened gene expression databases from humans and vocal-learning birds as well as the nonvocal-learning birds and primate. Their study indicates molecularly analogous regions that are homologous for song and speech for vocal-learning birds and humans. They also demonstrate that nonvocal-learning birds and macaques do not share in any significant way these traits. The idea that humans and songbirds have homologous regions that are not found in non-singing birds and also in macaques raises an interesting question. Are these shared regions due to convergent evolution or to a common progenitor? Convergent evolutions are sometimes known to occur among unrelated living beings, as similar solutions may arise for similar problems (Gould 1976), as in the example of the emergence of the eye in unrelated organisms (Ogura et al. 2004; Fernald 2006). On the convergent evolution view, the shared regions are analogies of each other. The other possibility is that the regions with similar functions in vocal-learning birds and in humans descended from a common ancestor. Given that the ancestors of present-day birds and mammals split 300 million years ago (Benton 1990), this suggests that, on the common progenitor view, the genetic resources for singing in primates existed over 300 million years ago. Darwin probably did not imagine that the song protolanguage stage lasted that long, and our ancestors may not have sung for all of that time, but just as we saw for Broca's and Wernicke's areas, what we see is that the genetic resources may have been present long before language emerged. On the convergent view, we are looking at a time depth of at least around nine million years if the existence of regions similar to the Broca's and Wernicke's areas in the brain of some great apes is an indication of the age of these regions. On the common progenitor view, the brain centers for songbirds imply a much older evolutionary history.

But what does singing have to do with human language? While birds still sing today, we see on the primate side only four species that sing: titi monkeys, indris, tarsiers, and gibbons. Of these four, gibbons are the only apes (Hylobatidae) and thus the most closely related to humans. Their songs have been studied for decades (Marshall and Marshall 1976; Haimoff 1984) but have only recently been compared to human language (Clarke et al. 2006). I will present a hypothesis for how language emerged in evolution that considers the system underlying singing to play a critical role, thus inheriting Darwin's idea of a singing prehuman, and linking it to the regions of the brain for singing that may have existed for million years and perhaps for over 300 million years.[4]

11.4 Basics of Human Language

According to the UNESCO report, *Endangered Languages*, there are approximately 6000 languages in the world. Some may question this number as an underestimation—for example, just in Papua New Guinea, we find 820 languages, and there are

[4] Samuels (2015) argues that songbirds have capacity to generate sound structure analogous to human language phonology.

many other regions of the world that hold a large number of languages. The precise number is hard to pin down because sometimes it is difficult to distinguish between dialects and languages. However many languages there are, one thing is certain: every language has the function, as Darwin described it, to connect "definite sounds with definite ideas" (Darwin 1871: 54). How does language do this? In the linguistics literature, the proposal is made that each sentence is composed of two layers of meaning, called "duality of semantics" (Chomsky 2005). We can use the following example to demonstrate this duality.

Did You Eat Bread?

This sentence contains three content words, *you*, *eat*, and *bread*. These words may occur in a variety of contexts with a fixed meaning, such as *I want you to eat bread, you shouldn't eat bread, and you seem to eat bread everyday*. This is the *lexical* layer, containing the meaning of individual words of a sentence. The sentence also contains the word *did*, a function word that has not just one, but two functions. By inflecting for tense, it indicates that the event represented in the sentence took place in the past, and by occurring at the head of the sentence, it signals that the sentence is interrogative in form. Tense and interrogative are two elements in the other layer of meaning, which I call the *expression* layer (Miyagawa 2010). The expression layer constitutes a chunk of the expression, typically a sentence, and its purpose is to give shape to the expression, such as interrogative, and any meaning that is associated with the entire expression, such as tense. This is in sharp contrast to the *lexical* layer, in which the meaning is represented at strictly the local level of each individual word. The two layers of meaning differ in two other significant ways. The lexical layer is composed of content words, and an adult English speaker knows around 60,000 such words. In contrast, the expression layer is composed of function words, and these number in the single digits to less than 15 depending on the theory one adopts. Also, the words of the lexical layer connects to specific meaning, often, though not always, referring to an entity in the real word, such as *table, school, and horse*. The expression layer provides the shape of the utterance, and it commonly communicates the intention of the speaker, such as the intention to ask a question, make a statement, issue a command, and so forth (Austin 1962). The two layers of meaning are represented as follows for the earlier example.

11.4.1 Duality of Semantics (Chomsky 2005; Miyagawa 2010)

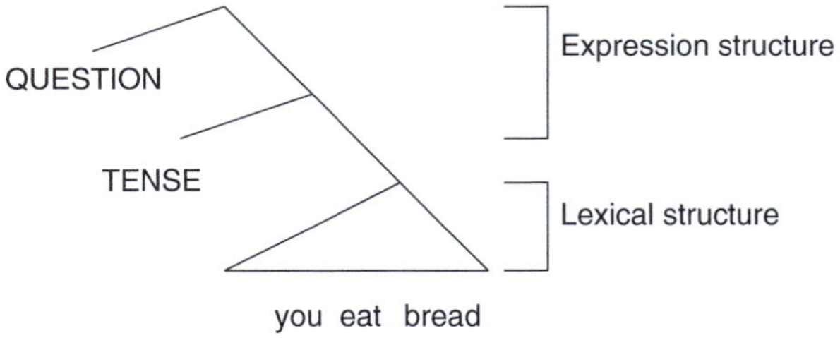

QUESTION indicates that the entire expression has the form of interrogative and represents the speaker's intention to ask a question, while TENSE contributes to the overall expression by indicating the tense of the utterance. Each individual word in the lexical layer is represented in the lower structure.

In Miyagawa et al. (2013), I argued, based on the work of a number of linguists, that although the expression and lexical layers are intricately intertwined in human language, they can be viewed as separate components each with its own specific properties, as we saw. The two components are in fact so fundamentally different that it appears that they just happen to occur together in language, but they could just as well function as separate systems. In Miyagawa et al. (2013, 2014), I in fact propose that the two layers correspond to two distinct systems that occur independently in nature. I call it the Integration Hypothesis of language evolution because the two layers happened to get integrated uniquely in humans and gave rise to language as we know it today. By separating language into these two components, we are able to entertain a range of possibilities that are difficult to consider in the other proposals for language in evolution. Given the richness and complexity of human language, it is certainly challenging to imagine that it developed rapidly within the last 100,000 years, yet there is no evidence of language prior to that time, as far as we can see (e.g., Tattersall 2016). How do we make sense of this enormously complex system that just "popped up" 100,000 years ago? By the Integration Hypothesis, the process of integration of the two independent systems occurred in recent evolutionary time. But the two systems themselves are older, much older, their progenitors being millions or, in one case, possibly hundreds of million years old. Below, I turn to the Integration Hypothesis.

11.5 Lexical and Expression Layers: Separate Systems in Nature

If the two layers of language, *lexical* and *expression*, have correlates in independent systems in nature, what are these systems? The calls of monkeys and apes are a natural candidate for the *lexical* system. There is a large body of literature on this topic (Seed and Tomasello 2010), an earlier work of which is on the Kenyan vervet monkeys (Seyfarth et al. 1980), which possess alarm calls for pythons, eagles, and leopards. Sticking to these simplest lexically based systems, what we see is an uttered object that correlates with a particular real-world state of affairs. An important point about this system is that each uttered object is coextensive with a real-world state ("holistic" in the sense of Wray 1998). A simple, and possibly a simple-minded, way to view this system is that it is a collection of vocal gestures that have a specific referent in the real world (Seyfarth and Cheney 1986; Miyagawa et al. 2013). It is a "pure" L(exical) system in that each unit is an isolated entity with a definite "meaning." The lexical protolanguage view points to this L system in our ancestors as forming the progenitors to language.[5]

[5]Although I consider the L system to consist of independent vocal gestures, there are studies that suggest that primates can use multiple calls to construe novel meaning (Dessalles 2007; Arnold and Zuberbühler 2006; Quattara et al. 2009).

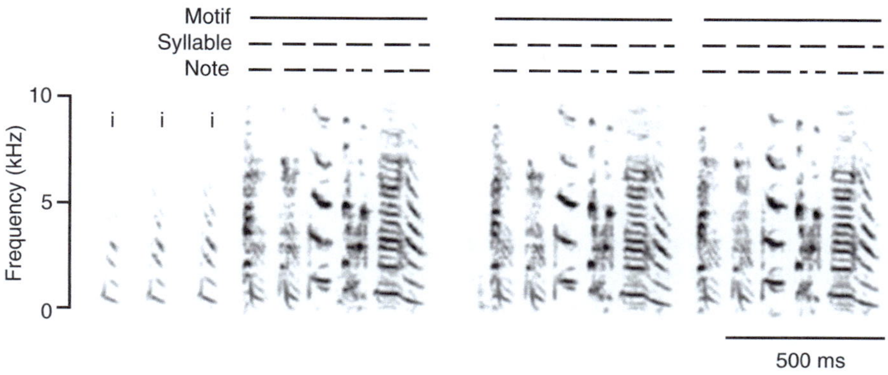

Fig. 11.3 The song of zebra finch (Berwick et al. 2011)

Miyagawa et al. (2013, 2014) link the E(xpression) layer of human language to the system underlying birdsong.[6] As Darwin first noted, there is a list of striking resemblances between birdsong and language, a list to which we can also add: a need for external input, sensitive developmental periods ending at sexual maturity, and hemispheric lateralization; and motor-auditory rehearsal systems (Bolhuis et al. 2010). There is one striking difference: lexical items in the sense of human language are absent from every variety of birdsong that has been studied, so that a birdsong appears to be a phonological syntax without a lexicon (Marler 2000). Every birdsong is a system that generates particular patterns, as in the case of the well-studied song of the zebra finch, which has a highly restricted set of "notes" that combine to form syllables, and syllables into motifs, and motifs into a complete song "bouts" (Berwick et al. 2011) (Fig. 11.3).

The zebra finch song is one of the simplest; other vocal-learning bird species possess more complex patterns with branches, loops, and repetitions (Berwick et al. 2011) (Fig. 11.4).

A unique property of human language is that a regular grammar (type 3) is inadequate for modeling it (Chomsky 1956). But this is not the case for the two proposed antecedents of human language. The lexical systems that employ isolated uttered units that correlate with real-world references, such as the alarm calls of vervet monkeys, are clearly couched in a simple regular grammar. The other layer, the expression layer, finds its antecedent in the kind of system underlying birdsong. Birdsongs have specific patterns, and these patterns can be complex, as in the example of the Bengalese finch, which loops back to various positions in the song and leads to considerable variation (Fig. 11.4). Nevertheless, all known birdsongs can be described as a *k-reversible* finite-state automaton, a form of regular grammar (Berwick et al. 2011); for example, we do not see center embedding, which was an example Chomsky used to prove the inadequacy of regular grammar for describing human language. While suggestions have been

[6] See Okanoya (2002) for an earlier proposal that links birdsong syntax and human language.

Fig. 11.4 Bengalese finch (Okanoya 2004)

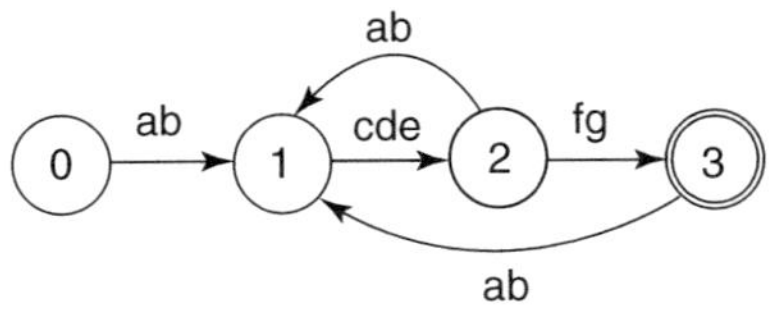

made that certain bird species can acquire recursive syntactic structures reminiscent of human language either through conditioning (Gentner et al. 2006) or spontaneously (Abe and Watanabe 2011), these results remain controversial (Beckers et al. 2012) and unconfirmed.

The Integration Hypothesis proposes that these two major systems in nature that underlie communication, L(exical) and E(xpression), integrated uniquely in humans to give rise to language. One challenge to the Integration Hypothesis is that, as noted above, contemporary languages are known to be based on a system more powerful than a regular grammar. As evidence for the Integration Hypothesis, in Miyagawa et al. (2013, 2014), we demonstrate that when we consider the two components of contemporary languages separately, L and E, each component may be characterized by a regular grammar, thus reflecting their antecedents in nature. It is only when the two systems integrated that gave rise to a system more powerful than what we see elsewhere in the animal kingdom.

Let us begin with the L layer. A trait of L units is that they don't combine directly, just as the discrete units of the vervet alarm calls do not combine to form new calls (Miyagawa et al. 2013).

To make these combinations possible, something from the E layer must intervene: the D(eterminer) represented by apostrophe *'s* (*John's book*), the copula that carries tense, an E layer phenomenon (*book is long*), and *to*, which helps to form the E layer of a clause (*I want to eat pizza*). Later, we will look at compounds, which ostensibly pose a challenge to this view that L units do not directly combine (Fig. 11.5).

To see the regular-grammar status of the E layer of human language, let us begin with a unique feature of human language not seen in other systems in the animal world, the feature commonly called "labeling" (Chomsky 1995). Given a word, its lexical category (noun, verb, etc.) forms the "label" of the larger phrase that contains it. For example, for the pair *eat* and *bread*, the verb *eat* labels the larger phrase, *eat bread*, as a V(erb), forming a verb phrase.

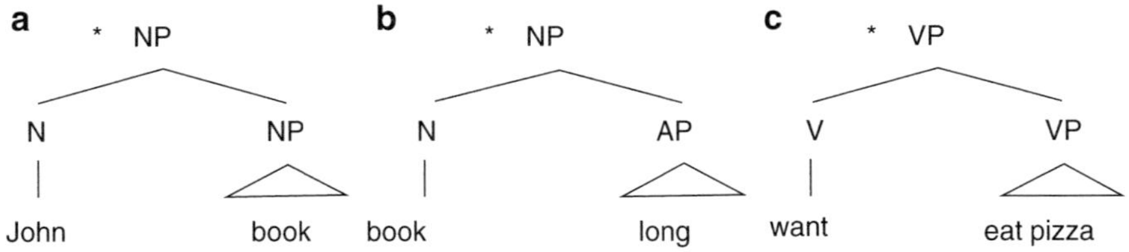

Fig. 11.5 Impossible Lexical Structures

Fig. 11.6 Labeling

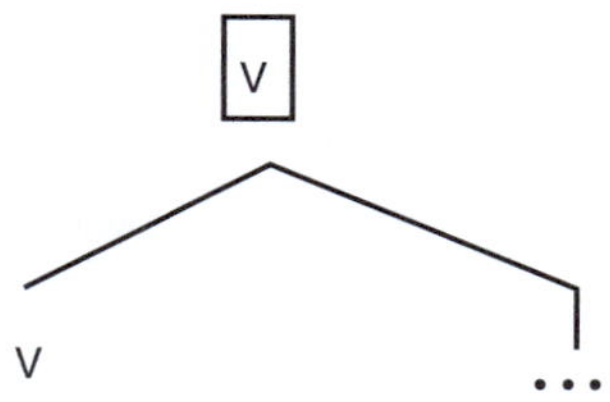

Fig. 11.7 Expression Structure

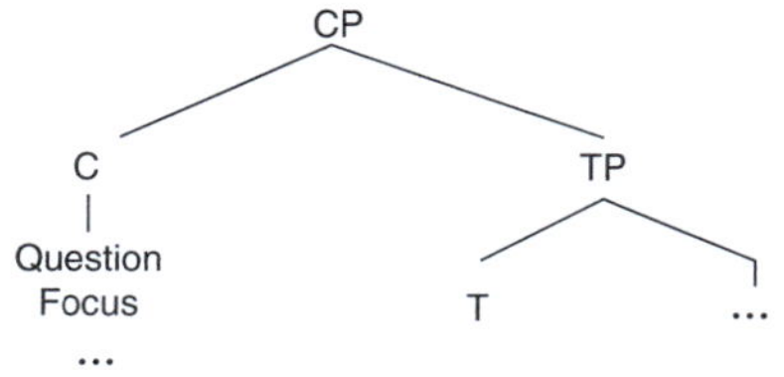

This property of labeling gives human language its unique ability to form hierarchical structures (Chomsky 1995, 2008; Hornstein 2009) and gives human syntactic structure the property of "discrete infinity" (Chomsky 2000) through recursively merging and labeling structures. However, as noted in Miyagawa et al. (2013), there is a severe limitation on the depth of the hierarchy when we isolate structures in the E layer. As we saw earlier, the E layer can contain tense; there is a second item, conventionally labeled "C(omplementizer)" that contains a range of expressive phrases such as Q(uestion), F(ocus), and so forth (Fig. 11.6).

C and T are the two most frequently cited labels within the E layer. Strikingly, these labels cannot occur as hierarchical structures of arbitrary depth; rather the CP-TP structure can only be one layer deep, a limitation consistent with the E layer being characterizable as a regular grammar (Fig. 11.7).

The limitation is that the E layer is restricted to a depth-one hierarchical structure, similar to the systems we see in the Bengalese finch and nightingale songs.[7] This suggests that the E layer of human language closely reflects the birdsong structure, a system of regular grammar. While there are theories of linguistics that posit multilayer within E (Rizzi 1997), there are alternatives that do not assume such a multilayer (e.g., Miyagawa 2010) (Fig. 11.8).[8]

---

[7]Arsenijevic and Hinzen (2012) also notice this limitation in the E layer and attribute this limitation to meaning. We believe that characterizing this limitation in terms of finite-state grammar captures the restriction, which we believe holds all kinds of E layers and not just the CP-TP structure, which is what Arsenijevic and Hinzen account for.

[8]Tallerman (2016) criticizes the idea that the expressive layer of human language parallels the structure of birdsong. She points out that while the elements of birdsongs always occur in an ordered sequence, in human language elements may undergo permutations—movement as called in linguistics—that alter the ordering sequence of items. This criticism is based on a fundamental misunderstanding of the expressive layer. As noted in Miyagawa et al. (2013, 2014), an operation such as movement is only possible after the two layers, E and L, integrated. The point is quite simple: what is moved are words in phrases; hence L as well as E must be involved. Tallerman

Fig. 11.8 An Impossible E layer Structure

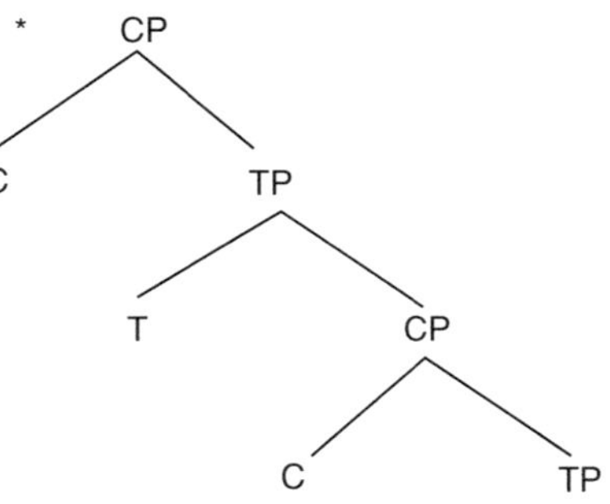

11.6 Progenitors of Human Language

The previous proposals about how human language developed all have some "beginning" that is a simpler stage of today's language, and through a progression of stages, each more complex than the earlier one, we arrive at today's human language. For example, and as already mentioned, lexical protolanguage, proposed by linguists (e.g., Bickerton 1990, 2000, 2014; Jackendoff 1999, 2002), views human language as having undergone a sequence of stages, from a one-word stage to a more complex system of expression; in the case of Jackendoff, and more recently Progovac (Progovac and Locke 2009; Progovac 2012, 2015), this second stage is characterized by the combination of single words into compounds through a rudimentary recursive *n*-ary operation that generates flat structures (Progovac and Locke 2009; Progovac 2012, 2015). Once equipped with this protolanguage that can form a two-word sequence, a primate's brain becomes ready at some point to create structures that ultimately lead to sentence formation. Why did syntax emerge? If it is the case that in protolanguage, any two words could be strung together, there would be no structure to define the relationship between the two words, so that there is a great deal of burden on the context to give meaning to the combination. For example, if one combines *child* and *picture,* as *child-picture*, we cannot tell if this means a picture drawn by a child, a picture that depicts a child, a picture for children, and so forth. One way to think about the reason why the protolanguage developed structure is to clearly mark the relationship between the combined items (Jackendoff 1999). This reduces burden on the brain to compute the meaning. From this, the system could have developed structures that define such relationships as modifier modified as in *blue sky*; verb complement, such as *eat bread*; and an agent of an act, such as *horse rider*. Linguists such as Jackendoff and Progovac point to the existence of certain compounds (*daredevil*) in contemporary languages as living fossils of an earlier, non-syntax stage of language.

---

incorrectly notes that movement is strictly handled by the E component. Rather, a nonfinite-state operation such as movement does not occur in E layer in isolation, but is the outcome of integration (Miyagawa et al. 2013, 2014).

Let us consider the lexical protolanguage proposal by looking at the most recent work on the topic—Progovac (2015). This work, by a scholar well versed in modern linguistic theory and with the major literature on evolution, has amassed together considerable data from contemporary languages to argue that from its initial one-word stage, language evolved through progressively complex stages. A unique feature of this work is that Progovac specifically argues that each of these stages reflects the syntactic structure postulated in the modern linguistic theory called minimalism (e.g., Chomsky 1995). Thus, from a one-word stage, language evolved into a two-word stage, and eventually it evolved to stages that progressively took on layers of functional structures, as in vP > TP > CP. Her interesting claim is that we find each of these stages as living fossils or in some related form in contemporary languages. Whether one agrees with her conclusion or not, the extensive data she presents is fascinating in and of themselves; I found particularly interesting the numerous "exocentric" compounds in Chap. 6 from Serbian and other languages. Independent of issues of language in evolution, these data, and other data in other chapters, present a body of empirical material for carrying out research on topics such as compounding and phrase structure. Progovac also presents the core of the recent linguistic theory (e.g., Chomsky 1995) in a way that is comprehensible to nonlinguists, which is a valuable service to the field.

There are some issues with her work. Progovac considers the emergence of language as an evolutionary process, but as already noted, living entities evolve, and language is not a biological entity. It is telling that all the examples Progovac gives as examples of evolutionary change are biological in nature except language—lungfish, body hair, and the eye. So, if she is correct in her description of how language developed, one must wonder precisely what this development is telling us about the organism that produced each of these stages.

Furthermore, if we consider "one-word" and "two-word" systems, they are fundamentally different. In "one-word" systems, such as the vervet alarm calls and the emotional interjections in contemporary human languages (*ouch*, *wow*), the units do not combine with other units to form a new unit. This is because each unit of utterance is coextensive with the referent or the emotion it represents; since they don't have any lexical category, these single units are better called "roots" (Nóbrega and Miyagawa 2015). It is simply not possible to combine the alarm calls such as *leopard* and *eagle* because each has an independent reference, and there is no sense in which one unit can combine with another unit to form a third unit with some other reference, unless these units undergo fundamental change. Same goes for human language interjections: there is no sense in which *ouch* and *wow* could combine to give a third emotional interjection simply because each interjection is a fully independent and isolated unit of utterance that expresses a specific emotional state. In contemporary languages, combining words into compounds or sentences minimally requires that each word ("root" to be more precise) be assigned a category (noun, verb, etc.), which automatically gives the word more structure than an unstructured "root" form of alarm calls or interjections; thus a word in a compound is complex (see Di Sciullo 2011, 2013, 2014).[9] As noted in Nóbrega and Miyagawa (2015), if one looks at compounds beyond those used as evidence for “living fossils,” one finds structure that is just as complex as a full sentence, thereby questioning the idea that there was an unstructured two-word protolanguage stage. From this viewpoint, the four stages that Progovac suggests should be questioned (she refers to her proposal as a three-stage model, but I am including the “zero” stage as the first stage).

11.6.1 Four Stages of Language Evolution (Progovac 2015: 15)

(a) One-word stage
(b) Paratactic stage: conjoin two words without creating hierarchy or headedness
(c) Proto-coordination stage: emergence of “conjunction/linker” that provides “segmental glue to hold the utterance together” (p. 13), presumably making multiple-word utterances possible
(d) Specific-functional category stage: “specific functional categories become available” (p. 13). “it is only at this stage that hierarchical structure, Move, and recursion become available.” (p. 14)

Based on what we have said, there are really only two crucial stages, the one-word stage, which is made up of isolated roots, and a stage that Progovac calls the “specific-functional category stage,” where “functional categories”—what I am calling the E system—become available. Once the E layer becomes associated with the isolated units of the L layer, combining the L units under Merge is possible, with labeling of the newly created structure. From this perspective, there would be no need to postulate the paratactic and proto-coordination stages. That also makes the picture simpler for creating multi-word expressions. Instead of having to postulate “proto-Merge” which presumably joined roots, and which arguably does not occur in contemporary languages (Di Sciullo 2013, 2014),[10] we would simply need to postulate the same Merge that exits today that takes words associated with lexical category and other E properties.

Despite these issues, there are a number of points I agree with, including Progovac’s suggestion that the progenitors of human language may have been around millions of years ago. I will return to this point below.

Turning to Darwin, a crucial feature of his view of how language emerged is that he separates cognitive abilities from the initial emergence of language-like behavior: “The mental powers in some early progenitor of man must have been more highly developed than in any existing ape, before even the most imperfect form of

---

[9] This is similar, and possibly the same, as the notion of “lexical envelop” that allows roots to merge with other “enveloped” units (Spelke 2003; Ott 2009; Boeckx 2011).

[10] See Boeckx (2015) for an alternate view.

speech could have come into use" (Darwin 1871: 57). The protolanguage Darwin conjectured is musical in nature, and involved singing with the intention, same as songbirds and gibbons, of courtship and territoriality. This eventually developed into a full-fledge language, according to him, through imitations aided by signs and gestures that became associated with definite meaning.[11] Darwin goes on to add that the emergence of language aided in the further development of the mind, very much in the spirit of the German philosopher and linguist Wilhelm von Humboldt, whose work Darwin studied closely.

The idea that prehumans sang appears to be particularly plausible given the recent genetic finding that humans with their speech, and birds with their song, share regions that are homologous for speech and song (Pfenning 2014). Where Darwin's conjecture becomes questionable is how could a musical system, which lacks lexical units, come to take on such units built into the song?

The Integration Hypothesis differs from all other hypotheses of language development in that it does not postulate a singular "proto" system that developed progressively into ever-complex systems that ultimately ended up as language as we know it today. Any "proto" system runs into the kinds of problems I noted above. If one begins with a word (or more accurately, root, which has no structure), how does syntax develop out of it to make it possible to combine these words into two-word utterances and beyond to sentences? If one starts with a musical form of protolanguage, how do words get introduced into this system? What the Integration Hypothesis suggests is that the sources for words and syntax existed independently as the L system underlying alarm calls and similar modes of communication and the E system underlying birdsong (and possibly gibbon song). It is reasonable to assume that these two systems have been around for a long time, millions of years or even longer. At some point in recent evolutionary time, the two systems integrated uniquely in humans to give rise to a system that generates patterns as an E system and contain lexical items as an L system. Why this happened is not clear, but we could imagine some things that had to have happened to make this integration possible. As noted earlier, there is a fundamental difference between one-word and two-word systems. While a one-word utterance can be coextensive with the referent or an emotion, a two-word (or more) combination requires each word to be nonreferential—underspecified for meaning, in other words—so that the two can combine to form a novel utterance that has its own unique reference. This means that each word must have undergone some process of abstraction. Thus, in *child-picture*, *child* does not refer to a definite entity, but it is a *kind* of an entity that shares essential properties (human, youth, etc.). The same goes for *picture*. What could have shifted the isolated utterances of the L system into these abstract entities that can be combined into patterns furnished by the E system? It is not clear, but again, we can see what must have occurred: each item, as a process of abstraction, took on category, such as noun, verb, and so forth. As a noun, for example, it no longer referred to a specific entity, but stood for a *kind* of an entity (*bread*). Furthermore, by taking

[11] See Jespersen (1922) for a different model of how song developed into language.

on categorical identity, it is able to merge with another unit and, crucially, to label the newly formed unit (V label for {eat bread}).

What about the E system? Birdsong is a pure E system that generates finite-state patterns without any specific reference to entities in the real world. It is, as already noted, a phonological syntax without a lexicon (Marler 2000). Just as some change must have taken place to the L system, from its "one-word" units that are coextensive with the meaning they embody, the E system must have undergone some change to accommodate the members of the L system in order for integration to have taken place. What could that change be? One possibility is found in Darwin's proposal of musical protolanguage. He conjectures that for the singing of prehumans to have transitioned to articulate language, they began to incorporate imitations of natural sounds that represented some entity, aided by gestures (Fitch 2010: 472), something also suggested by Farrar (1870) and Müller (1861). One way to think about this change is that the E system changed from a system of nonreferential patterns to one in which the patterns contained placeholders for items that refer to the real world. These placeholders are represented by things like imitation sounds, and imitation can represent an entity or an event. It then becomes natural to start inserting the members of the L system, which themselves have undergone change to be underspecified for meaning. If an underspecified root with the core meaning of dog is inserted into an "entity" (N) placeholder, it gets interpreted as DOG, but if it is inserted into the placeholder for an event (V), it is interpreted as HARASS.[12] As these members of the L system began to combine within the E system, they start to build structure in the way that lexical protolanguage proponents such as Jackendoff have conjectured. One crucial difference is that the building blocks of the structure are furnished by the E system, which comes with a system to link nodes, hence a potential to build structure, as opposed to the proposal of the protolanguage proponents, who conjecture that the structure somehow arose directly from L combinations. For that to happen, something (structure) must emerge literally from nothing, something that isn't out of the question in principle, but highly unlikely.

Once the integrated system achieves this state of having underspecified L units that can be inserted into the E system to take on category, we have the essence of the human language at work. The idea that human language emerged within the past 100,000 years does not mean that all the pieces of language developed within this time frame. All the essential pieces have existed for millions of years and developed gradually over a long evolutionary time span, and what happened to give rise to language was that these independently developed pieces integrated uniquely in humans to give rise to the kind of rich and complex system we know to be language. This gives the appearance of language having rapidly developed in recent evolutionary time, and while it is true that there is no evidence for language earlier in evolutionary history, we have ample evidence for the existence of L and E systems long before language emerged. It is the adventitious integration of these two highly developed, preadapted systems that gave rise to the remarkable system of human language.

[12] Thanks to Vitor Nóbrega for these examples and the general line of thinking outlined.

Among all the necessary conditions for integration, it appears that abstraction (or underspecification) of the L units is one crucial missing link. Other pieces were essentially there as independent systems: the units of L and the pattern-generating capacity of E. For the units of L to combine by tapping the pattern-generating E system, the L units must undergo abstraction, possibly by being assigned a category such as noun and verb. It is not clear how this happened. One possibility, and I only mention it as a conjecture, is the notion of displacement that Bickerton incorporates into his proposal of protolanguage. Note that displacement requires abstraction. A unit of L that refers, for example, to leopard, can do so if the referent is present in the physical proximity. For an organism to convey the existence of an entity in the absence of this entity, the organism must convey the *idea* of the entity. By necessity, this involves some ability to conceptualize the *kind* of the entity involved, hence abstraction, which is required for integration. Even if this turns out to be true, how this abstraction leads to the uttered unit becoming associated with a lexical category remains a mystery.

11.7 Concluding Remarks

Did language emerge rapidly or gradually? Given the rich and complex nature of language, it surely makes sense that the key components of language took a long evolutionary time to emerge. Yet, there is no evidence for such gradual development, in which language first appeared, say, as a one-word system followed by a two-word system and so forth. The Integration Hypothesis suggests a parallel development of independent systems in nature that underlie communication, the E and L systems, which allows us to capture the essence of both the emergent and gradualist views of language in evolution. Each system developed over a long span of time. At some point in recent evolutionary history, the two systems integrated uniquely in *Homo sapiens* to give rise to the kind of system we see today as language.

Acknowledgement I am grateful to Esther Clarke, Takashi Morita, Vitor Nóbrega, Kaz Okanoya, and Bridget Samuels for numerous suggestions on an earlier version of this chapter.

References

Abe K, Watanabe D (2011) Songbirds possess the spontaneous ability to discriminate syntactic rules. Nat Neurosci 14(8):1067–1074. doi:10.1038/nn.2869

Arnold K, Zuberbühler K (2006) Language evolution: semantic combinations in primate calls. Nature 441:303

Arsenijevic B, Hinzen W (2012) On the absence of X-within-X recursion in human grammar. Linguist Inq Summer 43(3):423–440

Austin JL (1962) How to do things with words. Harvard University Press, Cambridge

Beckers GJ, Bolhuis JJ, Okanoya K, Berwick RC (2012) Birdsong neurolinguistics: songbird context-free grammar claim is premature. Neuroreport 23:139–145. doi:10.1097/WNR.0b013e32834f1765

Benton MJ (1990) Phylogeny of the major tetrapod groups - morphological data and divergence dates. J Mol Evol 30:409–424
Berwick R (1998) Language evolution and the minimalist program: the origins of syntax. In: Hurford JR, Studdert-Kennedy M, Knight C (eds) Approaches to the evolution of language: social and cognitive bases. Cambridge University Press, Cambridge, pp 320–340
Berwick RC, Chomsky N (2011) The biolinguistic program: the current state of its development. In: The biolinguistic enterprise: new perspectives on the evolution and nature of human language faculty. Oxford University Press, Oxford
Berwick RC, Okanoya K, Beckers GJL, Bolhuis JJ (2011) Songs to syntax: the linguistics of birdsong. Trends Cogn Sci 15:113–121
Bickerton D (1990) Species and language. Chicago University Press, Chicago, IL
Bickerton D (1995) Language and human behavior. University College London Press, London
Bickerton D (1998) Catastrophic evolution: the case for a single step from protolanguage to full human language. In: Hurford JR, Kennedy MS, Knight C (eds) Approaches to the evolution of language: social and cognitive bases. Cambridge University Press, Cambridge, pp 341–358
Bickerton D (2000) How protolanguage became language. In: Knight C, Studdert-Kennedy M, Hurford JR (eds) The evolutionary emergence of language: social function and the origins of linguistic form. Cambridge University Press, Cambridge, pp 264–284
Bickerton D (2014) More than nature needs: language, mind and evolution. Harvard University Press, Cambridge, MA
Boeckx C (2011) Some reflections on Darwin's problem in the context of Cartesian biolinguistics. In: Di Sciullo A-M, Boeckx C (eds) The biolinguistic enterprise: new perspectives on the evolution and nature of the human language faculty. OUP, Oxford, pp 42–64
Boeckx C (2015) Elementary syntactic structures—prospects of a feature-free syntax. Cambridge University Press, Cambridge
Bolhuis JJ, Okanoya K, Scharff C (2010) Twitter evolution: converging mechanisms in birdsong and human speech. Nat Rev Neurosci 11:747–759
Bolhuis JJ, Tattersall I, Chomsky N, Berwick RC (2014) How could language have evolved? PLoS Biol 12(8):e1001934. doi:10.1371/journal.pbio.1001934
Bruner E, Iriki A (2016) Extending mind, visuospatial integration, and the evolution of the parietal lobes in the human genus. Quat Int 405:98–110
Calabrese A, Woolley SM (2015) Coding principles of the canonical cortical microcircuit in the avian brain. Proc Natl Acad Sci U S A 112:3517–3522. doi:10.1073/pnas.1408545112
Cantalupo C, Hopkins WD (2001) Asymmetric Broca's area in great apes. Nature 414(6863):505–505. doi:10.1038/35107134.
Chomsky N (1956) Three models for the description of language. IRE Trans Inf Theory 2:113–124
Chomsky N (1995) The minimalist program. MIT Press, Cambridge, MA
Chomsky N (2000) Minimalist inquiries: The framework. In: Martin R, Michaels D, Uriagereka J (eds), Step by step: Essays on minimalist syntax in honor to Howard Lasnik. MIT Press, Cambridge, MA, pp 89–155
Chomsky N (2005) Three factors in human language design. Linguist Inq 36:1–22
Chomsky N (2008) On phases. In: Freidin R, Otero C, Zubizarreta M-L (eds) Foundational issues in linguistic theory: essays in honor of Jean-Roger Vergnaud. MIT Press, Cambridge, MA, pp 133–166
Chomsky N (2010) Some simple evo-devo thesis: how true might them be for language? In: Larson RK, Deprez V, Yamakido H (eds) The evolution of human language: biolinguistic perspectives. Cambridge University Press, Cambridge, pp 45–62
Chomsky N (2012) The poverty of stimulus: unfinished business. Stud Chin Linguist 33–1:3–16
Chomsky N (2013) Problems of projection. Lingua 130-1:33–49
Chomsky N (2017) Language architecture and its import for evolution. Neurosci Biobehav Rev. doi: 10.1016/j.neubiorev.2017.01.053

Christiansen MH, Kirby S (2003) Language evolution: consensus and controversies. Trends Cogn Sci 7:300–307
Clarke E, Reichard UH, Zuberbühler K (2006) The syntax and meaning in wild gibbon songs. PLoS One 1:e73. doi:10.1371/journal.pone.0000073
Darwin C (1859) On the origin of species. John Murray, London
Darwin C (1871) The descent of man, and selection in relation to sex. John Murray, London
Deacon TW (1997) The symbolic species. W. W. Norton, New York
Dessalles JL (2007) Why we talk: the evolutionary origins of language. Oxford University Press, Oxford
Di Sciullo A-M (2011) A biolinguistic approach to variation. In: Di Sciullo A-M, Boeckx C (eds) The biolinguistic enterprise: new perspectives on the evolution and nature of human language faculty. Oxford University Press, Oxford, pp 305–326
Di Sciullo A-M (2013) Exocentric compounds, language and proto-language. Lang Inf Soc 20:1–26
Di Sciullo A-M (2014) Minimalism and I-morphology. In: Kosta P, Schurcks L, Radev-Bork T (eds) Minimalism and beyond: radicalizing the interfaces. John Benjamins, Amsterdam, pp 267–286
Farrar FW (1870) Philology & Darwinism. Nature 1:527–529
Fernald R (2006) Casting a genetic light on the evolution of eyes. Science 313:1914–1918. doi:10.1126/science.1127889
Fitch TW (2010) The evolution of language. Cambridge University Press, Cambridge
Fonseca-Azevedo K, Herculano-Houzel S (2012) Metabolic constraint imposes tradeoff between body size and number of brain neurons in human evolution. Proc Natl Acad Sci U S A 109(45):18571–18576
Gannon PJ, Holloway RL, Broadfield DC, Braun AR (1998) Asymmetry of chimpanzee planum temporale: humanlike pattern of Wernicke's brain language area homolog. Science 279(5348):220–222. doi:10.1126/science.279.5348.220
Gentner TQ, Fenn KM, Margoliash D, Nusbaum HC (2006) Recursive syntactic pattern learning by songbirds. Nature 440:1204–1207. doi:10.1038/nature04675
Gould SJ (1976) In defense of the analog: a commentary to N. Hotton. In: Masterson R, Hodos W, Jerison H (eds) Evolution, brain, & behavior: persistent problems. Wiley, New York, pp 175–179
Haimoff EH (1984) Acoustic and organizational features of gibbon songs. In: Preuschoft H, Chivers DJ, Brockelman WY, Creel N (eds) The lesser Apes. Evolutionary and behavioural biology. Edinburgh University Press, Edinburgh, pp 333–353
Harris KD (2015) Cortical computation in mammals and birds. Proc Natl Acad Sci U S A 112:3184–3185
Hauser MD, Chomsky N, Fitch WT (2002) The faculty of language: what is it, who has it, and how did it evolve? Science 298:1569–1579
Hewes GW (1973) Primate communication and the gestural origins of language. Curr Anthropol 14:5
Hornstein N (2009) A theory of syntax. Cambridge University Press, Cambridge
Hurford JR (2012) The origins of grammar: language in the light of evolution II. Oxford University Press, Oxford
Jackendoff R (1999) Possible stages in the evolution of the language capacity. Trends Cogn Sci 3:272–279. doi:10.1016/S1364-6613(99)01333-9
Jackendoff R (2002) Foundations of language: brain, meaning and evolution. Oxford University Press, Oxford
Jespersen O (1922) Language: its nature, development, and origin. George Allen and Unwin, London
Lipkind D, Marcus GF, Bemis DK, Sasahara K, Jacoby N, Takahasi M et al (2013) Stepwise acquisition of vocal combinatorial capacity in songbirds and human infants. Nature 498(7452):104–108

Liu WL, Gardner TJ, Nottebohm F (2004) Juvenile zebra finches can use multiple strategies to learn the song. Proc Natl Acad Sci U S A 101(52):18177–18182. doi:10.1073/pnas.0408065101.
Marler P (2000) Origins of music and speech: insights from animals. In: Wallin NL, Merker B, Brown S (eds) The origins of music. MIT Press, Cambridge, MA, pp 31–48
Marshall JT, Marshall ER (1976) Gibbons and their territorial songs. Science 193:235–237
Mithen S (2014) The cognition of Homo neanderthalensis and H. sapiens: does the use of pigment necessarily imply symbolic thought? In: Akazawa T, Ogihara N, Tanabe HC, Terashima H (eds) The dynamics of learning in neanderthals and modern humans. Springer, London, pp 7–15
Miyagawa S (2010) Why agree? Why move?: unifying agreement-based and discourse-configurational languages. MIT Press, Cambridge, MA
Miyagawa S, Berwick RC, Okanoya K (2013) The emergence of hierarchical structure in human language. Front Psychol 4:71. Article 71
Miyagawa S, Ojima S, Berwick RC, Okanoya K (2014) The integration hypothesis of human language evolution and the nature of contemporary languages. Front Psychol 5:564. doi:10.3389/fpsyg.2014.00564
Müller FM (1861) The theoretical stage, and the origin of language. In: Lectures on the science of language. Longman, Green, Longman, and Roberts, London
Newmeyer FJ (1991) Functional explanation in linguistics and the origins of language. Lang Commun 11:3–28
Newmeyer FJ (1998) On the supposed 'counterfunctionality' of universal grammar: some evolutionary implications. In: Hurford JR, Studdert-Kennedy M, Knight C (eds) Approaches to the evolution of language. Cambridge University Press, Cambridge, pp 305–319
Nóbrega V, Miyagawa S (2015) The precedence of syntax in the rapid emergence of human language in evolution as defined by the integration hypothesis. Front Psychol 6:271
O'Grady W (2005) How children learn language. Cambridge University Press, Cambridge
Ogura A, Ikeo K, Gojobori T (2004) Comparative analysis of gene expression for convergent evolution of camera eye between octopus and human. Genome Res 14:1555–1561. doi:10.1101/gr.2268104
Okanoya K (2002) Sexual display as a syntactical vehicle: the evolution of syntax in birdsong and human language through sexual selection. In: Wray A (ed) The transition to language. Oxford University Press, Oxford, pp 46–63
Okanoya K (2004) Song syntax in Bengalese finches: proximate and ultimate analyses. Adv Study Behav 34:297–346
Ott D (2009) The evolution of i-language: lexicalization as the key evolutionary novelty. Biolinguistics 3(2):255–269
Pfenning AR, Hara E, Whitney O, Rivas MV, Wang R, Roulhac PR et al (2014) Convergent transcriptional specializations in the brains of humans and song-learning birds. Science 346 (6215):1256846. doi:10.1126/science.1256846
Pinker S (1994) The language instinct. Harper Collins, New York
Pinker S, Bloom P (1990) Natural language and natural selection. Behav Brain Sci 13:707–784
Progovac L (2006) Fossilized imperative in compounds and other expressions. Online proceedings of the inaugural meeting of the Slavic Linguistics Society. http://www.indiana.edu/~sls2006/page6/page6.html
Progovac L (2008) What use is half a clause? In: Smith ADM, Smith K, Ferrer i Cancho R (eds) Evolution of language: proceedings of the 7th international EVOLANG conference. World Scientific, Hackensack, NJ, pp 259–266
Progovac L (2009) Layering of grammar: vestiges of evolutionary development of syntax in present-day languages. In: Sampsom G, Gil D, Trudgill P (eds) Language complexity as an evolving variable. Oxford University Press, Oxford, pp 203–212
Progovac L (2010) Syntax: its evolution and its representation in the brain. Biolinguistics 4(2–3):1–22
Progovac L (2012) Compounds and commands in the evolution of human language. Theor Historia Sci 9:49–70
Progovac L (2015) Evolutionary syntax. Oxford University Press, Oxford

Progovac L, Locke J (2009) The urge to merge: insult and the evolution of syntax. Biolinguistics 3(2–3):337–354
Quattara K, Lemasson A, Züberbuhler K (2009) Campbell's monkeys use affixation to alter call meaning. PLoS One 4(11):e7808
Rizzi L (1997) The fine structure of the left periphery. In: Haegeman L (ed) Elements of grammar: a handbook of generative syntax. Kluwer, Dordrecht
Samuels C (2015) Can a bird brain do phology? Front Psychol 6:1082. doi:10.3389/fpsyg.2015.01082
Seed A, Tomasello M (2010) Primate cognition. Top Cogn Sci 2:407–419
Seyfarth RM, Cheney DL (1986) Vocal development in vervet monkeys. Anim Behav 34:1640–1658
Seyfarth RM, Cheney DL, Marler P (1980) Monkey responses to three different alarm calls: evidence of predator classification and semantic communication. Science 210:801–803. doi:10.1126/science.7433999
Spelke E (2003) What makes us smart? Core knowledge & natural language. In: Dedre Gentner D, Goldin–Meadow S (eds) Language in mind: advances in the study of language & thought. MIT Press, Cambridge, MA, pp 277–311
Tallerman M (2007) Did our ancestors speak a holistic protolanguage? Lingua 117:579–604
Tallerman M (2016) Against the emergent view of language evolution. Proceedings of EvoLang XI
Tallerman M, Gibson KR (2012) Introduction: the evolution of language. In: Tallerman M, Gibson KR (eds) The Oxford handbook of language evolution. Oxford University Press, Oxford, pp 1–35
Tattersall I (2009) Language and the origin of symbolic thought. In: De Beaune SA, Coolidge FL, Wynn TG (eds) Cognitive archaeology and human evolution. Cambridge University Press, New York, pp 109–116
Tattersall (2016) Language origins: an evolutionary framework. Topoi 35:1
Tobias PV (1998) Evidence for the early beginnings of spoken language. Camb Archaeol J 8:72–78
Wrangham RW, Jones JH, Laden G, Conklin-Brittain NL (2000) The raw and the stolen: cooking and the ecology of human origins. Curr Anthropol 40(5):567–594
Wynn T (1998) Did *Homo erectus* speak? Camb Archaeol J 8:78–81

Part III
Evolution of Emotion

Chapter 12
Evolution of the Emotional Brain

Edmund T. Rolls

Abstract The brain systems and processing involved in emotion in vertebrates have evolved considerably. The way in which the primate orbitofrontal cortex has undergone great evolutionary development in primates and comes to overshadow the much evolutionarily older amygdala for many functions related to emotion is described. Indeed there may be no cortical area in rodents that is homologous to most of the primate including human orbitofrontal cortex. The primate including human orbitofrontal cortex (OFC) implements reward value. Value is not represented at earlier stages of processing in primates including humans. Invariant visual object recognition is used for many functions including memory formation, so perception is kept separate from emotion. In contrast, in rodents, value is represented even in the first taste relay in the brain, the nucleus of the solitary tract: there is no clear separation between perception and emotion. In rodents, even the taste pathways are connected differently, with subcortical connections bypassing the cortex (including orbitofrontal cortex) and making connections via a pontine taste area directly to the hypothalamus and amygdala. Goal value-directed choice is usual in primates and humans, whereas fixed action patterns, such as pecking in birds, are more common elsewhere. In humans, and perhaps some primates, syntactic reasoning and thereby planning allows selfish gene-specified (emotion-related) rewards to be rejected in favour of the long-term interests of the individual, the phenotype.

Keywords Emotion • Evolution of emotion • Orbitofrontal cortex • Amygdala • Value • Reward

E.T. Rolls
Oxford Centre for Computational Neuroscience, Oxford, UK

Department of Computer Science, University of Warwick, Coventry CV4 7AL, UK
e-mail: Edmund.Rolls@oxcns.org; http://www.oxcns.org

S. Watanabe et al. (eds.), *Evolution of the Brain, Cognition, and Emotion in Vertebrates*, Brain Science, DOI 10.1007/978-4-431-56559-8_12

12.1 Introduction

The brain systems and processing involved in emotion in vertebrates have evolved considerably. Some of the principles that I elucidate in this chapter include the following (Rolls 2014):

1. The primate orbitofrontal cortex has undergone great evolutionary development in primates and comes to overshadow the much evolutionarily older amygdala for many functions related to emotion. Indeed there may be no cortical area in rodents that is homologous to most of the primates including human orbitofrontal cortex.
2. The primate including human orbitofrontal cortex (OFC) implements reward value, as shown by devaluation experiments such as feeding to satiety.
3. Value is not represented at earlier stages of processing in primates including humans. Invariant visual object recognition is used for many functions including memory formation, so perception is kept separate from emotion.
4. In contrast, in rodents, value is represented even in the first taste relay in the brain, the nucleus of the solitary tract: there is no clear separation between perception and emotion. In rodents, even the taste pathways are connected differently, with subcortical connections bypassing the cortex (including orbitofrontal cortex) and making connections via a pontine taste area directly to the hypothalamus and amygdala.
5. In primates and humans, the orbitofrontal cortex implements one-trial rule-based reversal learning, and this is important in rapidly updating social behaviour. This is rapid updating of value-based representations. Maintaining the current rule in short-term memory and using this to bias neurons in the orbitofrontal cortex may be one computation that granular prefrontal cortex facilitates. Rodents may not be able to perform this.
6. The value representation in the primate and human orbitofrontal cortex is domain general, in that the amount and value of goods, and temporal discounting, operate transitively (as shown by trade-offs), providing a basis for economic decision-making. There is evidence that this is not the case in rodents.
7. Goal-directed choice may be the best measure of value and emotion, for there are many partly separate neural circuits for different emotion-related responses, e.g. autonomic output, freezing, fixed action patterns, and unconditioned approach or withdrawal.
8. Goal value-directed choice is usual in primates and humans, whereas fixed action patterns, such as pecking in birds, are more common elsewhere.
9. In humans, and perhaps some primates, syntactic reasoning and thereby planning allows selfish gene-specified (emotion-related) rewards to be rejected in favour of the long-term interests of the individual, the phenotype.

These principles are now elucidated.

12.2 An Anatomical and Functional Framework for Understanding the Neural Basis of Emotion

Emotions can be defined as states elicited by rewards and punishers, that is, by instrumental reinforcers, which are the goals for action (Rolls 2013, 2014, 2015b, 2016a). Motivational states can be defined as states in which an instrumental reinforcer is the goal for action (Rolls 2016b). The principle of operation is that genes can specify goals for actions that are in the selfish interests of the genes. By specifying the rewards (e.g. a sweet taste) and punishers (e.g. painful touch), the specification is simpler than trying to specify detailed behavioural responses to stimuli and allows much greater flexibility of the actions, which can be learned instead of prespecified by the genes. Emotions are states that can continue after the eliciting stimulus is no longer present, for example, when an expected reward is not obtained, and this is adaptive, for the state can influence ongoing goal directed behaviour, for example, to obtain a missing reward. The approach to emotions that I have described, as states elicited by (instrumental) rewards and punishers, relates emotions to goals and is therefore different from measuring emotion by respondents such as autonomic responses to unconditioned approach or flight. Indeed, a rich set of mechanisms are brought into play when rewards and punishers are delivered, and one must be very careful to distinguish the different types of mechanism involved, as set out in section 4.6.1 (pp. 159–165) of Emotion and Decision-Making Explained (Rolls 2014) and by Cardinal et al. (2002).

I now provide a framework for understanding some of the brain structures involved in emotion in primates including humans and at the same time contrast them with the structures that in terms of connectivity and function precede them and succeed them in the anatomical and functional hierarchy moving from left to right in Fig. 12.1 (Rolls 2014). This provides a framework within which to consider the evolution of these systems involved in emotion. In line with the definition of emotion provided above, the interest is in the brain systems that compute and represent reward value and then provide this as an input to decision and action systems.

In Tier 1 of Fig. 12.1, information is processed to a level at which the neurons represent ‘what’ the stimulus is, independently of the reward or punishment value of the stimulus. Thus neurons in the primary taste cortex represent what the taste is, and its intensity, but not its reward value (Rolls 2014). In the inferior temporal visual cortex, the representation is of objects, invariantly with respect to the exact position on the retina, size, and even view. Forming invariant representations involves a great deal of cortical computation in the hierarchy of visual cortical areas from the primary visual cortex V1 to the inferior temporal visual cortex (Rolls 2012a, 2016a). The fundamental advantage of this separation of ‘what’ processing in Tier 1 from reward value processing in Tier 2 is that any learning in Tier 2 of the value of an object or face seen in one location on the retina, size, and view will generalize to other views, etc. In rodents, there is no such clear separation of ‘what’ from ‘value’ representations. For example, in the taste system, satiety

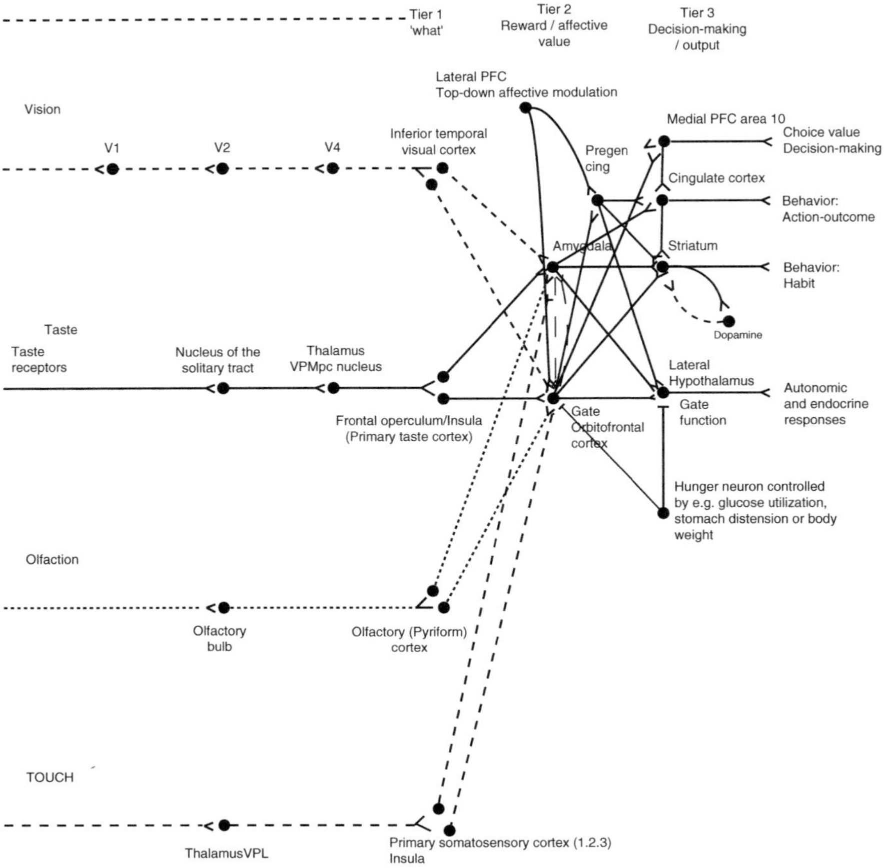

Fig. 12.1 Schematic diagram showing some of the gustatory, olfactory, visual, and somatosensory pathways to the orbitofrontal cortex, and some of the outputs of the orbitofrontal cortex, in primates. The secondary taste cortex and the secondary olfactory cortex are within the orbitofrontal cortex. V1—primary visual cortex. V4—visual cortical area V4. PreGen Cing—pregenual cingulate cortex. 'Gate' refers to the finding that inputs such as the taste, smell, and sight of food in some brain regions only produce effects when hunger is present (Rolls 2014). Tier 1: the column of brain regions including and below the inferior temporal visual cortex represents brain regions in which 'what' stimulus is present is made explicit in the neuronal representation, but not its reward or affective value which are represented in the next tier of brain regions (Tier 2), the orbitofrontal cortex and amygdala, and in the anterior cingulate cortex. In Tier 3 areas beyond these such as medial prefrontal cortex area 10, choices, or decisions about reward value are taken (Rolls 2008b, 2014; Rolls and Deco 2010). Top-down control of affective response systems by cognition and by selective attention from the dorsolateral prefrontal cortex is also indicated. Medial PFC area 10—medial prefrontal cortex area 10; VPMpc—ventralposteromedial thalamic nucleus, the thalamic nucleus for taste

influences taste processing at the first central synapse in the taste system (Rolls and Scott 2003), and this property makes the processing in rodents not only different from that in primates including humans but also much more difficult to analyse (Rolls 2014, 2015a).

In Tier 2 of Fig. 12.1, there are brain mechanisms in the orbitofrontal cortex that are involved in computing the reward value of primary (unlearned) reinforcers, as shown by devaluation experiments in which, for example, a food is fed to satiety (Rolls et al. 1989; Critchley and Rolls 1996a; Kringelbach et al. 2003; Rolls and Grabenhorst 2008), and by neuroeconomics experiments which show that the amount and quality of each commodity is encoded by orbitofrontal cortex neurons (Padoa-Schioppa 2011; Padoa-Schioppa and Assad 2008; Grabenhorst and Rolls 2011). The primary reinforcers include taste, touch (both pleasant touch and pain), and to some extent smell, and perhaps certain visual stimuli such as face expression. There is evidence that there is a representation of the (reward/punishment) value of many primary reinforcers in the orbitofrontal cortex, including taste, positive touch and pain, face expression, face beauty, and auditory consonance/dissonance. In neuroeconomics, these are termed 'outcome value' representations (Rolls 2014). Further evidence for value representations is that orbitofrontal cortex activations in humans to these stimuli are linearly related to the subjectively reported pleasantness of stimuli (medially) or to their unpleasantness (laterally) (Rolls 2014).

Brain regions in Tier 2 are also concerned with learning associations between previously neutral stimuli, such as the sight of objects or of individuals' faces, with primary reinforcers. These brain regions include the amygdala and orbitofrontal cortex, with the orbitofrontal cortex being especially important in the rapid, one-trial learning and reversal of stimulus-reinforcer associations. In neuroeconomics, these are termed 'expected value' representations. Once the Tier 2 brain regions have determined whether the input is reinforcing, whether primary or secondary, the signal is passed directly to output regions of the brain, with no need to produce and then feedback peripheral body or autonomic responses to the brain.

In Tier 2 in the orbitofrontal cortex, the representation is of the value of stimuli, and actions are not represented. The value of very many different types of stimuli, events, or goals is represented separately at the neuronal level, providing the basis for choice between stimuli and the selection at later stages of processing of an appropriate action to obtain the chosen goal.

In Tier 3, the medial prefrontal cortex area 10/ventromedial prefrontal cortex is implicated in decision-making between stimuli, in which a selection or choice must be made, moving beyond a representation of value on a continuous scale towards a decision between goods based on their value (Rolls 2014; Grabenhorst et al. 2011; Rolls et al. 2008).

The Tier 2 brain regions in which the reinforcing, and hence emotional, value of stimuli is represented in primates include mainly the orbitofrontal cortex and amygdala, which interface to three other main types of output system in Tier 3 (Figs. 12.1 and 12.2):

In Tier 3, the first is the autonomic and endocrine system, for producing such changes as increased heart rate and release of adrenaline, which prepare the body for action. Structures receiving from the orbitofrontal cortex, amygdala, and anterior cingulate cortex that provide a route for these autonomic effects include the hypothalamus and parts of the anterior insula close to the insular taste cortex (Rolls 2014; Critchley and Harrison 2013).

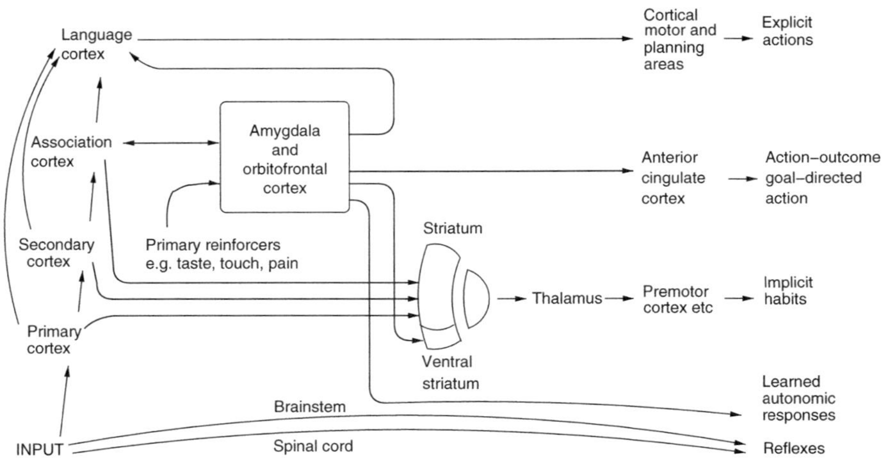

Fig. 12.2 Dual routes to the initiation of actions in response to rewarding and punishing stimuli. The inputs from different sensory systems to brain structures such as the orbitofrontal cortex and amygdala allow these brain structures to evaluate the reward- or punishment-related value of incoming stimuli, or of remembered stimuli. One type of route is via the language systems of the brain, which allow explicit (verbalizable) decisions involving multistep syntactic planning to be implemented. The other type of route may be implicit and includes the anterior cingulate cortex for action-outcome, goal-dependent learning and the striatum and rest of the basal ganglia for stimulus-response habits. The basal ganglia may be involved in selecting only one system for output. Outputs for autonomic responses can also be produced using outputs from the orbitofrontal cortex and anterior cingulate cortex (some of which are routed via the anterior insular cortex) and amygdala

The second type of output is to brain systems concerned with performing actions unconsciously or implicitly, in order to obtain rewards or avoid punishers. One of these brain systems is the basal ganglia for habit ('stimulus-response') behaviour, in which the behaviour becomes no longer under the control of the goal as shown by devaluation procedures but is a stimulus-to-motor-response association, which are necessary strong emotional states (Rolls 2014). A second brain system is the anterior cingulate cortex for goal-directed, action-outcome learning (Rolls 2014). (The 'outcome' is the reward or punisher that is or is not obtained when the action is performed to obtain the goal.) The anterior cingulate cortex contains representations of reward and punisher value, and thus of outcome, which are essential for learning associations between actions and the outcomes that follow actions. The mid-cingulate area contains representations of actions.

The third type of output in humans and perhaps related animals is to a system capable of planning many steps ahead and, for example, deferring short-term rewards in order to execute a long-term plan. This system may use syntactic processing to perform the planning and is therefore part of a linguistic system which performs explicit (conscious) processing, as described more fully elsewhere (Rolls 2014).

12.3 Evolution of the Primate Orbitofrontal Cortex

Many of the brain systems that are involved in emotion have undergone considerable development in primates (e.g. monkeys and humans) (Rolls 2014), as summarized next.

First, the temporal lobe has undergone great development in primates, and several systems in the temporal lobe are either involved in emotion (e.g. the amygdala) or provide some of the main sensory inputs to brain systems involved in emotion and motivation. For example, the amygdala and the orbitofrontal cortex, key brain structures in emotion, both receive inputs from the highly developed primate temporal lobe cortical areas, including those involved in invariant visual object recognition and face identity and expression processing (Rolls 2000, 2011, 2012a, 2014).

Second, the prefrontal cortex has undergone great development in primates, and one part of the prefrontal cortex, the orbitofrontal cortex, is very little developed in rodents yet is one of the major brain areas involved in emotion, motivation, and reward value processing including for taste, olfactory, and visual inputs in primates including humans. With this great development of the orbitofrontal cortex in primates, there may be division of functionality, with the primate taste insula not performing taste-related hedonic functions (Rolls 2015c). Indeed, it has been argued (on the basis of cytoarchitecture, connections, and functions) that the granular prefrontal cortex is a primate innovation (Preuss 1995; Wise 2008; Passingham and Wise 2012; Rolls 2014, 2015c), and the implication of the argument is that any areas that might be termed orbitofrontal cortex in rats (Schoenbaum et al. 2009) are homologous only to the agranular parts of the primate orbitofrontal cortex, that is, to areas 13a, 14c, and the agranular insular areas Ia (Passingham and Wise 2012) (shaded mid grey in Fig. 12.3). Indeed, there may be no cortical area in rodents that is homologous to most of the primates' including human orbitofrontal cortex (Preuss 1995; Wise 2008; Passingham and Wise 2012; Rolls 2014, 2015c). It follows from that argument that for most areas of the orbitofrontal and medial prefrontal cortex in humans and macaques (those shaded light grey in Fig. 12.3), special consideration must be given to research in macaques and humans.

Third, even the taste system (which might have been supposed to be phylogenetically old and preserved) of primates and rodents may be different, with obligatory processing from the nucleus of the solitary tract via the thalamus to the cortex in primates, but a subcortical pathway in rodents via a pontine taste area to the amygdala, and differences in where satiety influences taste-responsive neurons in primates and rodents (Norgren 1984; Rolls and Scott 2003; Small and Scott 2009; Rolls 2014, 2015a).

Fourth, with the great development of the orbitofrontal cortex in primates, the amygdala may become relatively less important in humans in emotion than in other vertebrates (Rolls 2014).

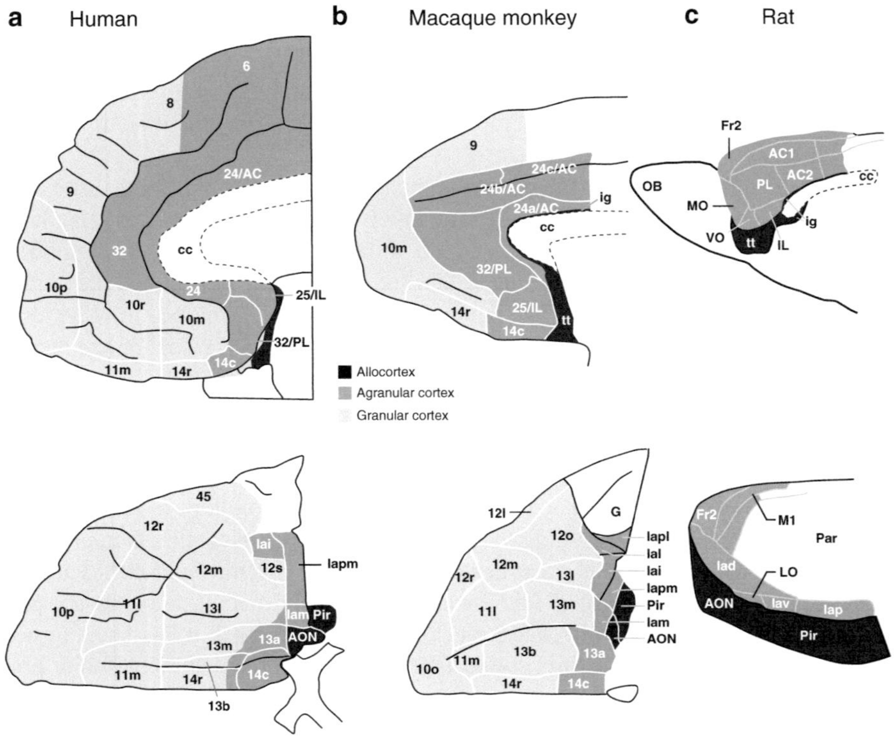

Fig. 12.3 Comparison of the orbitofrontal (below) and medial prefrontal (above) cortical areas in humans, macaque monkeys, and rats. (**a**) Medial (*top*) and orbital (*bottom*) areas of the human frontal codex (Ongur et al. 2003). (**b**) Medial (*top*) and orbital (*bottom*) areas of the macaque frontal cortex (Carmichael and Price 1994). (**c**) Medial (*top*) and lateral (*bottom*) areas of rat frontal cortex (Palomero-Gallagher and Zilles 2004). Rostral is to the *left* in all drawings. *Top row*: dorsal is up in all drawings. *Bottom row*: in (**a**, **b**), lateral is up; in (**c**), dorsal is up. Not to scale. Abbreviations: *AC* anterior cingulate cortex, *AON* anterior olfactory 'nucleus', *cc* corpus callosum, *Fr2* second frontal area, *Ia* agranular insular cortex, *ig* indusium griseum, *IL* infralimbic cortex, *LO* lateral orbital cortex, *MO* medial orbital cortex, *OB* olfactory bulb, *Pr* piriform (olfactory) cortex, *PL* prelimbic cortex, *tt* tenia tecta, *VO* ventral orbital cortex; Subdivisions of areas are labelled caudal (c), inferior (i), lateral (l), medial (m), orbital (o), posterior or polar (p), rostral(r), or by arbitrary designation (a, b) (After Passingham and Wise (2012). (**a**) Adapted from Dost Ongur, Amon T. Ferry, and Joseph L. Price, Architectonic subdivision of the human orbital and medial prefrontal cortex, Journal of Comparative Neurology, 460 (3), pp. 425–49 Copyright 2003 Wiley-Liss, Inc. (**b**) Adapted from S. T. Carmichael and J. L. Price, Architectonic subdivision of the orbital and medial prefrontal cortex in the macaque monkey, Journal of Comparative Neurology, 346 (3), pp. 366–402 Copyright 1994 Wiley-Liss, Inc. (**c**) Adapted from Palomero-Gallagher N. and Zilles K., Isocortex, in The Rat Nervous System 3rd edn. G. Paxinos, pp. 729–57 Copyright 2004, Elsevier Academic Press

12.4 The Primate Including Human Orbitofrontal Cortex (OFC) Implements Reward Value

Let us start with the reward value of taste, which can be measured in devaluation experiments such as feeding to satiety, which decreases food reward value, and in humans the pleasantness of food, to zero. In the macaque orbitofrontal cortex throughout its mediolateral extent, almost all neurons show a decrease to zero of the response to taste, that is, the neurons do not alter from their spontaneous firing rate, after feeding to satiety (Rolls et al. 1989; Critchley and Rolls 1996a; Pritchard et al. 2008; Rolls 2015c). Similar effects are found for fat texture (Rolls et al. 1999; Verhagen et al. 2003). In the human orbitofrontal cortex, we found a large decrease in the BOLD signal to a complex food (tomato juice vs. chocolate) fed to satiety, but not in the insula (Kringelbach et al. 2003). Moreover, this was a sensory-specific decrease in the BOLD signal, a useful indication that this was a response related to real satiety, which is to a considerable extent sensory-specific, and not for every food (Rolls 2016d). Moreover, this sensory-specific decrease was related to the decrease in the subjective pleasantness of the food eaten to satiety. Further, we are looking for a brain region not just where there may be small changes to the response to a taste fed to satiety, but a region where the response decreases to zero, for this is what happens to the pleasantness of food after it is fed to satiety, with little effect on its intensity (Rolls et al. 1983; Rolls and Grabenhorst 2008; Rolls 2014). A detailed analysis of the functions of different parts of the primate including human anterior insula in taste and related functions is provided elsewhere (Rolls 2015a).

Odours are also represented in the macaque orbitofrontal cortex (Critchley and Rolls 1996b; Rolls et al. 1996b). It was shown that the majority of orbitofrontal olfactory neurons decrease their responses to the odour of the food with which the monkey was fed to satiety (Critchley and Rolls 1996a). The subjective pleasantness or reward or affective value of odour is represented in the orbitofrontal cortex, in that feeding humans to satiety decreases the activation found to the odour of that food, and this effect is relatively specific to the food eaten in the meal (Francis et al. 1999; O'Doherty et al. 2000; cf. Morris and Dolan 2001). Further, the human medial orbitofrontal cortex has activation that is related to the subjective pleasantness of a set of odours, and a more lateral area has activation that is related to the degree of subjective unpleasantness of odours (Rolls et al. 2003). An fMRI investigation in humans showed that whereas in the orbitofrontal cortex the pleasantness vs. unpleasantness of odours is represented, this was not the case in primary olfactory cortical areas, where instead the activations reflected the intensity of the odours (Rolls et al. 2003).

There is a major visual input to many neurons in the orbitofrontal cortex, and what is represented by these neurons is in many cases the reinforcement association of visual stimuli, i.e. their reward/punishment value. The visual input is from the ventral, temporal lobe, visual stream concerned with 'what' object is being seen (see Rolls 2000, 2012a, 2016a). Many neurons in these temporal cortex visual areas have responses to objects or faces that are invariant with respect to size, position on the retina, and even view (Rolls 2000, 2007, 2008a, b, 2009, 2012a, 2016a), making these neurons ideal as an input to a system that may learn about the reinforcement association properties of objects and faces, for after a single learning trial, the learning then generalizes correctly to other views, etc. (see Rolls 2000, 2008b, 2012a, 2014, 2016a). Using this object-related information, orbitofrontal cortex visual neurons frequently respond differentially to objects or images depending on their reward association (Thorpe et al. 1983; Rolls et al. 1996b). The primary reinforcer that has been used is taste, and correlates of visual to taste association learning have been demonstrated in the human orbitofrontal cortex with fMRI (O'Doherty et al. 2002). Many of these neurons show visual-taste reversal in one or a very few trials. (In a visual discrimination task, they will reverse the stimulus to which they respond, from e.g. a triangle to a square, in one trial when the taste delivered for a behavioural response to that stimulus is reversed (Thorpe et al. 1983).) This reversal learning probably occurs in the orbitofrontal cortex, for it does not occur one synapse earlier in the visual inferior temporal cortex (Rolls et al. 1977), and it is in the orbitofrontal cortex that there is convergence of visual and taste pathways onto the same single neurons (Thorpe et al. 1983; Rolls and Baylis 1994; Rolls et al. 1996b). Moreover the majority of orbitofrontal visual food-related neurons decrease their responses to the sight of the food with which the monkey was fed to satiety. Thus for these neurons, the expected reward value of the sight of food is what is represented in the orbitofrontal cortex (Critchley and Rolls 1996a).

Another type of visual information represented in the orbitofrontal cortex that is relevant to emotion is information about faces. There is a population of orbitofrontal cortex neurons that respond in many ways similarly to those in the temporal cortical visual areas (Rolls 1984, 1992, 1996, 2000, 2007; Rolls and Deco 2002). The orbitofrontal cortex face-responsive neurons, first observed by Thorpe et al. (1983), then by Rolls et al. (2006), tend to respond with longer latencies than temporal lobe neurons (140–200 ms typically, compared to 80–100 ms); also convey information about which face is being seen, by having different responses to different faces; and are typically rather harder to activate strongly than temporal cortical face-selective neurons, in that many of them respond much better to real faces than to two-dimensional images of faces on a video monitor (Rolls et al. 2006; Rolls 2011) (cf. Rolls and Baylis 1986). Some of the orbitofrontal cortex face-selective neurons are responsive to face expression, gesture, or movement (Rolls et al. 2006). The findings are consistent with the likelihood that these neurons are activated via the inputs from the temporal cortical visual areas in which face-selective neurons are found (see Fig. 12.1). The significance of the neurons is likely to be related to the fact that faces convey information that is important in social reinforcement in at least two ways that could be implemented by these neurons. The first is that some may encode face expression (Rolls et al. 2006) (cf. Hasselmo et al. 1989), which can indicate reinforcement. The second way is that they encode information about which individual is present (Rolls et al. 2006), which by stimulus-reinforcement association learning is important in evaluating and utilizing learned reinforcing inputs in social situations, e.g. about the current reinforcement value as decoded by stimulus-reinforcement association, to a particular individual. Between them, these neurons represent whose face has a particular expression, and this is important in social situations. This system is likely to be a primate specialization, made possible by the great development of the temporal lobes, which compute invariant representations of faces, which make this functionality in the orbitofrontal cortex possible (Rolls 2012a, 2016a).

This system has also been shown to be present in humans. For example, Kringelbach and Rolls (2003) showed that activation of a part of the human orbitofrontal cortex occurs during a face discrimination reversal task. In the task, the faces of two different individuals are shown, and when the correct face is selected, the expression turns into a smile. (The expression turns to angry if the wrong face is selected.) After a period of correct performance, the contingencies reverse, and the other face must be selected to obtain a smile expression as a reinforcer. It was found that activation of a part of the orbitofrontal cortex occurred specifically in relation to the reversal, that is, when a formerly correct face was chosen, but an angry face expression was obtained. In a control task, it was shown that the activations were not related just to showing an angry face expression. Thus in humans, there is a part of the orbitofrontal cortex that responds selectively in relation to face expression specifically when it indicates that behaviour should change, and this activation is error-related (Kringelbach and Rolls 2003) and occurs when the error neurons in the orbitofrontal cortex become active (Thorpe et al. 1983).

Value is not represented at earlier stages of processing than the orbitofrontal cortex in primates including humans.

Rolls, Scott, and colleagues have shown that the primary taste cortex in the primate anterior insula and adjoining frontal operculum contains not only taste neurons tuned to sweet, salt, bitter, sour (Scott et al. 1986a; Yaxley et al. 1990; Rolls and Scott 2003; Scott and Plata-Salaman 1999), and umami as exemplified by monosodium glutamate (Baylis and Rolls 1991; Rolls et al. 1996a) but also other neurons that encode oral somatosensory stimuli including viscosity, fat texture, temperature, and capsaicin (Verhagen et al. 2004). None of the insular taste cortex neurons had responses to olfactory stimuli, and none could be shown to have responses to visual stimuli that were clearly not just related to mouth movements and the accompanying somatosensory input (Verhagen et al. 2004), in contrast to the orbitofrontal cortex where responses to olfactory and visual stimuli associated with food are common (Thorpe et al. 1983; Rolls et al. 1996b, c, 2010; Critchley and Rolls 1996a, b; Rolls 2015c). Water can activate some neurons in cortical taste areas (Rolls et al. 1990; Yaxley et al. 1990), and this has also been found in the rodent insula (MacDonald et al. 2012). Whether this is by mouth feel relative to saliva, or by ionic content relative to saliva, or by some other mechanism, is not known.

Neurons in the macaque primary taste cortex do not represent the reward value of taste, that is, the appetite for a food, in that their firing is not decreased to zero by feeding the taste to satiety (Yaxley et al. 1988; Rolls et al. 1988). This was confirmed in 17 separate experiments on neurons in the insular and frontal opercular primary taste cortex, using anatomical confirmation that these neurons were in the primary taste cortex by the use of X-ray localization and then histological reconstruction. The neurons showed no reduction in their firing to the taste (typically glucose) after it had been fed to satiety (Yaxley et al. 1988; Rolls et al. 1988).

In macaques, neural processing peripheral to the primary taste cortex is consistent with this, with taste responses found in the rostral part of the nucleus of the solitary tract (Scott et al. 1986b) that are not influenced by feeding to satiety (Yaxley et al. 1985).

Consistently, in humans, BOLD activations in the insular taste cortical area were linearly related to the intensity but not the pleasantness of the tastes (Grabenhorst and Rolls 2008). The converse was found for the orbitofrontal cortex: the BOLD activations in the orbitofrontal cortex but not the anterior and mid-insular taste cortical areas were linearly related to the pleasantness of the tastes (Grabenhorst and Rolls 2008).

For odour, there are similar findings, with activations in the pyriform cortex correlated with the intensity of odours and not their pleasantness, whereas in the orbitofrontal cortex activations are correlated with the pleasantness of odours, but not with their intensity (Rolls et al. 2003).

Consistently, for visual stimuli, the reward value of objects including the sight of food are not represented in the inferior temporal visual cortex in that there is no effect of feeding to satiety, and reversal of reward value does not reverse neuronal responses in the inferior temporal visual cortex (Rolls et al. 1977).

In rodents, reward value is represented even in the first taste relay in the brain, and in the olfactory bulb; and there are direct subcortical pathways.

First, there are major anatomical differences in the neural processing of taste in rodents and primates (Rolls and Scott 2003; Small and Scott 2009; Scott and Small 2009; Rolls 2014, 2015c). In primates, the rostral part of the nucleus of the solitary tract (NTS, the first central taste relay) projects to the taste thalamus and thus to the cortex (Figs. 12.1 and 12.4); whereas in rodents the majority of NTS taste neurons project to the pontine parabrachial nucleus (PbN), referred to as the rodent 'pontine taste area' (Small and Scott 2009; Cho et al. 2002) (Fig. 12.4). From the PbN, the rodent gustatory pathway bifurcates into two pathways: (1) a ventral 'affective' projection to the hypothalamus, central grey, ventral striatum, bed nucleus of the stria terminalis, and amygdala; and (2) a dorsal 'sensory' pathway, which first synapses in the thalamus and then the agranular and dysgranular insular gustatory cortex (Norgren and Leonard 1971; Norgren 1974, 1976, 1990) (Fig. 12.4). In primates (including humans) there is strong evidence to indicate that the PbN gustatory relay is absent (Small and Scott 2009).

Second, a functional difference of rodent taste processing from that of primates is that physical and chemical signals of satiety have been shown to reduce the taste responsiveness of neurons in the nucleus in the solitary tract, and the pontine taste

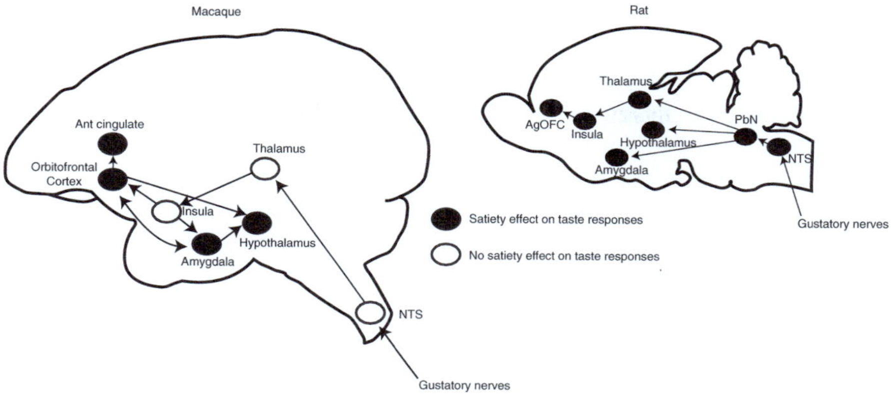

Fig. 12.4 Taste pathways in the macaque and rat. In the *macaque*, gustatory information reaches the nucleus of the solitary tract (NTS), which projects directly to the taste thalamus (ventral posteromedial nucleus, pars parvocellularis, VPMpc) which then projects to the taste cortex in the anterior insula (insula). The insular taste cortex then projects to the orbitofrontal cortex and amygdala. The orbitofrontal cortex projects taste information to the anterior cingulate cortex. Both the orbitofrontal cortex and the amygdala project to the hypothalamus (and to the ventral striatum). In macaques, feeding to normal self-induced satiety does not decrease the responses of taste neurons in the NTS or taste insula (and by inference not VPMpc) (see text). In the *rat*, in contrast, the NTS projects to a pontine taste area, the parabrachial nucleus (PbN). The PbN then has projections directly to a number of subcortical structures, including the hypothalamus, amygdala, and ventral striatum, thus bypassing thalamocortical processing. The PbN in the rat also projects to the taste thalamus (VPMpc), which projects to the rat taste insula. The taste insula in the rat then projects to an agranular orbitofrontal cortex (AgOFC), which probably corresponds to the most posterior part of the primate OFC, which is agranular. (In primates, most of the orbitofrontal cortex is granular cortex, and the rat may have no equivalent to this (Small and Scott 2009; Passingham and Wise 2012; Rolls 2014; Wise 2008; Rolls 2015c.) In the rat, satiety signals such as gastric distension and satiety-related hormones decrease neuronal responses in the NTS (see text), and by inference therefore in the other brain areas with taste-related responses, as indicated in the figure

area, of the rat, with decreases in the order of 30% (Scott and Small 2009; Rolls and Scott 2003; Glenn and Erickson 1976; Giza and Scott 1983, 1987; Giza et al. 1993; Hajnal et al. 1999). (Given this evidence, as expected, neuronal responses in many areas of the rat brain including the insula and amygdala are decreased by satiety (de Araujo et al. 2006).) The implication of this whole body of evidence is that in rodents, sensory (perceptual) and reward (hedonic) processing are not independent. In contrast, in primates, the reward value of tastants is represented in the orbitofrontal cortex in that the responses of orbitofrontal cortex taste neurons are modulated by hunger in just the same way as is the reward value or palatability of a taste, and this is not found in the taste insula (Rolls 2015c). Thus in the primary taste cortex of nonhuman primates (and at earlier stages of taste processing including the nucleus of the solitary tract (Yaxley et al. 1985)), the reward value of taste is not represented, and instead the identity and intensity of the taste are represented (Rolls 2014). A perceptual correlate of this is that when humans feed to satiety, the intensity of the flavour changes very little, whereas the pleasantness of the flavour decreases to zero (Rolls et al. 1983), showing that in humans perceptual representations of taste and olfaction are kept separate from hedonic representations. This is adaptive, in that we do not go blind to the sight, taste, and smell of food after eating it to satiety and can therefore still learn about where food is located in the environment even when we are not hungry (Rolls 2014). Moreover, and consistently, activations in the human insular primary taste cortex are related to the intensity and not to the pleasantness of taste (Grabenhorst et al. 2008; Grabenhorst and Rolls 2008).

The importance of cortical processing of taste in primates, first for identity and intensity in the primary taste cortex, and then for reward value in the orbitofrontal cortex, is that both types of representation need to be interfaced to visual and other processing that requires cortical computation. For example, it may have adaptive value to be able to represent exactly what taste is present and to link it by learning to the sight and location of the source of the taste, even when hunger is not present and reward is not being produced, so that the source of that taste can be found in future, when it may have reward value. More generally, when we see and taste a food (perhaps in a particular place) when hunger is not present and the food has no reward value, it is still important to be able to learn associations between these representations, including for semantic memory. In line with cortical processing to dominate the processing of taste in primates, there is no modulation in primates of taste responsiveness at or before the primary taste cortex, and the pathways for taste are directly from the nucleus of the solitary tract in the brainstem to the taste thalamus and then to the taste cortex (Figs. 12.1 and 12.4) (Rolls 2014, 2015a).

12.5 Rapid, Rule-Based Reward Reversal Learning in Primates: Orbitofrontal Cortex vs. Amygdala

In primates and humans, the orbitofrontal cortex implements one-trial rule-based reversal learning, and this is important in rapidly updating social behaviour. This is rapid updating of value-based representations.

Visual-to-taste association learning and its reversal by neurons in the orbitofrontal cortex can take place in as little as one trial (Thorpe et al. 1983; Rolls et al. 1996b; Deco and Rolls 2005; Rolls and Deco 2016). This is rule-based, in that if on one trial the expected reward is not obtained, on the very next trial a previously punished visual stimulus is shown, it will be chosen on the basis that the rule for which stimulus is associated with reward has changed. This has clear adaptive value in enabling particular foods with a good or bad taste to be learned and recognized quickly, important in foraging and in food selection for ingestion. The visual inputs reach the orbitofrontal cortex from the inferior temporal visual cortex, where neurons respond to visual objects independently of their reward value (e.g. taste) as shown by satiety and reversal learning tests (Rolls et al. 1977; Rolls 2008b, 2012a). The visual-to-taste associations are thus learned in the orbitofrontal cortex (Rolls 2014). These visual-taste neurons thus respond to expected value (Rolls 2014). Consistent evidence is available in humans, in that the lateral orbitofrontal cortex is activated on reversal trials, when an error is detected (Kringelbach and Rolls 2003), consistent with the presence of error neurons in the primate orbitofrontal cortex (Thorpe et al. 1983; Rolls 2016a). Further, patients with damage to the orbitofrontal cortex are impaired on rapid stimulus-reward reversal learning (Hornak et al. 2004; Rolls et al. 1994). Maintaining the current rule in short-term memory and using this to bias neurons in the orbitofrontal cortex may be one computation that granular prefrontal cortex facilitates, because of its highly developed local recurrent collateral system which can form an attractor network and hold the current rule in short-term memory (Rolls 2016a; Rolls and Deco 2016). Rodents may not be able to perform one-trial rule-based stimulus-reward reversal.

The amygdala is a structure in the temporal lobe with somewhat similar connections to the orbitofrontal cortex (see Fig. 12.1). The amygdala has been present in evolution for much longer than the primate orbitofrontal cortex and appears to differ from the orbitofrontal cortex in that it cannot implement one-trial, rule-based, visual discrimination reversal when the taste or flavour associated with the visual stimulus is reversed (Rolls 2014). The primate amygdala contains neurons that respond to taste and oral texture (Sanghera et al. 1979; Scott et al. 1993; Kadohisa et al. 2005a, b). Some neurons respond to visual stimuli associated with reinforcers such as taste but do not reflect the reinforcing properties very specifically, do not rapidly learn and reverse visual-to-taste associations, and are much less affected by reward devaluation by feeding to satiety than are orbitofrontal cortex neurons (Sanghera et al. 1979; Yan and Scott 1996; Kadohisa et al. 2005a, b; Wilson and Rolls 2005; Rolls 2014). The primate orbitofrontal cortex appears to be much more closely involved in flexible (rapidly learned, and affected by reward devaluation) reward representations than in the primate amygdala (Rolls 2014), although both amygdala and orbitofrontal cortex lesions can impair the preference for an object on the first trial after devaluation by feeding to satiety of the food-related value of that object (Murray and Izquierdo 2007). In addition, in humans, amygdala lesions appear to have less profound effects on emotion and emotion-related learning than do orbitofrontal cortex lesions (Rolls 2014).

12.6 Neuroeconomic Representation of Value in the Primate Including Human Orbitofrontal Cortex

The reward value representations in the primate orbitofrontal cortex of taste, olfactory, and flavour stimuli are appropriate for economic decision-making in a number of ways (Rolls 2014, 2015c). First, the responses of orbitofrontal cortex neurons reflect the quality of the commodity or 'good' (e.g. the sight or taste of food) multiplied by the amount available (Padoa-Schioppa and Assad 2006; Padoa-Schioppa 2011). In humans, activations in the orbitofrontal cortex reflect the 'subjective value' of foods (where 'subjective value' in economics refers strictly to what is chosen by a subject rather than to conscious subjective pleasantness (Rolls 2014, 2015c)), assessed in a task in which the value is measured by choices between different foods and different amounts of money (Plassmann et al. 2007). Moreover these neurons reflect the value of reward stimuli, and not actions made to obtain them (Rolls 2014; Thorpe et al. 1983; Rolls et al. 1990; Verhagen et al. 2003; Padoa-Schioppa and Assad 2006).

The value representation in the primate and human orbitofrontal cortex is domain general, in that the amount and value of goods, and temporal discounting, operate transitively (as shown by trade-offs), providing a basis for economic decision-making (Padoa-Schioppa 2011; Rolls 2014). There is evidence that this is not the case in rodents (Padoa-Schioppa 2011; Rolls 2014).

Goal-directed choice may be the best measure of value and emotion, for there are many partly separate neural circuits for different emotion-related responses, e.g. autonomic output, freezing, fixed action patterns, and unconditioned approach or withdrawal. The functions of the amygdala.

Given the approach to emotions as states elicited by instrumental reinforcers, goal-directed value as a basis for action is a crucial system in emotion (Rolls 2014). However, some of the brain structures implicated in emotion produce other outputs that are adaptive, even if not fundamental to emotion and goal-directed behaviour [section 4.6.1 (pp. 159–165) of Emotion and Decision-Making Explained (Rolls 2014), and Cardinal et al. (2002)], though possibly of earlier evolutionary origin.

First, autonomic responses such as increased heart rate can be produced by brainstem pathways, the hypothalamus, amygdala, orbitofrontal cortex, and cingulate cortex (Figs. 12.1 and 12.2) (Critchley and Harrison 2013). Autonomic responses can become classically conditioned in structures such as the amygdala, as can freezing responses (LeDoux 2012; Phelps and LeDoux 2005), but of course these are conditioned reflexes, with no flexibility of response (Rolls 2014). Even approach to a food can become classically conditioned (Cardinal et al. 2002; Rolls 2014).

Second, the striatum/basal ganglia route, evolutionarily old, which receives from the amygdala as well as the cortex, is involved in learning stimulus-response habits, which tend to be overlearned and are not under the direct control of the goal value (Figs. 12.1 and 12.2) (Rolls 2014).

In this context, it is interesting to consider the role of the amygdala in these types of response and in emotion. Neurons in the primate amygdala do not show rapid, one-trial reversal nor are their responses very consistently reduced to zero by devaluation produced, for example, by feeding to satiety (Sanghera et al. 1979; Rolls and Scott 2003; Rolls 2014). This is in contrast to the orbitofrontal cortex (Rolls 2014), though amygdala neurons may be more involved when aversive stimuli are used (Morrison et al. 2011). Lesions of the macaque amygdala do not impair stimulus-reward reversal learning, whereas lesions of the orbitofrontal cortex do (Murray and Izquierdo 2007). Lesions of the rodent amygdala impair many classically conditioned responses such as autonomic responses and freezing (LeDoux 2012; Phelps and LeDoux 2005; Cardinal et al. 2002; Rolls 2014), and the importance of the rat amygdala, with its much less well-developed orbitofrontal cortex than primates, in olfactory reversal learning, has been emphasized (Schoenbaum et al. 1999). Further, the changes in emotion in patients with amygdala lesions are much less marked than those in patients with orbitofrontal cortex damage, and special tests, analogous in some cases to those developed in rodent studies, are necessary to reveal deficits (Phelps and LeDoux 2005; Whalen and Phelps 2009). For example, patients with amygdala lesions are impaired at learning classically conditioned skin conductance responses when a blue square is associated with a shock and are also impaired in acquiring the same autonomic response to fear by verbally instructed learning or by observational learning. The human amygdala appears to be important mainly for some fear responses to some stimuli, such as whether an individual backs off in a social encounter (Feinstein et al. 2011).

Taken together, these findings provide evidence that in primates including humans the amygdala becomes overshadowed by the orbitofrontal cortex. The orbitofrontal cortex has a much more important role in the computation of reward and punishment value, as measured by effects of devaluation and goal-directed one-trial rule-based learning and reversal, and thereby in emotion, which shows major changes after damage to the orbitofrontal cortex (Hornak et al. 1996, 2003, 2004,; Rolls et al. 1994; Rolls 2014). The computational bases are considered elsewhere of primate one-trial rule-based reversal (Deco and Rolls 2005), how non-reward neuronal activity is computed in the orbitofrontal cortex (Rolls and Deco 2016), and how these process may be related to depression considered as involving emotions produced by non-reward (Rolls et al. 2016; Rolls 2016c). The computational bases for the importance of the orbitofrontal cortex in emotion are described in *Cerebral Cortex: Principles of Operation* (Rolls 2016a), which also considers more widely the ways in which the cerebral cortex has evolved.

Goal value-directed choice is usual in primates and humans, whereas fixed action patterns, such as pecking in birds, are more common elsewhere.

The computation of reward value, and then its use as the target for goal-directed learning under the control of the goal value, is a flexible way for genes to influence behaviour and appears to be at the heart of primate including human emotion. However, genes may also encode stimulus-response reflexes, and this is seen, for example, in the pecking of birds at grain-like objects (Rolls 2014; Brown and Jenkins 1968).

In humans, and perhaps some primates, syntactic reasoning and thereby planning allow selfish gene-specified (emotion-related) rewards to be rejected in favour of the long-term interests of the individual, the phenotype.

The evolutionary adaptive value of emotions is that different genes specify different goals in their own self-interest, and actions can then be learned and performed by instrumental learning to obtain the goals. In addition, a rational thought system involved in multistep planning using syntax can allow gene-specified goals to be deferred or avoided in order to achieve longer-term types of goal that may be more advantageous to the individual than to the genes (Rolls 2012b, 2014). Decisions between these systems are likely to be taken by a probabilistic cortical attractor decision-making network (Rolls and Deco 2010; Rolls 2014).

References

Baylis LL, Rolls ET (1991) Responses of neurons in the primate taste cortex to glutamate. Physiol Behav 49:973–979

Brown PL, Jenkins HM (1968) Autoshaping of the pigeon's key-peck. J Exp Anal Behav 11:1–8

Cardinal N, Parkinson JA, Hall J, Everitt BJ (2002) Emotion and motivation: the role of the amygdala, ventral striatum, and prefrontal cortex. Neurosci Biobehav Rev 26:321–352

Cho YK, Li CS, Smith DV (2002) Gustatory projections from the nucleus of the solitary tract to the parabrachial nuclei in the hamster. Chem Senses 27(1):81–90

Critchley HD, Harrison NA (2013) Visceral influences on brain and behavior. Neuron 77:624–638. doi:10.1016/j.neuron.2013.02.008

Critchley HD, Rolls ET (1996a) Hunger and satiety modify the responses of olfactory and visual neurons in the primate orbitofrontal cortex. J Neurophysiol 75:1673–1686

Critchley HD, Rolls ET (1996b) Olfactory neuronal responses in the primate orbitofrontal cortex: analysis in an olfactory discrimination task. J Neurophysiol 75:1659–1672

de Araujo IE, Gutierrez R, Oliveira-Maia AJ, Pereira A Jr, Nicolelis MA, Simon SA (2006) Neural ensemble coding of satiety states. Neuron 51:483–494. doi:10.1016/j.neuron.2006.07.009

Deco G, Rolls ET (2005) Synaptic and spiking dynamics underlying reward reversal in orbitofrontal cortex. Cereb Cortex 15:15–30

Feinstein JS, Adolphs R, Damasio A, Tranel D (2011) The human amygdala and the induction and experience of fear. Curr Biol 21(1):34–38. doi:10.1016/j.cub.2010.11.042. S0960-9822(10)01508-3 [pii]

Francis S, Rolls ET, Bowtell R, McGlone F, O'Doherty J, Browning A, Clare S, Smith E (1999) The representation of pleasant touch in the brain and its relationship with taste and olfactory areas. Neuroreport 10:453–459

Giza BK, Scott TR (1983) Blood glucose selectively affects taste-evoked activity in rat nucleus tractus solitarius. Physiol Behav 31:643–650

Giza BK, Scott TR (1987) Intravenous insulin infusions in rats decrease gustatory-evoked responses to sugars. Am J Phys 252:R994–1002

Giza BK, Deems RO, Vanderweele DA, Scott TR (1993) Pancreatic glucagon suppresses gustatory responsiveness to glucose. Am J Phys 265:R1231–R1237

Glenn JF, Erickson RP (1976) Gastric modulation of gustatory afferent activity. Physiol Behav 16:561–568

Grabenhorst F, Rolls ET (2008) Selective attention to affective value alters how the brain processes taste stimuli. Eur J Neurosci 27:723–729

Grabenhorst F, Rolls ET (2011) Value, pleasure, and choice in the ventral prefrontal cortex. Trends Cogn Sci 15:56–67

Grabenhorst F, Rolls ET, Bilderbeck A (2008) How cognition modulates affective responses to taste and flavor: top down influences on the orbitofrontal and pregenual cingulate cortices. Cereb Cortex 18:1549–1559

Grabenhorst F, Rolls ET, Margot C (2011) A hedonically complex odor mixture captures the brain's attention. NeuroImage 55:832–843

Hajnal A, Takenouchi K, Norgren R (1999) Effect of intraduodenal lipid on parabrachial gustatory coding in awake rats. J Neurosci 19:7182–7190

Hasselmo ME, Rolls ET, Baylis GC (1989) The role of expression and identity in the face-selective responses of neurons in the temporal visual cortex of the monkey. Behav Brain Res 32:203–218

Hornak J, Rolls ET, Wade D (1996) Face and voice expression identification in patients with emotional and behavioural changes following ventral frontal lobe damage. Neuropsychologia 34:247–261

Hornak J, Bramham J, Rolls ET, Morris RG, O'Doherty J, Bullock PR, Polkey CE (2003) Changes in emotion after circumscribed surgical lesions of the orbitofrontal and cingulate cortices. Brain 126:1691–1712

Hornak J, O'Doherty J, Bramham J, Rolls ET, Morris RG, Bullock PR, Polkey CE (2004) Reward-related reversal learning after surgical excisions in orbitofrontal and dorsolateral prefrontal cortex in humans. J Cogn Neurosci 16:463–478

Kadohisa M, Rolls ET, Verhagen JV (2005a) Neuronal representations of stimuli in the mouth: the primate insular taste cortex, orbitofrontal cortex, and amygdala. Chem Senses 30(5):401–419. doi:10.1093/chemse/bji036

Kadohisa M, Rolls ET, Verhagen JV (2005b) The primate amygdala: neuronal representations of the viscosity, fat texture, temperature, grittiness and taste of foods. Neuroscience 132(1):33–48. doi:10.1016/j.neuroscience.2004.12.005

Kringelbach ML, Rolls ET (2003) Neural correlates of rapid reversal learning in a simple model of human social interaction. NeuroImage 20:1371–1383

Kringelbach ML, O'Doherty J, Rolls ET, Andrews C (2003) Activation of the human orbitofrontal cortex to a liquid food stimulus is correlated with its subjective pleasantness. Cereb Cortex 13:1064–1071

LeDoux J (2012) Rethinking the emotional brain. Neuron 73(4):653–676. doi:10.1016/j.neuron.2012.02.004

MacDonald CJ, Meck WH, Simon SA (2012) Distinct neural ensembles in the rat gustatory cortex encode salt and water tastes. J Physiol 590:3169–3184. doi:10.1113/jphysiol.2012.233486

Morris JS, Dolan RJ (2001) Involvement of human amygdala and orbitofrontal cortex in hunger-enhanced memory for food stimuli. J Neurosci 21:5304–5310

Morrison SE, Saez A, Lau B, Salzman CD (2011) Different time courses for learning-related changes in amygdala and orbitofrontal cortex. Neuron 71(6):1127–1140. doi:10.1016/j.neuron.2011.07.016

Murray EA, Izquierdo A (2007) Orbitofrontal cortex and amygdala contributions to affect and action in primates. Ann N Y Acad Sci 1121:273–296

Norgren R (1974) Gustatory afferents to ventral forebrain. Brain Res 81(2):285–295. doi:0006-8993(74)90942-1 [pii]

Norgren R (1976) Taste pathways to hypothalamus and amygdala. J Comp Neurol 166:17–30

Norgren R (1984) Central neural mechanisms of taste. In: Darien-Smith I (ed) Handbook of physiology - the Nervous system III. Sensory processes, vol 1. American Physiological Society, Washington, pp 1087–1128

Norgren R (1990) Gustatory system. In: Paxinos G (ed) The human nervous system. Academic, San Diego, pp 845–861

Norgren R, Leonard CM (1971) Taste pathways in rat brainstem. Science 173(4002):1136–1139

O'Doherty J, Rolls ET, Francis S, Bowtell R, McGlone F, Kobal G, Renner B, Ahne G (2000) Sensory-specific satiety related olfactory activation of the human orbitofrontal cortex. Neuroreport 11:893–897

O'Doherty JP, Deichmann R, Critchley HD, Dolan RJ (2002) Neural responses during anticipation of a primary taste reward. Neuron 33:815–826

Padoa-Schioppa C (2011) Neurobiology of economic choice: a good-based model. Annu Rev Neurosci 34:333–359. doi:10.1146/annurev-neuro-061010-113648

Padoa-Schioppa C, Assad JA (2006) Neurons in the orbitofrontal cortex encode economic value. Nature 441(7090):223–226

Padoa-Schioppa C, Assad JA (2008) The representation of economic value in the orbitofrontal cortex is invariant for changes of menu. Nat Neurosci 11(1):95–102

Passingham REP, Wise SP (2012) The neurobiology of the prefrontal Cortex. Oxford University Press, Oxford

Phelps EA, LeDoux JE (2005) Contributions of the amygdala to emotion processing: from animal models to human behavior. Neuron 48(2):175–187. doi:10.1016/j.neuron.2005.09.025. S0896-6273(05)00823-8 [pii]

Plassmann H, O'Doherty J, Rangel A (2007) Orbitofrontal cortex encodes willingness to pay in everyday economic transactions. J Neurosci 27(37):9984–9988. doi:10.1523/JNEUROSCI.2131-07.2007

Preuss TM (1995) Do rats have prefrontal cortex? The rose-woolsey-akert program reconsidered. J Cogn Neurosci 7:1–24
Pritchard TC, Nedderman EN, Edwards EM, Petticoffer AC, Schwartz GJ, Scott TR (2008) Satiety-responsive neurons in the medial orbitofrontal cortex of the macaque. Behav Neurosci 122:174–182. doi:10.1037/0735-7044.122.1.174. 2008-01943-018 [pii]
Rolls ET (1984) Neurons in the cortex of the temporal lobe and in the amygdala of the monkey with responses selective for faces. Hum Neurobiol 3:209–222
Rolls ET (1992) Neurophysiological mechanisms underlying face processing within and beyond the temporal cortical visual areas. Philos Trans R Soc Lond B 335:11–21
Rolls ET (1996) The orbitofrontal cortex. Philos Trans R Soc Lond B 351:1433–1444
Rolls ET (2000) Functions of the primate temporal lobe cortical visual areas in invariant visual object and face recognition. Neuron 27:205–218
Rolls ET (2007) The representation of information about faces in the temporal and frontal lobes. Neuropsychologia 45:125–143
Rolls ET (2008a) Face processing in different brain areas, and critical band masking. J Neuropsychol 2:325–360
Rolls ET (2008b) Memory, attention, and decision-making: a unifying computational neuroscience approach. Oxford University Press, Oxford
Rolls ET (2009) The neurophysiology and computational mechanisms of object representation. In: Dickinson S, Tarr M, Leonardis A, Schiele B (eds) Object categorization: computer and human vision perspectives. Cambridge University Press, Cambridge, pp 257–287
Rolls ET (2011) Face neurons. In: Calder AJ, Rhodes G, Johnson MH, Haxby JV (eds) The Oxford handbook of face perception. Oxford University Press, Oxford, pp 51–75
Rolls ET (2012a) Invariant visual object and face recognition: neural and computational bases, and a model, VisNet. Front Comput Neurosci 6(35):1–70. doi:10.3389/fncom.2012.00035
Rolls ET (2012b) Neuroculture. On the implications of Brain science. Oxford University Press, Oxford
Rolls ET (2013) What are emotional states, and why do we have them? Emot Rev 5:241–247
Rolls ET (2014) Emotion and decision-making explained. Oxford University Press, Oxford
Rolls ET (2015a) Functions of the anterior insula in taste, autonomic, and related functions. Brain Cogn 110:4–19. doi:10.1016/j.bandc.2015.1007.1002
Rolls ET (2015b) Limbic systems for emotion and for memory, but no single limbic system. Cortex 62:119–157
Rolls ET (2015c) Taste, olfactory, and food reward value processing in the brain. Prog Neurobiol 127–128:64–90. doi:10.1016/j.pneurobio.2015.03.002
Rolls ET (2016a) Cerebral cortex: principles of operation. Oxford University Press, Oxford
Rolls ET (2016b) Motivation Explained: ultimate and proximate accounts of hunger and appetite. Adv Motiv Sci 3:187–249
Rolls ET (2016c) A non-reward attractor theory of depression. Neurosci Biobehav Rev 68:47–58. doi:10.1016/j.neubiorev.2016.05.007
Rolls ET (2016d) Reward systems in the brain and nutrition. Annu Rev Nutr 36:435–470
Rolls ET, Baylis GC (1986) Size and contrast have only small effects on the responses to faces of neurons in the cortex of the superior temporal sulcus of the monkey. Exp Brain Res 65(1):38–48
Rolls ET, Baylis LL (1994) Gustatory, olfactory, and visual convergence within the primate orbitofrontal cortex. J Neurosci 14:5437–5452
Rolls ET, Deco G (2002) Computational neuroscience of vision. Oxford University Press, Oxford
Rolls ET, Deco G (2010) The noisy brain: stochastic dynamics as a principle of brain function. Oxford University Press, Oxford
Rolls ET, Deco G (2016) Non-reward neural mechanisms in the orbitofrontal cortex. Cortex 83:27–38. doi:10.1080/23273798.2016.1203443
Rolls ET, Grabenhorst F (2008) The orbitofrontal cortex and beyond: from affect to decision-making. Prog Neurobiol 86:216–244
Rolls ET, Judge SJ, Sanghera M (1977) Activity of neurones in the inferotemporal cortex of the alert monkey. Brain Res 130:229–238

Rolls ET, Rolls BJ, Rowe EA (1983) Sensory-specific and motivation-specific satiety for the sight and taste of food and water in man. Physiol Behav 30:185–192
Rolls ET, Scott TR, Sienkiewicz ZJ, Yaxley S (1988) The responsiveness of neurones in the frontal opercular gustatory cortex of the macaque monkey is independent of hunger. J Physiol 397:1–12
Rolls ET, Sienkiewicz ZJ, Yaxley S (1989) Hunger modulates the responses to gustatory stimuli of single neurons in the caudolateral orbitofrontal cortex of the macaque monkey. Eur J Neurosci 1:53–60
Rolls ET, Yaxley S, Sienkiewicz ZJ (1990) Gustatory responses of single neurons in the caudolateral orbitofrontal cortex of the macaque monkey. J Neurophysiol 64:1055–1066
Rolls ET, Hornak J, Wade D, McGrath J (1994) Emotion-related learning in patients with social and emotional changes associated with frontal lobe damage. J Neurol Neurosurg Psychiatry 57:1518–1524
Rolls ET, Critchley H, Wakeman EA, Mason R (1996a) Responses of neurons in the primate taste cortex to the glutamate ion and to inosine 5′-monophosphate. Physiol Behav 59:991–1000
Rolls ET, Critchley HD, Mason R, Wakeman EA (1996b) Orbitofrontal cortex neurons: role in olfactory and visual association learning. J Neurophysiol 75:1970–1981
Rolls ET, Critchley HD, Treves A (1996c) The representation of olfactory information in the primate orbitofrontal cortex. J Neurophysiol 75(5):1982–1996
Rolls ET, Critchley HD, Browning AS, Hernadi A, Lenard L (1999) Responses to the sensory properties of fat of neurons in the primate orbitofrontal cortex. J Neurosci 19:1532–1540
Rolls ET, Kringelbach ML, de Araujo IET (2003) Different representations of pleasant and unpleasant odors in the human brain. Eur J Neurosci 18:695–703
Rolls ET, Critchley HD, Browning AS, Inoue K (2006) Face-selective and auditory neurons in the primate orbitofrontal cortex. Exp Brain Res 170:74–87
Rolls ET, Grabenhorst F, Parris BA (2008) Warm pleasant feelings in the brain. NeuroImage 41:1504–1513
Rolls ET, Critchley HD, Verhagen JV, Kadohisa M (2010) The representation of information about taste and odor in the orbitofrontal cortex. Chemosens Percept 3:16–33. doi:10.1007/s12078-009-9054-4
Rolls ET, Cheng W, Qiu J, Liu W, Chang M, Huang C-C, Wang X-F, Zhang J, Pu J-C, Tsai S-J, Lin C-P, Wang F, Xie P, Feng J (2016) Medial reward and lateral non-reward orbitofrontal cortex circuits change in opposite directions in depression. Brain 139:3296–3309
Rolls ET, Scott TR (2003) Central taste anatomy and neurophysiology. In: Doty RL (ed) Handbook of olfaction and gustation, 2nd edn. Dekker, New York, pp 679–705
Sanghera MK, Rolls ET, Roper-Hall A (1979) Visual responses of neurons in the dorsolateral amygdala of the alert monkey. Exp Neurol 63:610–626
Schoenbaum G, Chiba AA, Gallagher M (1999) Neural encoding in orbitofrontal cortex and basolateral amygdala during olfactory discrimination learning. J Neurosci 19:1876–1884
Schoenbaum G, Roesch MR, Stalnaker TA, Takahashi YK (2009) A new perspective on the role of the orbitofrontal cortex in adaptive behaviour. Nat Rev Neurosci 10(12):885–892. doi:10.1038/nrn2753. nrn2753 [pii]
Scott TR, Karadi Z, Oomura Y, Nishino H, Plata-Salaman CR, Lenard L, Giza BK, Aou S (1993) Gustatory neural coding in the amygdala of the alert macaque monkey. J Neurophysiol 69(6):1810–1820
Scott TR, Plata-Salaman CR (1999) Taste in the monkey cortex. Physiol Behav 67:489–511
Scott TR, Small DM (2009) The role of the parabrachial nucleus in taste processing and feeding. Ann N Y Acad Sci 1170:372–377. doi:10.1111/j.1749-6632.2009.03906.x. NYAS03906 [pii]
Scott TR, Yaxley S, Sienkiewicz ZJ, Rolls ET (1986a) Gustatory responses in the frontal opercular cortex of the alert cynomolgus monkey. J Neurophysiol 56:876–890
Scott TR, Yaxley S, Sienkiewicz ZJ, Rolls ET (1986b) Taste responses in the nucleus tractus solitarius of the behaving monkey. J Neurophysiol 55:182–200
Small DM, Scott TR (2009) Symposium overview: what happens to the pontine processing? Repercussions of interspecies differences in pontine taste representation for tasting and feeding. Ann N Y Acad Sci 1170:343–346. doi:10.1111/j.1749-6632.2009.03918.x. NYAS03918 [pii]

Thorpe SJ, Rolls ET, Maddison S (1983) Neuronal activity in the orbitofrontal cortex of the behaving monkey. Exp Brain Res 49:93–115

Verhagen JV, Kadohisa M, Rolls ET (2004) The primate insular/opercular taste cortex: neuronal representations of the viscosity, fat texture, grittiness, temperature and taste of foods. J Neurophysiol 92(3):1685–1699. doi:10.1152/jn.00321.2004

Verhagen JV, Rolls ET, Kadohisa M (2003) Neurons in the primate orbitofrontal cortex respond to fat texture independently of viscosity. J Neurophysiol 90(3):1514–1525. doi:10.1152/jn.00320.2003

Whalen PJ, Phelps EA (eds) (2009) The human amygdala. Guilford, New York

Wilson FAW, Rolls ET (2005) The primate amygdala and reinforcement: a dissociation between rule-based and associatively-mediated memory revealed in amygdala neuronal activity. Neuroscience 133:1061–1072

Wise SP (2008) Forward frontal fields: phylogeny and fundamental function. Trends Neurosci 31(12):599–608. doi:10.1016/j.tins.2008.08.008. S0166-2236(08)00207-5 [pii]

Yan J, Scott TR (1996) The effect of satiety on responses of gustatory neurons in the amygdala of alert cynomolgus macaques. Brain Res 740(1–2):193–200

Yaxley S, Rolls ET, Sienkiewicz ZJ (1988) The responsiveness of neurons in the insular gustatory cortex of the macaque monkey is independent of hunger. Physiol Behav 42:223–229

Yaxley S, Rolls ET, Sienkiewicz ZJ (1990) Gustatory responses of single neurons in the insula of the macaque monkey. J Neurophysiol 63:689–700

Yaxley S, Rolls ET, Sienkiewicz ZJ, Scott TR (1985) Satiety does not affect gustatory activity in the nucleus of the solitary tract of the alert monkey. Brain Res 347:85–93. doi:0006-8993(85)90891-1 [pii]

Chapter 13
Evolutionary Origin of Empathy and Inequality Aversion

Shigeru Watanabe and Yutaka Kosaki

Abstract An important function of emotion is to enable individuals to adapt to the environment through induction of physiological and behavioral responses directly toward, or in anticipation of, biologically significant events such as food and predators. Another important function is to provide a social signal for other individuals in the group. This emotional signal often induces the same emotional state in the observer, a process called emotional contagion, which serves as a surrogate for others to learn through observation and provide cues to take actions in pro-social ways or, on occasions, in Machiavellian ways. Empathy, a term used to encompass these various social functions of emotion, is thus crucial for the survival of many species of animals including humans. In this chapter, we review literature concerning experimental studies of empathy in the laboratory animals, mostly rodents, which could provide a clue to understand the evolutionary origin of empathy. We first review basic findings concerning emotional contagion and introduce recent studies that examined the importance of social comparison in automatic empathetic responses, which indicate that the nonstandard forms of empathy such as envy, schadenfreude, as well as inequity aversion, may exist in rodents. We then discuss the functional significance of empathy by reviewing literature on observational learning and helping behavior. We then offer mechanistic analyses of empathy on the basis of the principles of associative learning. Finally, we discuss the evolutionary origin of social comparison.

Keywords Emotional contagion • Social comparison • Envy • Schadenfreude Pro-social behavior • Associative learning • Sexual selection

S. Watanabe
Department of Psychology, Keio University, 2-14-5 Mita, Minato-ku, Tokyo 108-8345, Japan
e-mail: swat@flet.keio.ac.jp

Y. Kosaki
Keio Advanced Research Centres, Keio University, 2-15-45 Mita, Minato-ku, Tokyo, Japan
e-mail: yutaka100@gmail.com

S. Watanabe et al. (eds.), *Evolution of the Brain, Cognition, and Emotion in Vertebrates*, Brain Science, DOI 10.1007/978-4-431-56559-8_13

13.1 Introduction

For many species of animals, social communication among conspecifics is as critical for survival as individual learning and memory. In particular, the acquisition of information through perception of emotional state in other individuals has immediate relevance to survival. Many animals display emotional expressions including autonomic and somatic responses upon involvement in and in expectation of biologically significant events, such as predator detection and avoidance, procuring food, and attracting a mate. A primary function of these emotional responses is to allow animals to cope with and prepare for such events. On the other hand, among the spectrum of emotional expressions are those that are apparently less likely to promote adaptation in as direct a way as defensive/appetitive responses; these include facial expressions, vocalizations, and emission of pheromones. These overt emotional expressions are thought to function as social signals for conspecifics. Although there is no agreed definition of emotion (Frigida 2007), empathy can be broadly defined as a phenomenon in which the detection of such an emotional expression provokes an emotional experience in the observer. Based on the evoked emotional state, animals may also be engaged in voluntary actions to affect the emotional state of others, which could be regarded as pro-social and sympathetic behavior.

This chapter reviews the major empirical findings concerning the expression of empathy in nonhuman animals—especially rodents—and offers mechanistic and functional analyses of this phenomenon. We first review basic findings concerning emotional contagion, which is a fundamental and automatic form of empathy and then introduce recent studies that have examined the importance of social comparison in empathetic responses. We then discuss the functional significance of basic empathy by reviewing the literature on social learning and voluntary empathetic behavior. We then offer analyses on behavioral mechanisms underlying basic empathy on the basis of the principles of associative learning. Finally, we discuss the evolutionary origin of social comparison.

13.2 Emotional Contagion or State-Matching of Emotion: Automatic Positive and Negative Empathy

Many researchers agree that at the most fundamental level of empathy lies an automatic contagion of emotional states between conspecifics (e.g., Preston and de Waal 2002). Historically, it is a modern version of "affectum imitatio" by the seventeenth-century philosopher Spinoza. In this section, we briefly review basic findings on emotional or the state-matching form of basic empathy, in which the perception of other's positive or negative emotional expression by an observer induces a similar emotional state in that observer.

13.2.1 Negative Emotional Contagion

In one of the earliest studies of empathy in animals, Church (1959) demonstrated negative emotional contagion in rats. He showed that electric shocks applied to a demonstrator rat suppressed instrumental behavior (lever pressing to get food reward) in other rats that observed the conspecific being shocked, which was interpreted as evidence for a transfer of fear from the demonstrator to the observer. Importantly, the suppression was stronger if the observers themselves had previously experienced such a shock (i.e., a shared experience), and even stronger if the previous shock experience had been contingent upon shocks to the demonstrator, suggesting the involvement of associative learning mechanisms in emotional contagions (see Sect. 13.4.1 for detailed analysis). Watanabe and Ono (1986) later replicated the findings in pigeons. More recently, Langford et al. (2006) demonstrated negative emotional contagion in mice by measuring enhancement of pain reaction to noxious stimuli in mice that observed similar expressions of pain in the demonstrators. Knapska et al. (2006) observed increased fear reaction, measured in an acoustic startle paradigm, in mice that interacted with a cage mate that had been shocked immediately before a social interaction. Other more indirect demonstrations of emotional contagion involve facilitation of subsequent fear conditioning upon observing fear in conspecifics as shown in monkeys (e.g., Mineka et al. 1984), rats (e.g., Bruchey et al. 2010), and mice (e.g., Jeon et al. 2010), which will be discussed further in Sect. 13.5.1.1.

13.2.2 Positive Emotional Contagion

In contrast to the numerous demonstrations of negative emotional contagion, it is less common to observe positive emotional contagion in animals. One such example is the so-called social facilitation of the positively reinforcing effect of a drug (e.g., Watanabe 2011a), which will be discussed in Sect. 13.3.1. Apart from this social facilitation, one study showed both positive and negative emotional contagion in pigs (Reimert et al. 2013). Other potential cases for positive empathy have been reported in studies demonstrating the social transmission of food preference (e.g., Galef and Wigmore 1983). The socially transmitted food preference, however, can be acquired even when the demonstrator is anesthetized (Galef and Wigmore 1983; Galef and Stein 1985), suggesting that the food preference can be acquired in the absence of positive emotion exhibited by the demonstrator. Galef et al. (1988) suggested that the smell of novel food together with carbon disulfides in the demonstrator's breath suffices to establish the social transmission of food preference.

13.3 Social Comparison and Empathy

As important as the emotional contagion might be in affecting an individual's emotional state, we also know that such an empathetic response is substantially modulated by comparison of our own emotional states with those of others. For example, people suffering stress may feel less stress when they know that others experience the same stress. Also, people suffering stress may feel more stress when surrounded by individuals who are not experiencing stress. Thus, the value of any event—positive or negative—can be modulated by social comparisons with reference to the states of others. In this section, we discuss how perception of social equality/inequality modulates emotional experiences in animals.

13.3.1 Social Equality

Social equality is a situation in which demonstrators and observers experience a similar emotion, and it facilitates both positive and negative emotions. In some cases, however, social equality can reduce negative emotion in an observer.

13.3.1.1 Social Facilitation by Equality in Negative Emotion

There are many examples of social facilitation of drug effects in humans. For example, people tend to consume more alcoholic beverages when they are with other drinkers, and they often start smoking cigarettes if they have friends who smoke (Glynn 1989). Thus, social facilitation may require not only the presence of other individuals but also state-matching between them. For example, an adolescent rat will consume more ethanol following interactions with a social partner injected with ethanol than with a social partner injected with water (Hunt et al. 2001; Maldonado et al. 2008). However, Gauvin et al. (1994) found that the presence of a cage mate with or without alcohol consumption reduced the conditioned aversion to ethanol. To clarify whether the state-matching of a pharmacological effect influences the reinforcing properties of methamphetamine, Watanabe (2011a) examined conditioned place preference in three groups of mice: One group was subjected to conventional conditioned place preference training in the absence of other mice (single group), a second group was conditioned in the presence of cage mates who experienced the same schedule of drug/saline injections (paired group), and the third group was conditioned in the presence of differentially treated cage mates (i.e., one mouse was injected with saline whereas the other was injected with methamphetamine—control-paired group). The paired group demonstrated facilitation of conditioned place preference as compared with the single group, but the control-paired group did not. Thus, the social facilitation effect does not represent just the simple summation of drug and social rewards but rather critically depends on whether the two individuals experience the matching emotional state. Functionally, it may be argued that social facilitation based on positive empathy can serve to strengthen bonding in social groups. In human societies, drinking, eating, and singing together (e.g., military or college songs) are common practices aimed to facilitate social bonding.

13.3.1.2 Stress-Reducing Effects by Equality in Negative Emotion

Langford et al. (2006) reported that a pain response induced by injection of formalin into the hind leg of mice was enhanced when a cage mate was also injected with formalin, as already mentioned in Section x. It is interesting that the relative magnitude of the induced pain in the demonstrator and the observer influenced both the strength and direction of emotional contagion exhibited by the observer. When the observer was injected with a low dose of formalin (1%) and the cage mate was injected with a high dose of formalin (5%), the observer showed an augmented pain response, whereas when the observer was injected with a high dose of formalin and the cage mate was injected with a low dose, the pain response of the observer decreased. Social equality also has stress-reducing effects.

A variety of stressors have been reported to raise body temperature (stress-induced hyperthermia, SIH; for review, see Bouwknecht et al. 2007), including a novel cage (Houtepen et al. 2011), social threat (Pardon et al. 2004), social defeat (Keeney et al. 2001), and restraint (Thornhill et al. 1979). The infrared thermograph is a suitable tool to measure SIH in social contexts because it is completely noninvasive and allows the assessment of several individuals simultaneously. Watanabe (2015) compared SIH in different social conditions using the thermograph. In the single-stress condition, the mouse was inserted into a restraint device (a holder for blood sampling) to induce SIH. In the equality condition, each of four animals in each cage was restrained individually at the same time, and the holders were placed in a radial arrangement facing the center of the home cage so that each animal could see all other animals. The social equality condition caused less SIH than the single condition, indicating that social equality reduced restraint stress.

Watanabe (2011b) employed a similar experimental design to study social modification of stress with two more indices, namely, corticosterone level and enhancement of aversive memory (Roozendaa et al. 2009). Corticosterone level was lower after restraint stress was applied with restrained cage mates in comparison to single-stress condition, consistent with the thermograph results (Fig. 13.1b). Another index of stress is the enhancement of aversive memory. The retention of the step-down passive-avoidance learning was weaker in mice that had experienced restraint stress under the equality condition than those that experienced the stress alone, indicating that the mice in the equality condition suffered less stress (Fig. 13.1c).

The effect of social equality on negative emotion may depend on familiarity between animals. Langford et al. (2006) reported that the presence of a free mouse resulted in fewer displays of pain reactions by a mouse injected with acetic acid, but only if the free mouse was unfamiliar. Watanabe (2016b) examined whether familiarity between animals affects the social equality induced by reduction of restraint

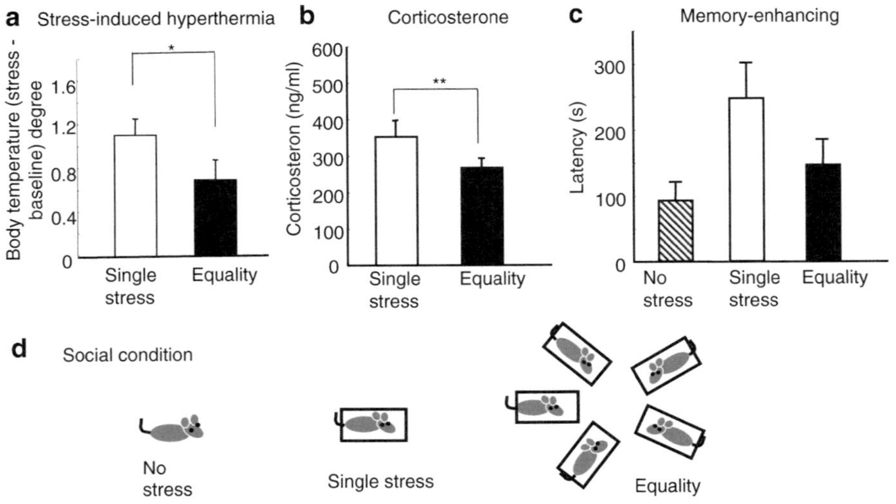

Fig. 13.1 Stress-reducing effects of social equality. (**a**) Stress-induced hyperthermia. The vertical axis indicates the change in body surface temperature from baseline (Watanabe 2015). (**b**) Level of corticosterone (Watanabe 2011b). (**c**) Aversive memory-enhancing effect. The vertical axis indicates the latency to step down to the floor (Watanabe 2011b). (**d**) Schematic representation of the social condition

SIH. The presence of similarly stressed but unfamiliar mice did not reduce SIH indicating that social equality can reduce stress only in the presence of familiar conspecific.

13.3.2 *Social Inequality*

The issue discussed in this section is whether and how such social inequality affects emotional experiences. One case of social inequality is when one person is happier than another, which can cause negative emotion in the latter. Envy, or reversed empathy, is a common emotion in humans (Smith 2013). Another case of inequality is schadenfreude, in which the misfortune of others causes positive emotion in the observer. Enjoying misfortune of others is considered by many to be a particularly human-unique emotion. In this section, we review studies in animals that tested whether envy and schadenfreude are indeed unique to humans.

13.3.2.1 Reversed Empathy or Disadvantageous Inequality Aversion

"Reversed empathy" is a non-state-matching form of empathy in which distress results from observing the fortune of others. It is generally considered to be non-adaptive or less adaptive because it does not offer an obvious benefit for the

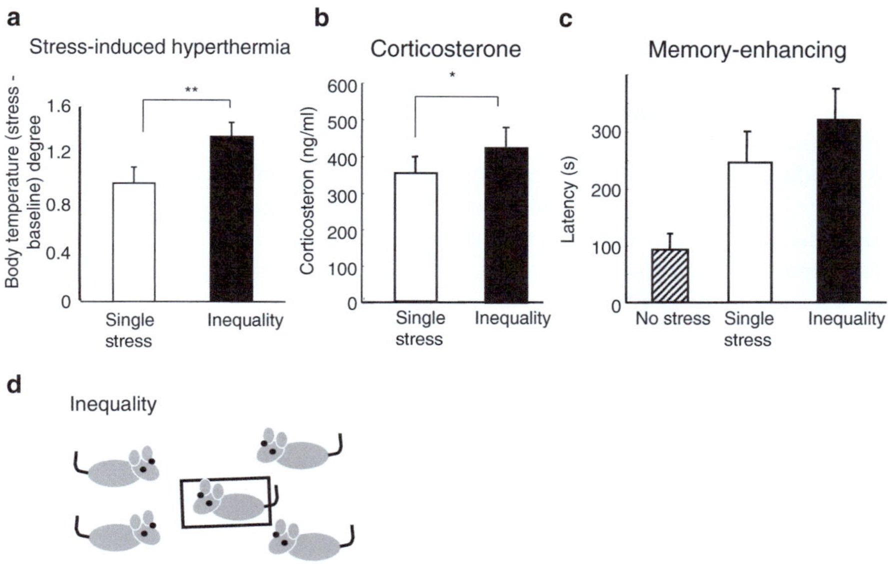

Fig. 13.2 Stress-reducing effects of social inequality. (**a**) Stress-induced hyperthermia. The vertical axis indicates the change in body surface temperature from baseline (Watanabe 2015). (**b**) Level of corticosterone (Watanabe 2011b). (**c**) Aversive memory-enhancing effect. The vertical axis indicates the latency to step down to the floor (Watanabe 2011b). (**d**) Schematic representation of the social condition

observer. One interesting example of reversed empathy is the stress-enhancing effect. Watanabe (2011b) examined the reversed empathy in a situation similar to that in which he studied the SIH-reducing effect of social equality. In this case, the subject mouse was restrained in the holder while four other cage mates freely moved around the holder. The SIH was higher in this inequality condition than in the single-stress condition, suggesting that the experimentally induced social inequality increased the stress (Fig. 13.2a). This conclusion was supported by results from two other measures. First, the level of corticosterone after the restraint was the highest in mice that experienced restraint with free-moving cage mates (Fig. 13.2b). Second, the memory-enhancing effect of restraint-induced stress was also higher in the social inequality condition than in the single-stress condition (Fig. 13.2). Therefore, all three indices of stress, namely, SIH, corticosterone level, and the memory-enhancing effect, consistently pointed to the idea that social inequality increased the restraint-induced stress, suggesting that the mice experienced reversed empathy.

The social inequality condition can be considered a type of social threat because restrained subjects cannot fight back against or escape the free cage mates. Social defeat-induced stress increases not only corticosterone level (Koolhaas et al. 1997) and immune response (Kollack-Walker et al. 1997) but also body temperature in rats (Meerlo et al. 1997). In the social inequality condition described above, physical contact was possible but only at a limited level; the free mice often climbed on the holder, but they could not bite the subject inside because the slit of the holder was so narrow. Thus it is unlikely that the increased stress in the social inequality condition was produced by social defeat.

To determine whether the actual movement of free mice is crucial for the stress-enhancing effect of free cage mates, Watanabe (2015) replicated the social inequality condition using anesthetized cage mates (i.e., the cage mates were not restrained but were immobile). In fact the anesthetized cage mates did not have a stress-enhancing effect, indicating that the movement of the free mice is the crucial factor for increasing the stress of restrained mice. Thus, the subjects could discriminate between immobile and freely moving cage mates.

Social inequality might also have more general aversive effects in nonhuman animals—a phenomenon called inequality aversion. Studies in nonhuman primates indicate that primates possess a sense of fairness in some extent. For example, capuchin monkeys refused to receive cucumbers when others obtained more valuable grapes (Brosnan and de Waal 2003). This behavior appears nonadaptive because receiving some food must be better than receiving no food, yet the aversive property of inequitable outcomes has been demonstrated in several nonhuman primates (Yamamoto and Takimoto 2012).

There is little evidence for inequality aversion in non-primate animals, but owners of dogs believe that their dogs experience jealousy (Morris et al. 2008) and some sense of fairness (Salovey and Rodin 1989). Range et al. (2012) trained dogs to "give the paw" and then extinguished this behavior. Dogs displayed faster extinction when given extinction training with another dog that was rewarded than when extinction-trained alone (Range et al. 2012). However, dogs preferred a trainer that over-rewarded other dogs for sitting on command to a trainer that provided fair rewards (Horowitz 2012), providing no evidence for inequality aversion. The dogs had no preference between an under-rewarding trainer and a fair trainer either, which again suggests no preference for the fair trainers. There have been several discussions about inequality aversion in nonhuman animals (Chen and Santos 2006; Sheskin et al. 2014).

Another type of inequality aversion is negative emotional response caused by less happier others. This aversion is called advantageous inequality. We, human, feel negative emotion in advantageous inequality situation but there is almost no report of the advantageous inequality aversion in animals. Even humans, the advantageous inequality is less aversive than the disadvantageous inequality. Ultimatum game and dictator game are experimental design often used for experiments of the inequality aversion in humans. In the ultimatum game, a proposer offers how split money between the proposer and the responder can reject the offer. If the responder rejects the offer, both the proposer and responder lose the money. In other words, the responder can punish the proposer. In the dictator game, the responder cannot refuse the offer. Typical proposers offer around 40% of money to responder in the ultimatum game (Güth and Tietz 1990) whereas the proposers offer quite small amount to the responder in the dictator game (Forsythe et al. 1994). This suggests the advantageous inequality aversion exists but does not so well control human behavior.

13.3.2.2 Schadenfreude

Schadenfreude, the mirror image of reversed empathy or envy, occurs when the failure or misfortune of demonstrators induces pleasure in the observer. This phenomenon is apparently irrational, because misfortune of others does not provide any value except in competitive situations. Schadenfreude is commonly observed in human society, however, suggesting that it is unique to humans.

Watanabe (2012) examined whether mice exhibit schadenfreude. The subject mouse was placed in the central compartment of a three-compartment chamber, and two familiar stimulus mice were placed in the side compartments. One stimulus mouse was injected with formalin at one paw just before being placed in the side compartment, whereas the other stimulus mouse did not receive an injection. The subject spent more time with the formalin-injected stimulus mouse. This preference for the distressed cage mate could represent schadenfreude-like emotion. Alternatively, this behavior may represent a predisposition to explore unusual states of familiar conspecifics. Yet another interpretation is that it reflects rescue behavior or pre-concern for others.

Virtually all human and nonhuman animal societies have stratified social classes or dominance hierarchies that are fundamental emergent properties of social groups (Ellis 1993). The adaptive basis of stratification is the intrinsic competition for limited resources such as food, water, territory, and mating opportunities. In human studies, subjects experienced schadenfreude when observing failure of a higher-status achiever (Feather 2008). Social status, or the dominant/subordinate relation, is known to play a crucial role in social preference in mice (Van Loo et al. 2001). Accordingly, Watanabe (2014) measured social rank among three mice using a competitive food retrieval test before replicating the social preference test. In the social preference test, the subordinate mice spent more time with a dominant cage mate that had received a formalin injection than with a non-injected mid-rank cage mate, but the preference was not evident when the dominant individual had not received an injection. The dominant mice themselves, in turn, did not display any preference for subordinate mice regardless of whether they were in pain. Moreover, the time spent by the subordinate mouse with the formalin-injected dominant mouse correlated negatively with the social dominance distance between them (Fig. 13.3).

The pain response of a slightly more dominant mate attracted the subordinate mice the most. This may reflect a greater chance for the subordinate to overturn the hierarchical positions, as the pain response of a higher-ranking animal would be perceived as indicating weakness (Fig. 13.3). Although there were no explicit attacks on the suffering demonstrators, the pain response may still have signaled the potential opportunity for an attack and thus led to the observed preference. These results suggest that the pain response of a dominant cage mate has a rewarding property for a subordinate; in other words, it is under these social conditions that a schadenfreude-like phenomenon could occur. As already discussed in Section xx, the presence of an unfamiliar male has an analgesic effect that reduces the pain response in male mice (Langford et al. 2006, 2010), which has the potential effect of concealing the appearance of weakness.

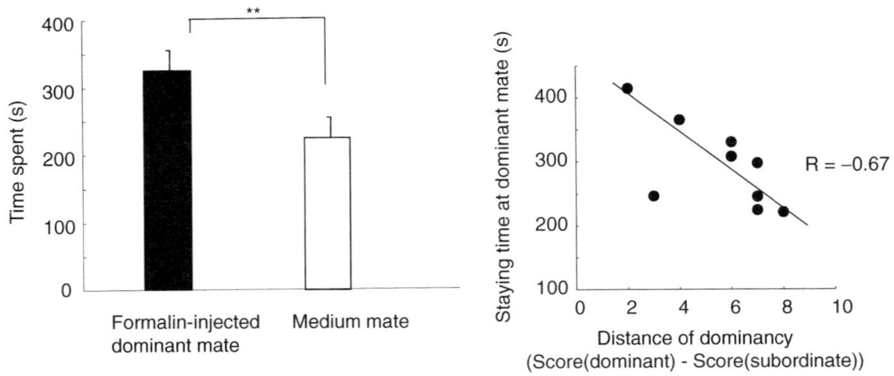

Fig. 13.3 Demonstration of schadenfreude in mice. *Left*: time spent close to the formalin-injected dominant cage mate and the medium cage mate without the formalin injection. *Right*: correlation between time spent close to the formalin-injected dominant cage mate and social distance between the subjects and the dominant cage mate

Humans may also be more likely to feel schadenfreude when the demonstrator is of slightly higher social position (Smith 2013). Schadenfreude in humans, however, has one particular feature—namely, its expression is concealed at least in some cases. In many casual settings, it may be socially acceptable to explicitly enjoy the minor misfortune of close friends, but it is less acceptable to express joy for the misfortune of persons of higher position. Hiding the experience of schadenfreude is likely unique to humans, and hidden schadenfreude can be considered a social skill that has been developed to avoid conflict among humans. However, a kind of tactical deception has been observed in extra-pair copulation in primates (le Roux et al. 2013). Dominant males punished individuals involved in extra-pair copulation, and the individuals suppressed vocalization during the extra-pair copulation. These observations suggest inhibition of emotion might be possible in animals.

13.3.3 Sense of Fairness

Several studies have suggested that human social intelligence involves a strong sense of fairness and that humans have a desire to punish others who act unfairly, even if there is a cost associated with the punishment (Fehr and Gachter 2002). In the ultimatum game, the responder rejects unequal offers by the proposer even if both of the players lose money by the rejection. The rejection is illogical behavior because small amount of money is better than nothing, but it functions as punishment of unfair behavior of the proposer. Desire to punish the unfair behavior has been suggested to contribute to schadenfreude. Brain imaging studies have shown that observing pain responses in others activates brain areas involved in one's own sensation of pain, but this empathetic response was not apparent when participants observed the pain response of individuals who had played unfairly in a game before the scan (Singer et al. 2006). Furthermore, male participants showed activation of the nucleus accumbens, which is usually activated by reward, when they saw a pain response of the unfair player (Singer et al. 2006), suggesting that witnessing "just deserts" for an offender is rewarding.

It is likely that an aversion toward inequality and a sense of fairness has helped to sustain human societies and form the basic foundations for principles of social exchange, reciprocity, and law. A sense of unfairness could in principle emerge both when one is happier than others or when one is less happy than others, but humans appear to feel a much stronger sense of unfairness in the latter case than in the former. To test whether this is also the case for mice, Watanabe (unpublished data) measured SIH in a free mouse surrounded by restrained cage mates. In contrast to the robust SIH in the reversed empathy condition (Watanabe 2011b), the free mouse did not show SIH, indicating that the social inequality did not produce stress if the subject was happier than others. Our biased sensitivity for unfairness means that humans prefer everybody being unhappy to a mixture of happy and unhappy people. As described in section of reversed empathy, advantageous inequality is less aversive than advantageous inequality. According to de Waal (2009), "We're all for fair play so long as it helps us."

13.4 Mechanisms of Empathy from an Associative Learning Perspective

13.4.1 Emotional Contagion Through Pavlovian Conditioning and a Role for Stimulus Generalization

We start our mechanistic descriptions of empathy with its most fundamental form, namely, emotional contagion, by focusing on some of the earliest demonstrations in rats (Church 1959) and pigeons (Watanabe and Ono 1986), which were briefly mentioned in Section x. Church (1959) presented an electric shock to a rat while another rat was lever pressing for food reinforcement in an adjacent compartment. The shock to the conspecific suppressed lever pressing in the observer, but only transiently, so that repeated presentations of shocks resulted in progressively weaker suppression. If, however, the observer rats had previously been given separate training where a shock to the demonstrators predicted a contingent shock to the observers, the observers later showed a greater level of suppression in response to the shock to the demonstrator and for a longer period across sessions. Intriguingly, another group of observers, which previously received shocks but noncontingently to the shocks to the demonstrators, also exhibited suppression, but to a lesser degree than the group that received contingent shocks but more strongly than the other group that experienced no shock by themselves.

13.4.1.1 Two Basic Learning Processes Underlying Emotional Contagion

Church (1959) offered a straightforward explanation of the emotional contagion: it is acquired through concurrent experience of another individual's expression of an emotional state (e.g., squeaking) and the subject's own emotional state (e.g., pain). That is, emotional contagion is learned via Pavlovian conditioning. In this case, the demonstrator's emotional expression served as a conditioned stimulus (CS), signaling the pain to the observers, which constitutes an unconditioned stimulus (US). Repeated pairing of these events established an association between them and consequently induced in the observer a conditioned suppression of instrumental behavior—a typical conditioned response (CR) related to anticipation of noxious stimuli. Equally important is a process responsible for the conditioned suppression shown by the group that had prior experiences of shocks that were noncontingent upon shocks to the demonstrator. To explain the result, Church (1959) appealed to the notion of stimulus generalization and suggested that rats in this group initially formed an association between their own emotional reaction to the shock (squeaks; CS_{self}) and the emotional state induced by the shock (pain; US), which then generalized to the demonstrator's similar emotional expression (CS_{other}) during the test stage to cause the pain-related state in the observers.

Figure 13.4 summarizes the two mechanisms described above. First, the animals acquire emotional contagion by directly forming an association between conspecifics' emotional expressions (CS_{other}) and their own emotional states (US). Second, emotional contagion can also be acquired via an auto-association of their own emotional expression (CS_{self}) and emotional state (US), which generalizes to other individuals' similar emotional expression (CS_{other}).

13.4.2 Role of a Shared Experience, Familiarity, and Similarity Between Conspecifics

The stimulus generalization account offers a clear explanation for why a shared experience facilitates empathetic responses in many animals and in different experimental settings (e.g., Bruchey et al. 2010; Church 1959; Langford et al. 2006; Watanabe and Ono 1986). Also, the associative account can explain why the degree of empathy depends on familiarity as well as genetic proximity between individuals (e.g., Ben-Ami Bartal et al. 2014; Jeon et al. 2010; Langford et al. 2006, 2010). Familiarity is a direct function of the amount of time conspecifics live together, and the longer they live together the greater the chance for one's emotional expression to be directly associated with another's emotional experience through joint experiences of the same emotional events. Importantly, individuals may display slightly different emotional expressions to the same event. Therefore, emotional contagion acquired through interaction with one conspecific may be expressed to a lesser degree with an unfamiliar conspecific owing to generalization decrement.

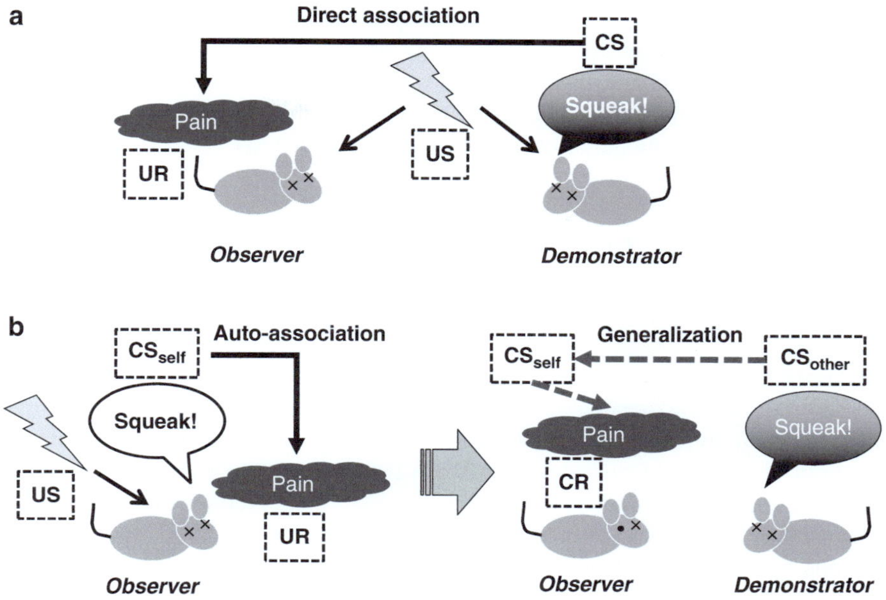

Fig. 13.4 Two associative mechanisms underlying emotional contagion. (**a**) When the demonstrator and the observer experience an electric shock (US) at the same time or in close temporal proximity, the pain reaction of the demonstrator can be directly associated with the state of pain in the observer. As a result, the former becomes a CS to induce the latter in the absence of the US for the observer, enabling emotional contagion. (**b**) The observer initially forms an association between its own pain reaction (CS_{self}) and the state of pain. Subsequently, the demonstrator's similar pain reaction (CS_{other}) induces a state of pain in the observer (conditioned response, CR) through stimulus generalization between CS_{self} and CS_{other}

Moreover, emotional contagion through stimulus generalization would be critically modulated by the similarities in the physical (visual), olfactory, and acoustic features of emotional expressions between the observer and the demonstrator animals; the more similar the two individuals, the stronger the generalization and hence stronger emotional contagion. This may contribute to the relatively greater degree of emotional contagion between siblings (e.g., Jeon et al. 2010; Langford et al. 2010). It is also possible, by extension, to account for the apparently graded degrees of empathy we can feel for other species based on stimulus generalization; we tend to empathize at ease with physically similar animals—other primates and certain other mammalian species such as dogs and cats—and possibly vice versa (e.g., Joly-Mascheroni et al. 2008)—whereas it is usually more difficult to empathize with phylogenetically more distant animals such as fish and reptiles, not to mention insects and other invertebrates. For example, monkeys pressed a bar to terminate electric shocks delivered to another monkey, but they did not exhibit the same behavior to an albino rat, suggesting potential limits on cross-species empathy (Mirsky et al. 1958). Finally, a recent human study of individuals with autism-spectrum disorder (ASD) indicated that those with ASD had no difficulty in empathizing with a story character who exhibited ASD-typical cognitive characteristics, whereas their empathy was impaired when the character exhibited cognitive features of typically developing individuals (Komeda et al. 2015). This result suggests that generalization decrement might contribute to empathy impairment in people with ASD.

13.4.3 Empathy, Imitation, and the Mirror Neuron System

13.4.3.1 Behavioral and Neural Mechanisms of Motor Imitation

So far we have discussed that empathy, at least in its most fundamental form, can be acquired through associative learning. It should be mentioned that a very similar idea has been put forward by Heyes and colleagues to account for motor imitation (Catmur et al. 2009; Cook et al. 2010, 2012; Heyes 2001, 2010; Heyes and Ray 2000). In their associative sequence learning (ASL) model (Heyes 2001; Heyes and Ray 2000), the sensory-motor correspondence of the observed and executed actions—a fundamental feature of motor imitations—is assumed to depend on associative learning mechanisms. Specifically, the ASL model assumes two associative mechanisms by which the sensory-motor correspondence is acquired.

First, organisms can form an association between a sensory representation of another's action (e.g., a sight of another individual's grasping hand) and their own motor command for the same action. Second, organisms can also form an association between the sensory representation of their own movement (e.g., a sight of one's own hand grasping) and the motor command for that particular action (e.g., grasping). Then, just as discussed for emotional contagion, the self-formed sensory-motor association can generalize to the sight of another individual's similar action, which then activates the motor command for the same action in the observer.

13.4.3.2 Neural Substrates of Empathy and Imitation

Imitation and emotional contagion, therefore, share a similar behavioral mechanism. Does this imply that these phenomena are supported by the same neural substrate? The ASL model assumes that the formation of sensory-motor association reflects the development of sensory-motor matching properties in the mirror neuron system (Heyes 2010). The mirror neurons were first discovered in area F5 of the ventral premotor area in macaques (Di Pellegrino et al. 1992; Gallese et al. 1996; Rizzolatti et al. 1996), where these neurons fired both when the monkey performed a certain action and upon seeing another individual performing the same action. Subsequently, neurons with similar properties were also found in the inferior parietal lobe (IPL; Gallese et al. 2002; Fogassi et al. 2005). Human functional magnetic resonance imaging (fMRI) studies found similar activations that reflect sensory-motor matching in the inferior frontal gyrus (IFG; Kilner et al. 2009), considered homologous to the F5 in macaques, and in the IPL (Chong et al. 2008). Evidence that the sensory-motor matching property of these mirror neurons is learned through experience comes from studies involving professional musicians and dancers and shows that the mirror-like neural activity is substantially modulated by prior experience of the same kind of action in the observer (Calvo-Merino et al. 2005; Haslinger et al. 2005). This is conceptually similar to the role of a shared experience in emotional contagion discussed earlier.

The critical issue is whether the same neural circuit for the sensory-motor mirror neurons also mediates emotional contagion. The associative account of emotional contagion predicts that emotional mirror-like neuronal properties could emerge anywhere in the brain where there exist converging inputs from areas responsible for the perception of emotional expressions and emotional experiences. Several fMRI studies have demonstrated activations in the IFG, a key area for sensory-motor matching, when participants observed or voluntarily imitated emotional facial expressions of others (Carr et al. 2003; Dapretto et al. 2006). Although these studies did not ensure that the participants actually experienced the same emotional state as that indicated by the demonstrators, it seems plausible that imitating emotional facial expression consequently affected subjects' emotional experiences (e.g., Darwin 1872; Ekman et al. 1983). This suggests a potential third mechanism of emotional contagion, that is, via automatic motor imitation.

Other studies demonstrated activations in the anterior cingulate cortex (ACC) and the anterior insular, both when participants experienced painful stimulations and when they observed pain-related facial expressions of others (Botvinick et al. 2005), observed another person's hand in a painful situation (Jackson et al. 2005), or even when they were instructed with an arbitrary cue that the painful stimulation is being inflicted upon their partner (Singer et al. 2004). Although these fMRI studies generally indicate that the neural substrates that partially overlap those for imitation mediate the self-other correspondence in the emotional domain, i.e., empathy, it remains to be tested whether changes in the BOLD signal come from exactly the same neurons within the implicated areas. This would require a technique such as MRI adaptation method, as employed in the demonstrations of sensory-motor matching (Chong et al. 2008; Kilner et al. 2009).

13.4.4 Mechanisms of Reversed Empathy and Schadenfreude

Owing mainly to the paucity of empirical data, it remains unclear whether reversed empathy and schadenfreude are also elicited automatically just as in the state-matching emotional contagion or whether they are instead second-order forms of empathy that require processing of additional information regarding social context and even the predicted consequences of another's emotional expressions. Although a study in mice mentioned earlier demonstrated social ranks to be an important factor in schadenfreude (Watanabe 2014), this does not necessarily imply that these forms of empathy are acquired by a mechanism different from that underlying the state-matching forms of empathy.

Based on the associative account, it may be argued that these non-state-matching forms of empathy, i.e., envy and schadenfreude, are also acquired through the same associative mechanisms but critically reflect the non-matching nature of environmental contingency between the emotional states of oneself and others. One particular situation in which such reversed contingency could exist is in the context of social competition. Animals living in a socially competitive environment may be consistently exposed to such a contingency in which injuries to higher-ranking individuals lead to a better social position of lower-ranking individuals, which should come with various rewards to the latter individuals. Thus, it may be predicted that animals exposed to a greater degree of such reversed social contingency would be more likely to display reversed empathy and schadenfreude. Indeed, schadenfreude-like behavior was displayed by mice of lower social ranks toward those of higher ranks, but not vice versa (Watanabe 2014). It will be possible in future studies to assess more directly the social competition hypothesis by manipulating the degree of reversed contingency between conspecifics.

It is worth noting that emotional contagion via the second process, i.e., through auto-association and stimulus generalization, should not work for reversed empathy and schadenfreude. This is for an obvious reason that within an individual a certain emotional expression is invariably linked to an experience of the same type of emotion. This in turn suggests that the emergence of envy and schadenfreude should usually be counteracted by the generalization-based mechanism that facilitates state-matching between conspecifics. An implication of this argument is as follows: whether the perception of another's emotional expression results in the state-matching or non-state-matching forms of empathy depends on the relative strength of the two processes—first on the strength of the direct process reflecting a given social contingency between emotional states of conspecifics and second on the strength of the generalization-based process that should facilitate state-matching between conspecifics. A reduction in the generalization process for any reason would favor the non-state-matching forms of empathy.

A straightforward prediction is that reversed empathy and schadenfreude would be stronger both when the conspecifics are physically more different and when they live in a socially competitive context, especially for a long time. Although the latter prediction is supported by at least one study in mice (Watanabe 2014), it has not been tested whether similarity between conspecifics also modulates the degree of schadenfreude and reversed empathy. In the human social psychology literature, it has been suggested that schadenfreude is more likely to occur between different groups of people in a rivalry context, for example, between supporters of different football clubs and males and females in an imaginary context of business competition (Spears 2013), which is consistent with the similarity hypothesis. However, in these studies the physical differences (e.g., gender, shirt color) are confounded by the presence of social competition, and it is not clear whether the similarity itself affects the degree of schadenfreude.

Although these considerations generally suggest that reversed empathy and schadenfreude may be acquired through fundamentally similar mechanisms as those for the state-matching emotional contagion, it is possible that they reflect more complex mechanisms that, for instance, require a clear distinction of and comparison between the emotional states of self and others.

13.4.5 Mechanisms of "Cognitive" Empathy

Thus far, this chapter has focused on so-called emotional or affective empathy. It is common practice to distinguish these forms of empathy from "cognitive" empathy, in which observers are said to understand the desires and intentions of other individuals (e.g., Baron-Cohen and Wheelwright 2004; Preston and de Waal 2002). It is also common to explain such "higher-order" empathy using cognitive rather than behavioral concepts such as mental state, belief, and desire. In this section, we briefly introduce the idea that cognitive empathy also could be explained in the framework of associative learning, in particular the learning theory that deals with goal-directed instrumental behavior of animals, which translates the notions of "belief" and "desire" into more objective, behaviorally observable terms.

One prominent example of cognitive empathy is the ability to understand other individuals' belief and desire through observation of their behavior, an ability often referred to as "theory of mind" (Premack and Woodruff 1978). A prerequisite to such cognitive feat is the ability to perform an action, first hand, based on a belief about its causal relationship to its outcome as well as to perform based on a desire to obtain that outcome. These are the two criteria for goal-directed action; a voluntary action is considered to be goal-directed if it is performed based on the causal relationship, or contingency, between the action and its outcome (i.e., belief) as well as based on an expectation about the value of the outcome in the current motivational state (i.e., desire; see Dickinson 1985, for a review). Studies with laboratory animals have convincingly suggested that they are capable of such purposive, goal-directed action. For example, rats are sensitive to action-outcome contingency (e.g., Dickinson and Charnock 1985) as well as to the expected value of an outcome associated with the action (e.g., Adams and Dickinson 1981; Balleine and Dickinson 1992).

Once an animal has acquired a goal-directed action based on a belief and a desire about its consequence, the associative knowledge may be later activated by observing a conspecific engaged in a similar action. For example, a rat trained to press a lever for food and pull a chain for water would readily learn to press the lever when hungry and pull the chain when thirsty. When this rat later observes a conspecific engaged in lever pressing, it may trigger the representation of contingency between lever pressing and the food outcome and the learned value of, or desire for, the food in the state of hunger. Thus, through a mechanism similar to that described for emotional contagion, animals may come to represent a particular goal associated with that action (belief) and a desire for that goal upon seeing a known action performed by others.

Importantly, for such retrieval of belief and desire through observation of another's action to be a full-fledged theory of mind, it is necessary for animals to distinguish between their own beliefs and desires and those of others based on self-other distinction (Frith and Frith 1999). In other words, belief and desire induced by the observation of another's action should be attributed to the actor rather than to the observer. To date, it has proved difficult to demonstrate this type of fully developed cognitive empathy in nonhuman animals, even chimpanzees, by typically testing understanding of a false belief in other individuals (for a review, see Call and Tomasello 2008). Maybe surprisingly, however, a recent study using Eurasian jays showed that male jays can change the type of food they offer to females by observing the females being pre-fed a particular type of food, suggesting that the males can understand the current desire of females (Ostojić et al. 2013). Importantly, the observation of females being pre-fed a particular type of food did not affect males' own preference for that type of food over another, suggesting that these male birds distinguished their own desire from that of the female's. It remains to be tested whether this is a specially adapted ability of these birds (thus reflecting a potential contingency between the quality of food sharing and partner choice in this species) or an ability that is shared by other species. Also, at present, it remains unclear exactly how such self-other distinction is made possible in terms of both behavioral and neural mechanisms (Mitchell 2009; Isoda and Noritake 2013).

13.5 Function of Empathy

13.5.1 Social Learning and Empathy

Animals living in groups are likely to experience common environmental contingencies. For instance, the shadow of a flying owl would predict the risk of predation not only for an individual shrew but also for all shrews living nearby. Thus, the animals would benefit from conspecifics' emotional expressions if they were able to use these signals for the purpose of preparing themselves for upcoming events. Such learning, of course, can be achieved using these social cues in the same way as other nonsocial cues, but the chance of survival would be much higher if the social cues were capable of automatically inducing the same emotional states in observers, which would exempt them from learning about potentially fatal consequences first hand. One major function of empathy, therefore, would be to facilitate such adaptive social learning.

13.5.1.1 Observational Fear Conditioning

The functional significance of empathy is highlighted by experimental studies showing the formation or modulation of fear-related learning through observation of fear in a demonstrator. In this paradigm, the subject is typically exposed to a demonstrator that expresses fear to a certain type of stimulus. Subsequently, the subjects come to show conditioned fear to that stimulus despite not experiencing the fearful event themselves. A classic example is a series of studies by Mineka and colleagues (e.g., Mineka et al. 1984; Cook and Mineka 1989) in which laboratory-reared rhesus monkeys, initially non-fearful of a snake, acquired fear of snakes by observing wild-reared monkeys expressing intense fear of the snake.

Recent studies in rodents showed similar results. Bruchey et al. (2010) initially trained demonstrator rats on direct fear conditioning between a tone CS and an electric shock US. Then, an observer rat was paired with one of the demonstrators, while both animals were presented with the CS only. Despite not experiencing the CS-US contingency directly, the observer subsequently showed conditioned freezing to the tone CS when tested alone. Rats in the control group, which had no opportunity to observe the fear expression of the demonstrators, did not express fear to the CS.

13.5.1.2 Roles of a Shared Experience and Familiarity in Observational Fear Conditioning

Kim et al. (2010) reported a similar result in rats and further demonstrated that this type of social fear learning was enhanced if the observers previously had an experience of being shocked. This result is consistent with the facilitation of emotional contagion by a shared experience (e.g., Church 1959; Watanabe and Ono 1986), thereby supporting the idea that this type of social conditioning is mediated by emotional contagion. Jeon et al. (2010) also reported a similar result in mice. Importantly, the social fear conditioning was more efficient if the demonstrator was a familiar individual such as a sibling or a long-time mating partner. As social proximity and familiarity comprise another set of modulatory variables in emotional contagion (e.g., Langford et al. 2006; see Section x), the result of Jeon et al. (2010) constitutes further evidence that emotional contagion mediates social fear conditioning. Jeon et al. (2010) further demonstrated that inactivation of the ACC or the midline and intralaminar thalamic nuclei that project to the ACC resulted in impaired social fear conditioning, in agreement with human fMRI studies showing involvement of the ACC in emotional contagion of fear (e.g., Botvinick et al. 2005; Singer et al. 2004; see Sect. 13.3.2).

Overall, these studies suggest that demonstrators' fear transfers to observers through emotional contagion and thereby serves as a proxy US for the observers to establish predictive learning about fearful events. Thus, once a certain emotional expression acquired the contagious property among conspecifics through learning (and possibly to some extent through evolution), the members of that group would no longer need to learn every stimulus-outcome contingency individually, and the merit of such function would be substantial when learning from scratch is too costly.

13.5.2 Voluntary Empathetic Behavior

Another important function of empathy, in particular emotional contagion, is to provide a motivational source to support animals' voluntary behavior. In other words, detection of conspecifics' emotional states through emotional contagion should act as a foundation on which appropriate instrumental behavior is performed.

13.5.2.1 Reducing the Distress of Conspecifics

There are now several demonstrations that rats can perform a voluntary action to reduce the distress of a conspecific. In one of the earliest studies, Rice and Gainer (1962) trained a rat under a contingency where lever pressing could lower a hoisted conspecific that expressed signs of distress. They found that the subject lever pressed faster than control animals for which an object, rather than a conspecific, was suspended. Giving electric shock to a conspecific yielded different results; Rice (1964) reported that rats often retreated to a corner of the chamber as if they tried to escape the distressed conspecific instead of pressing a lever to stop the shocks. Using a choice procedure, Aoyama and Okaichi (1994) demonstrated a similar aversion to a conspecific's distress in rats. They trained rats to lever press for food, where one lever was associated with food while another was associated with food and delivery of electric shock to a conspecific in an adjacent compartment. The rats showed a significant preference for the lever associated with the food only. This result supports early research with monkeys in a similar setting (Wechkin et al. 1964).

Another example of voluntary empathetic behavior can be found in what is called helping or rescuing behavior in rats. Ben-Ami Bartal et al. (2011, 2014) demonstrated that rats voluntarily opened a door of a restrainer to liberate a cage mate inside. The door opening was more frequent when a cage mate was trapped in the restrainer than when an object was inside the restrainer or when it was empty. Similarly, Sato et al. (2015) showed that rats opened a door to allow a cage mate to escape from water in an adjacent compartment. In studies done by Sato et al. (2015) as well as by Aoyama and Okaichi (1994), subject rats that had experienced being soaked in the water or shocked, i.e., those having a shared experience, showed a greater degree of empathetic behavior, suggesting the involvement of emotional contagion as a foundation for these instrumental behaviors.

13.5.2.2 Nature of Reinforcer in the Rescue Behavior

The door-opening behavior by rats in studies of both Ben-Ami Bartal et al. (2011, 2014) and Sato et al. (2015) emerged gradually across many trials, indicating that the behavior was acquired through instrumental learning reinforced by the consequence of that action. A critical issue is what exactly served as a reinforcer to support the acquisition of the rescue behavior.

Ben-Ami Bartal et al. (2011) stated that the rats could detect the distress of a cage mate trapped inside a restrainer and were thus motivated to free the conspecific to reduce the conspecific's distress. They went further to argue that such rescue behavior can be considered to reflect pro-sociality and altruism in these animals (see also Sato et al. 2015). It is important to note, however, that instrumental motivation can take two forms. First, the behavior could occur if the reduction of the conspecific's distress was inherently rewarding or led to a positive emotional state in the helper rats, that is, a positive reinforcement process. Second, the distress of the conspecific could cause a similar distress—through emotional contagion—in the helper rats that then opened the door for the purpose of reducing their own distress. Termination or omission of aversive events serves as a negative reinforcer to allow acquisition and maintenance of instrumental behavior (e.g., Sidman 1953; Thorndike 1911). Because the subjects were confined in a small experimental arena, they did not have a chance to escape from the distressed conspecific, except by rescuing them. Thus, one way to directly test whether rescue behavior is controlled by positive or negative reinforcement would be to offer animals choices and see if they choose or avoid the opportunity to rescue the distressed conspecific. To our knowledge, such a test has not been conducted.

There is also a third possibility that the helper rats simply sought a direct social contact with the trapped cage mate. In replicating the study of Ben-Ami Bartal et al. (2011), Silberberg et al. (2014) indeed showed that door-opening behavior was not acquired if the trapped rats were liberated to the other side of the divided compartment so that the helper rat was unable to have direct contact with the freed cage mate. Notably, once the rescue behavior was acquired, it was maintained in the absence of direct social contact (Ben-Ami Bartal et al. 2011; Silberberg et al. 2014), possibly mediated by one of the other reinforcement processes mentioned above. In this regard, it is important that Sato et al. (2015) demonstrated that rats did not learn to open the door if a cage mate was present in an adjacent compartment but not in the water and therefore not distressed, which argues against the social contact account. One way to explain this set of findings would be to assume that the social contact with the *distressed* conspecific was necessary, at least initially, to establish the door-opening behavior. It seems plausible that the unusual state of a cage mate has informational value for the observer and therefore triggers innate exploratory behavior toward it. Then, through several repeated experiences, the helper animal may incidentally learn the contingency between the door opening and the reduced distress (of both the helper and the helped animal), which would further strengthen and maintain the behavior via negative reinforcement.

Whether there is anything more than novelty seeking and negative reinforcement with respect to empathetic behavior in animals remains to be tested, but until this issue is addressed, it seems more appropriate to refrain from referring to these behaviors as pro-social or altruistic—at least when discussing basic processes rather than the functional significance of such behavior. A similar view has been expressed by Vasconcelos et al. (2012).

13.6 Evolutionary Origin of Social Comparison

It is not necessary to feel negative emotions in response to the negative emotions of others, but the negative emotions of others may signal our own danger or aversive experience. Thus, the aversive property of others' distress has adaptive value. We do not have to feel pleasure for the pleasure of others, but feeling pleasure together may improve social bonding within a group. Contagion mechanisms and the sharing of affective states may thus promote group cohesion and enhance fitness, survival, and

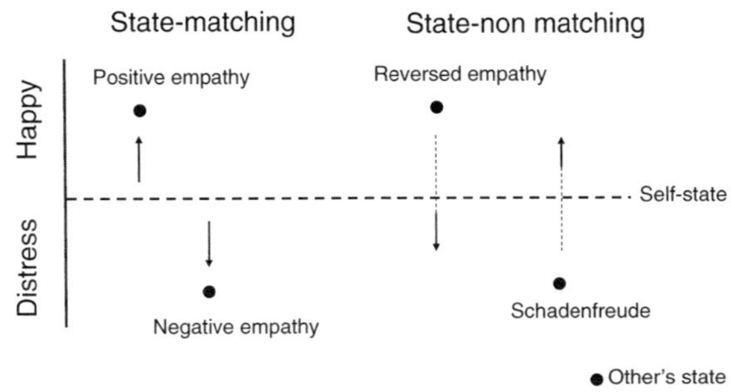

Fig. 13.5 Relationships between self-state (*horizontal dashed line*) and state of others (*black dots*). See text for explanation

reproductive opportunities within social groups. The empathetic response of humans to the emotional displays of others has an evolutionary origin and is clearly biologically protective and adaptive (see Watanabe 2016a).

Our happiness and distress are determined not only by the absolute value of happiness and distress but also by the relative values based on social comparison. As shown in Fig. 13.5, in the case of state-matching empathy, an observer has an upper limit for happiness—that is, the happiness of the demonstrator. Similarly, the distress of an observer has a limit, which coincides with that of the demonstrator. On the other hand, distress in non-state-matching empathy has no limit. Even a happy person in an ordinary standard could be unhappy in the presence of a happier individual. Thus, relative status, rather than absolute status, determines our emotional experiences. The economist J. M. Keynes (1931), for example, predicted that capitalism would reach a stationary state as we acquired sufficient wealth with less labor because of technological innovation. He predicted the increment of GDP and decrement of work time. Contrary to this prediction, however, we in fact continue to work long or even longer hours as technological innovation has increased. According to economists, "need" has a stationary state but "wish" does not (Sidelsky and Sidelsky 2009). A wealthy person who earns plenty of money—even too much to be spent during his/her lifetime—may still want more money because any wealthy person could indeed be unhappy in the presence of a wealthier person. This funny emotion is a consequence of social comparison. The reversed empathy is such case.

On the other hand, a poor person can be happy in the presence of a poorer person. Kobayashi Issa, a famous poet of eighteenth-century Japan, made a haiku as follows:

> "In cold autumn wind, a beggar compares himself with me."

Issa was very poor and shabby looking, so that the beggar felt comfortable observing Issa's appearance.

What is the biological basis for such irrational feelings induced by social comparison? The answer should lie in the process of sexual selection, an idea originally proposed by Darwin (1859, 1872) who tried to explain evolution of certain phenotypes that are apparently not adaptive, such as the tail of male peacocks and antlers of deer. He assumed that females of many species have a sense of "beauty" and thus a preference for beautiful males. This anthropomorphic explanation has been criticized, and modern evolutionary biologists propose that females prefer some physical characters that signal health or "good genes." A female's choice of mating partner is based on comparisons among male competitors. To be chosen, therefore, males of nonhuman animal species, for instance, must have a larger body size, a greater repertoire of songs, a larger territory, display more complex dances, etc. Social rank is one such adaptive index. Savanna baboons show a positive correlation between male dominance rank and ability to monopolize females (Alberts et al. 2003). A recent genetic analysis revealed that Genghis Khan—probably the most politically powerful male in human history—was a highly **(probably the highest)** reproductively successful male (Nasidze et al. 2005). A male with resources enough to maintain his life and his possible family may not be able to attract a female when a male with more resources appears. Because choice by females is based on comparison, males always have pressure to be larger, smarter, wealthier, etc. In other words, the presence of relatively larger, smarter, or wealthier competitors induces an aversive feeling (reversed empathy). On the other hand, if females choose a male within a group, even a weak male may have a chance to attract a female in comparison with other weaker males in the group. Thus, the presence of weaker competitors leads to the possibility of attracting females and induces positive emotion. Of course, we must consider cultural evolution in addition to biological evolution in order to understand human behavior, but our strong tendency to consider social comparison should be based on sexual selection during evolution, and indeed it may explain many irrational social behaviors of humans.

Acknowledgments We want to express our gratitude to Toru Shimizu and Tedd Patton for their comments on the MS of this chapter. This research was supported by a Grant-in-Aid for Scientific Research on Innovative Areas, Japan Society for the Promotion of Science (25118001).

References

Alberts SC, Wattas HE, Altmann J (2003) Queuing and queue-jumping; long-term patterns of reproductive skew in male savannah baboons. Papio Cynocephalus. Anim Behav 65:821–840

Adams CD, Dickinson A (1981) Instrumental responding following reinforcer devaluation. Q J Exp Psychol 33:109–121

Aoyama K, Okaichi H (1994) The influence of conspecific distress responses on the lever choice behavior in the rat. Shinrigaku Kenkyu 65:286–294

Balleine B, Dickinson A (1992) Signalling and incentive processes in instrumental reinforcer devaluation. Q J Exp Psychol 45:285–301

Baron-Cohen S, Wheelwright S (2004) The empathy quotient: an investigation of adults with Asperger syndrome or high functioning autism, and normal sex differences. J Autism Dev Disord 34:163–175

Ben-Ami Bartal I, Decety J, Mason P (2011) Empathy and pro-social behavior in rats. Science 334:1427–1430

Ben-Ami Bartal I, Rodgers DA, Sarria MSB, Decety J, Mason P (2014) Pro-social behavior in rats is modulated by social experience. eLife 3:e01385

Botvinick M, Jha AP, Bylsma LM, Fabian SA, Solomon PE, Prkachin KM (2005) Viewing facial expressions of pain engages cortical areas involved in the direct experience of pain. NeuroImage 25:312–319
Bouwknecht JA, Olivier B, Paylor RE (2007) The stress-induced hyperthermia paradigm as a physiological animal model for anxiety: a review of pharmacological and genetic studies in the mouse. Neurosci Biobehav Rev 31:41–59
Brosnan SF, de Waal BM (2003) Monkeys reject unequal pay. Nature 425:297–299
Bruchey AK, Jones CE, Monfils MH (2010) Fear conditioning by-proxy: social transmission of fear during memory retrieval. Behav Brain Res 214:80–84
Call J, Tomasello M (2008) Does the chimpanzee have a theory of mind? 30 years later. Trends Cogn Sci 12:187–192
Calvo-Merino B, Glaser DE, Grezes J, Passingham RE, Haggard P (2005) Action observation and acquired motor skills: an FMRI study with expert dancers. Cereb Cortex 15:1243–1249
Carr L, Iacoboni M, Dubeau MC, Mazziotta JC, Lenzi GL (2003) Neural mechanisms of empathy in humans: a relay from neural systems for imitation to limbic areas. Proc Natl Acad Sci 100:5497–5502
Catmur C, Walsh V, Heyes C (2009) Associative sequence learning: the role of experience in the development of imitation and the mirror system. Philos Trans R Soc B Biol Sci 364:2369–2380
Chen MK, Santos LR (2006) Some thoughts on the adaptive function of inequality aversion: an alternative to Brosnan's social hypothesis. Soc Justice Res 19:201–207
Chong TTJ, Cunnington R, Williams MA, Kanwisher N, Mattingley JB (2008) fMRI adaptation reveals mirror neurons in human inferior parietal cortex. Curr Biol 18:1576–1580
Church RM (1959) Emotional reactions of rats to the pain of others. J Comp Physiol Psychol 52:132
Cook M, Mineka S (1989) Observational conditioning of fear to fear-relevant versus fear-irrelevant stimuli in rhesus monkeys. J Abnorm Psychol 98:448–459
Cook R, Dickinson A, Heyes C (2012) Contextual modulation of mirror and countermirror sensorimotor associations. J Exp Psychol Gen 141:774–787
Cook R, Press C, Dickinson A, Heyes C (2010) Acquisition of automatic imitation is sensitive to sensorimotor contingency. J Exp Psychol Hum Percept Perform 36:840–852
Dapretto M, Davies MS, Pfeifer JH, Scott AA, Sigman M, Bookheimer SY, Iacoboni M (2006) Understanding emotions in others: mirror neuron dysfunction in children with autism spectrum disorders. Nat Neurosci 9:28–30
Darwin C (1859) On the origin of species by means of natural selection. John Murray, London
Darwin C (1872) The expression of the emotions in man and animals. John Murray, London
De Waal F (2009) The age of empathy. Harmony Book, New York, p 183
Dickinson A (1985) Actions and habits: the development of behavioural autonomy. Philos Trans R Soc B Biol Sci 308:67–78
Dickinson A, Charnock DJ (1985) Contingency effects with maintained instrumental reinforcement. Q J Exp Psychol 37:397–416
Di Pellegrino G, Fadiga L, Fogassi L, Gallese V, Rizzolatti G (1992) Understanding motor events: a neurophysiological study. Exp Brain Res 91:176–180
Ekman P, Levenson RW, Friesen WV (1983) Autonomic nervous system activity distinguishes among emotions. Science 221:1208–1210
Ellis L (ed) (1993) Social stratification and socioeconomic inequality, A comparative biosocial analysis, vol vol. 1. Praeger, Westport
Feather NT (2008) Effects of observer's own status on reactions to a high achiever's failure: deservingness, resentment, schadenfreude and sympathy. Aust J Psychol 60:31–43
Forsythe JLH, Savin NE, Sefton M (1994) Fairness in simple bargaining experiments. Games Econ Behav 6:347–369
Fehr E, Gachter S (2002) Altruistic punishment. Nature 415:137–140
Fogassi L, Ferrari PF, Gesierich B, Rozzi S, Chersi F, Rizzolatti G (2005) Parietal lobe: from action organization to intention understanding. Science 308:662–667
Frigida NH (2007) The laws of emotion. Lawrence, Erlbaum Associates, Mahwah

Frith CD, Frith U (1999) Interacting minds—a biological basis. Science 286:1692–1695

Galef BG, Mason JR, Preti G, Bean NJ (1988) Carbon disulfide: a semiochemical mediating socially-induced diet choice in rats. Physiol Behav 42:119–124

Galef BG, Stein M (1985) Demonstrator influence on observer diet preference: analyses of critical social interactions and olfactory signals. Anim Learn Behav 13:31–38

Galef BG, Wigmore SW (1983) Transfer of information concerning distant foods: a laboratory investigation of the 'information-centre' hypothesis. Anim Behav 31:748–758

Gallese V, Fadiga L, Fogassi L, Rizzolatti G (1996) Action recognition in the premotor cortex. Brain 119:593–609

Gallese V, Fogassi L, Fadiga L, Rizzolatti G (2002) Action representation and the inferior parietal lobule. In: Prinz W, Hommel B (eds) Common mechanisms in perception and action: attention and performance, vol 19. Oxford University Press, Oxford, pp 247–266

Gauvin D, Briscoe RJ, Goulden KL, Holloway FA (1994) Aversive attributes of ethanol can be attenuated by dyadic social interaction in the rat. Alcoholism 11:247–251

Glynn T (1989) Essential elements of school-based smoking prevention programs. J Sch Health 59:181–188

Güth W, Tietz R (1990) Ultimatum bargaining behavior: a survey and comparison of experimental results. J Econ Psychol 11:417–449

Haslinger B, Erhard P, Altenmüller E, Schroeder U, Boecker H, Ceballos-Baumann AO (2005) Transmodal sensorimotor networks during action observation in professional pianists. J Cogn Neurosci 17:282–293

Heyes C (2001) Causes and consequences of imitation. Trends Cogn Sci 5:253–261

Heyes C (2010) Where do mirror neurons come from? Neurosci Biobehav Rev 34:575–583

Heyes C, Ray E (2000) What is the significance of imitation in animals? Adv Study Behav 29:215–245

Horowitz A (2012) Fair is fine, but more is better: limits to inequity aversion in the domestic dog. Soc Justice Res 25:195–212

Houtepen LC, Peterse DP, Westphal KGC, Olivier B, Christiaan H, Vinkers CH (2011) The autonomic stress-induced hyperthermia response is not enhanced by several anxiogenic drugs. Physiol Behav 102:105–109

Hunt PS, Holloway JL, Scordalakes EM (2001) Social interaction with an intoxicated sibling can result in increased intake of ethanol by periadolescent rats. Dev Psychobiol 38:101–109

Isoda M, Noritake A (2013) What makes the dorsomedial frontal cortex active during reading the mental states of others? Front Neurosci 7:232

Jackson PL, Meltzoff AN, Decety J (2005) How do we perceive the pain of others? A window into the neural processes involved in empathy. NeuroImage 24:771–779

Jeon D, Kim S, Chetana M, Jo D, Ruley HE, Lin SY, Rabah D, Kinet JP, Shin HS (2010) Observational fear learning involves affective pain system and Cav1.2 Ca^{2+} channels in ACC. Nat Neurosci 13:482–488

Joly-Mascheroni RM, Senju A, Shepherd AJ (2008) Dogs catch human yawns. Biol Lett 4: 446–448

Keeney AJ, Hogg S, Marsden CA (2001) Alterations in core body temperature, locomotor activity, and corticosterone following acute and repeated social defeat of male NMRI mice. Physiol Behav 74:177–184

Keynes JM (1931) The collected writings of John Maynard Keynes, Essays in persuasion, vol 9. Cambridge, Cambridge University Press

Kilner JM, Neal A, Weiskopf N, Friston KJ, Frith CD (2009) Evidence of mirror neurons in human inferior frontal gyrus. J Neurosci 29:10153–10159

Kim EJ, Kim ES, Covey E, Kim JJ (2010) Social transmission of fear in rats: the role of 22-kHz ultrasonic distress vocalization. PLoS One 5:e15077

Knapska E, Nikolaev E, Boguszewski P, Walasek G, Blaszczyk J, Kaczmarek L, Werka T (2006) Between-subject transfer of emotional information evokes specific pattern of amygdala activation. Proc Natl Acad Sci U S A 103:3858–3862

Kollack-Walker S, Watson SJ, Akil H (1997) Social stress in hamsters: defeat activates specific neurocircuits within the brain. J Neurosci 17:8842–8855
Koolhaas JM, Meerlo P, De Boer SF, Strubbe JH, Bohus B (1997) The temporal dynamics of the stress response. Neurosci Biobehav Rev 21:775–782
Komeda H, Kosaka H, Saito DN, Mano Y, Jung M, Fujii T et al (2015) Autistic empathy toward autistic others. Soc Cogn Affect Neurosci 10:145–152
Langford DJ, Crager SE, Shehzad Z, Smith SB, Sotocinal SG, Levenstadt JS et al (2006) Social modulation of pain as evidence for empathy in mice. Science 312:1967–1970
Langford DJ, Tuttle AH, Brown K, Deschenes S, Fischer DB, Mutso A et al (2010) Social approach to pain in laboratory mice. Soc Neurosci 5:163–170
le Roux A, Snyder-Mackler N, Roberts EK, Beehner JC, Bergman TJ (2013) Evidence for tactical concealment in a wild primate. Nat Commun 4:1462. doi:10.1038/ncomms2468
Maldonado AM, Finkbeiner LM, Kirstein CL (2008) Social interaction and partner familiarity differentially alter voluntary ethanol intake in adolescent male and female rats. Alcoholism 42:641–648
Meerlo P, Overkamp GJF, Koolhaas JM (1997) Behavioural and physiological consequences of a single social defeat in roman high- and low-avoidance rats. Psychoneuroendocrinology 22:155–168
Mineka S, Davidson M, Cook M, Keir R (1984) Observational conditioning of snake fear in rhesus monkeys. J Abnorm Psychol 93:355
Mirsky IA, Miller RE, Murphy JV (1958) The communication of affect in rhesus monkeys. J Am Psychoanal Assoc 6:433–441
Mitchell JP (2009) Inferences about mental states. Philos Trans R Soc Lond B Biol Sci 364:1309–1316
Morris PH, Doe C, Godsell E (2008) Secondary emotions in non-primate species? Behavioural reports and subjective claims by animal owners. Cognit Emot 22:3–20
Nasidze I, Quinque D, Dupanloup I, Cordaux R, Kokshunova L, Stoneking M (2005) Genetic evidence for the Mongolian ancestry of Kalmyks. Am J Phys Anthropol 128:846–854
Ostojić L, Shaw RC, Cheke LG, Clayton NS (2013) Evidence suggesting that desire-state attribution may govern food sharing in Eurasian jays. Proc Natl Acad Sci 110:4123–4128
Pardon MC, Kendall DA, Perez-Diaz F, Duxon MS, Marsden CA (2004) Repeated sensory contact with aggressive mice rapidly leads to an anticipatory increase in core body temperature and physical activity that precedes the onset of aversive responding. Eur J Neurosci 20:1033–1050
Premack D, Woodruff G (1978) Does the chimpanzee have a theory of mind? Behav Brain Sci 1:515–526
Preston SD, de Waal FB (2002) Empathy: its ultimate and proximate bases. Behav Brain Sci 25:1–20
Range F, Leitner K, Viranyi Z (2012) The influence of the relationship and motivation on inequity aversion in dogs. Soc Justice Res 25:170–194
Reimert I, Bolhuis JE, Kemp B, Rodenburg TB (2013) Indicators of positive and negative emotions and emotional contagion in pigs. Physiol Behav 109:42–50
Rice GE (1964) Aiding behavior vs fear in the albino rat. Psychol Rec 14:165–170
Rice GE, Gainer P (1962) "altruism" in the albino rat. J Comp Physiol Psychol 55:123–125
Rizzolatti G, Fadiga L, Gallese V, Fogassi L (1996) Premotor cortex and the recognition of motor actions. Cogn Brain Res 3:131–141
Roozendaa B, McEwen BS, Chattarji S (2009) Stress, memory and the amygdala. Nat Rev Neurosci 10:423–433
Salovey P, Rodin J (1989) Envy and jealousy in close relationships. Rev Pers Soc Psychol 10:221–246
Sato N, Tan I,Tate K, Okada M (2015) Rats demonstrate helping behavior toward a soaked conspecific. Anim Cogn 18:1039–1047
Sheskin M, Ashayeri K, Skerry A, Santos LR (2014) Capuchin monkeys (*Cebus apella*) fail to show inequality aversion in a no-cost situation. Evol Hum Behav 35:80–88
Sidelsky R, Sidelsky E (2009) How much is enough? Other Press, New York

Sidman M (1953) Avoidance conditioning with brief shock and no exteroceptive warning signal. Science 118:157–158
Silberberg A, Allouch C, Sandfort S, Kearns D, Karpel H, Slotnick B (2014) Desire for social contact, not empathy, may explain "rescue" behavior in rats. Anim Cogn 17:609–618
Singer T, Seymour B, O'Doherty J, Kaube H, Dolan RJ, Frith CD (2004) Empathy for pain involves the affective but not sensory components of pain. Science 303:1157–1162
Singer T, Seymour B, O'Doherty JP, Stephan KE, Dolan RJ, Frith CD (2006) Empathic neural responses are modulated by the perceived fairness of others. Nature 439:466–469
Smith RH (2013) The joy of pain: Schadenfreude and the dark side of human nature. Oxford University Press, New York
Spears R (2013) Evidence for group-based schadenfreude from the gender domain: from intragroup to intergroup rivalry. TPM Test Psychom Methodol Appl Psychol 20:301–311
Thorndike EL (1911) Animal intelligence: experimental studies. Macmillan, New York
Thornhill JA, Cooper KE, Veale WL (1979) The effect of repeated periods of restraint and continued exposure to various ambient temperatures on the core temperature responses of drug-naive rats. Can J Physiol Pharmacol 57:1011–1017
Van Loo PL, de Groot AC, Van Zutphen BFM, Bauemans V (2001) Do male mice prefer or avoid each other's company? Influence of hierarchy, kinship and familiarity. J Appl Anim Welf Sci 4:91–103
Vasconcelos M, Hollis K, Nowbahari E, Kacelnik A (2012) Pro-sociality without empathy. Biol Lett 8:910–912
Yamamoto S, Takimoto A (2012) Empathy and fairness: psychological mechanisms for eliciting and maintaining prosociality and cooperation in primates. Soc Justice Res 25:233–255
Watanabe S (2011a) Drug-social interactions in the reinforcing property of methamphetamine in mice. Behav Pharmacol 22:203–206
Watanabe S (2011b) Empathy and reversed empathy of stress in mice. PLoS One 6:e23357
Watanabe S (2012) Distress of mice induces approach behavior but has an aversive property for conspecifics. Behav Process 91:141–144
Watanabe S (2014) The dominant/subordinate relation between mice modifies the approach behavior toward a cage mate experiencing pain. Behav Process 103:1–4
Watanabe S (2015) Social factors modulate restraint stress induced hyperthermia in mice. Brain Res 1624:134–139
Watanabe S (2016a) Evolutionary origins of empathy. In: Watt DJ, Panksepp J (eds) Neurobiology and psychology of empathy. Nova Science Publishers, Hauppauge, New York, 37–62
Watanabe S (2016b) Mirror perception in mice: preference for and stress reduction by mirrors. Int J Comp Psychol 29, http://escholarship.org/uc/item/1388v4pg
Watanabe S, Ono K (1986) An experimental analysis of "empathic" response: effects of pain reactions of pigeon upon other pigeon's operant behavior. Behav Process 13:269–277
Wechkin S, Masserman JH, Terris W Jr (1964) Shock to a conspecific as an aversive stimulus. Psychon Sci 1:47–48

MIX
Papier aus verantwortungsvollen Quellen
Paper from responsible sources
FSC® C105338

If you have any concerns about our products,
you can contact us on
ProductSafety@springernature.com

In case Publisher is established outside the EU,
the EU authorized representative is:
Springer Nature Customer Service Center GmbH
Europaplatz 3, 69115 Heidelberg, Germany

Printed by Libri Plureos GmbH
in Hamburg, Germany